Cognition and the Brain

The Philosophy and Neuroscience Movement

This volume provides an up-to-date and comprehensive overview of the philosophy and neuroscience movement, which applies the methods of neuroscience to traditional philosophical problems and uses philosophical methods to illuminate issues in neuroscience. At the heart of the movement is the conviction that basic questions about human cognition, many of which have been studied for millennia, can be answered only by a philosophically sophisticated grasp of neuroscience's insights into the processing of information by the human brain. Essays in this volume are clustered around five major themes: data and theory in neuroscience; neural representation and computation; visuomotor transformations; color vision; and consciousness.

Andrew Brook is Director of the Institute of Cognitive Science and Professor of Philosophy at Carleton University in Ottawa, Ontario. He is the author of a number of books and articles, including *Kant and the Mind* and, with Paul Raymont, *A Unified Theory of Consciousness.*

Kathleen Akins is Director of the McDonnell Project in Philosophy and Neuroscience and Associate Professor of Philosophy at Simon Fraser University in Burnaby, British Columbia.

Cognition and the Brain

The Philosophy and Neuroscience Movement

Edited by

ANDREW BROOK

Carleton University

KATHLEEN AKINS

Simon Fraser University

CAMBRIDGE UNIVERSITY PRESS
Cambridge, New York, Melbourne, Madrid, Cape Town, Singapore, São Paulo

Cambridge University Press
40 West 20th Street, New York, NY 10011-4211, USA

www.cambridge.org
Information on this title: www.cambridge.org/9780521836425

© Cambridge University Press 2005

First published 2005

Printed in the United States of America

A catalog record for this publication is available from the British Library.

Library of Congress Cataloging in Publication Data

Cognition and the brain : the philosophy and neuroscience movement / edited by
Andrew Brook, Kathleen Akins.
p. cm.
Includes bibliographical references and index.
ISBN 0-521-83642-5 (hardback : alk. paper)
1. Cognitive neuroscience – Philosophy. I. Brook, Andrew. II. Akins, Kathleen.
III. Title.
QP360.5.c6326 2005
612.8′233 – dc22   2004022551

ISBN-13 978-0-521-83642-5 hardback
ISBN-10 0-521-83642-5 hardback

Contents

Contributors

With a few noted exceptions, the contributors' doctoral degrees are in philosophy. Nearly all of them have doctoral-level training in neuroscience, too. The list focusses on that aspect of their background.

Kathleen Akins (PhD Michigan), Associate Professor of Philosophy and Director, James D. McDonnell Project in Philosophy and Neuroscience, Simon Fraser University. Her education in neuroscience includes coursework in neuropsychology and sensory processing as a PhD student and clinical rounds at Massachusetts General Hospital with Dr. Marcel Kinsbourne while a research associate at Tufts. She was a member of the Neuroscience Faculty working in the laboratory of Joseph Malpeli on mammalian vision at the University of Illinois, Urbana-Champaign.

Andrew Brook (DPhil Oxon.), Professor of Philosophy and Director, Institute of Cognitive Science, Carleton University. Author or editor of seven books including *A Unified Theory of Consciousness* (with Paul Raymont, MIT Press, forthcoming) and numerous papers centred on consciousness and the Kantian model of the mind, Brook is the reason for the qualification 'nearly all' in the opening comment.

Paul M. Churchland (PhD Pittsburgh), Professor of Philosophy, UC San Diego. Churchland began learning neuroscience in the lab of Professor Larry Jordan at the University of Manitoba in 1978. After he moved to UCSD in 1984, Francis Crick and Terry Sejnowski (both at the Salk Institute) had a lasting influence on his grasp of both experimental and theoretical neuroscience. However, he says that he has learned more neuroscience from his wife and colleague, Patricia Churchland, than

vii

from anyone else. Both of their children are experimental neuroscientists, so, as he puts it, "I am now learning to sit respectfully at *their* feet."

Diego Cosmelli has just finished his PhD in Cognitive Neuroscience, Ecole Polytechnique, Paris, where he worked under the direction of the late Francisco Varela. His main interests centre on the relationship between brain dynamics and human experience.

Chris Eliasmith (PhD Philosophy, Neuroscience and Psychology [PNP] programme, Washington University, St. Louis), Assistant Professor of Philosophy and Engineering and Director, Computational Neuroscience Research Group, University of Waterloo. While at WUSTL, Eliasmith collaborated with Charles Anderson in computational neuroscience for six years. Their recent book was the result. After completing his PhD in the PNP programme, he was a Postdoctoral Fellow at the Washington University Medical School in Dr. David van Essen's lab.

Rick Grush (PhD Cognitive Science and Philosophy, UC San Diego), Associate Professor of Philosophy, UCSD. As a graduate student he did coursework in neuroscience, worked in Vilayanur Ramachandran's lab for several years, and studied the mathematical underpinnings of neuro-computing with Robert Hecht-Nielsen. His work in this area is in theoretical cognitive neuroscience, the attempt to formulate theories of how cognitive phenomena (such as imagery, thought, temporal perception, spatial representation, etc.) are implemented in neural mechanisms.

Valerie Gray Hardcastle (PhD Philosophy and Cognitive Science, UC San Diego), Professor of Philosophy, Virginia Polytechnic. Hardcastle studied neuroscience while a graduate student at UCSD and as a member of the James S. McDonnell Project in Philosophy and the Neurosciences for the last five years. More recently, she has collaborated with C. Matthew Stewart, Johns Hopkins University, on issues of localization in the brain. Her area in the philosophy of mind and cognitive science is how (and whether) psychological phenomena relate to underlying neurophysiology. She is the author of *The Myth of Pain* (1999), four other books, and numerous articles and book chapters.

Pierre Jacob (PhD Harvard), Professor of Philosophy and Director, Jean Nicod Institute, Paris. A member of the world-famous CREA in Paris for many years and founder of the Institute of Cognitive Science in Lyon, he has no formal training in neuroscience. He learned neuro-science in the course of collaborations with cognitive neuroscientists of the Lyon Institute: Driss Boussaoud, Jean-René Duhamel, Angela Sirigu, and especially Marc Jeannerod, with whom he has written several papers and a book entitled *Ways of Seeing: The Scope and Limits of Visual Cognition* (Oxford University Press, 2003). His collaboration with Jeannerod has focussed on the two-visual-systems model of human vision.

Zoltán Jakab (Dipl. Psych, Loránd Eötvös University, Budapest; PhD Cognitive Science, Carleton), Békésy Postdoctoral Fellow, Budapest University of Technology and Economics. He studied neuroscience in the 1990s with Péter Érdi at the Central Research Institute for Physics in Budapest. The focus of his research was building a detailed, physiologically faithful neural network model of the hippocampal system. His current work is centred on colour vision.

Sean D. Kelly (PhD UC Berkeley), Assistant Professor of Philosophy, Princeton. He began work in neuroscience while a student in the MA programme in Cognitive and Linguistic Sciences at Brown. His thesis, written under the direction of neuroscientist Jim Anderson, was on the computational properties of neural nets. At UC Berkeley, he worked not only with the philosophers Hubert Dreyfus and John Searle but also with the neuroscientist Walter Freeman. He has also learned a lot of neuroscience during his affiliation with the James S. McDonnell Project in Philosophy and the Neurosciences over the last five years.

Antoine Lutz (PhD Cognitive Neuroscience, University of Jussieu, Paris VI), Postdoctoral Fellow, W. M. Keck Laboratory for Functional Brain Imaging and Behavior, Waisman Center, University of Wisconsin–Madison. His PhD research was done under the supervision of the late Francisco J. Varela. His current focus is the neurofunctional and neurodynamical characterization of meditative states in a group of highly trained Buddhist practitioners.

Pete Mandik (PhD PNP programme, Washington University, St. Louis), Assistant Professor of Philosophy and Coordinator, Cognitive Science Lab, William Paterson University. Mandik learned his neuroscience in the PNP programme and as a member of the James S. McDonnell Project in Philosophy and the Neurosciences over the last five years. In addition to his many other publications, he is one of the editors (with William Bechtel, Jennifer Mundale, and Robert Stufflebeam) of *Philosophy and the Neurosciences: A Reader* (Basil Blackwell, 2001). His primary areas of interest concern neurophilosophical approaches to both consciousness and representational content.

Victoria McGeer (PhD Toronto), Research Associate, Center for Human Values, Princeton. McGeer was a Postdoctoral Fellow in Alison Gopnik's developmental psychology lab at Berkeley. Most of the rest of her training in neuroscience has been through the James S. McDonnell Project in Philosophy and the Neurosciences over the last five years. Her work is informed by cognitive neuroscience and examines the philosophical preconceptions that underlie certain research programmes in cognitive psychology (mostly in connection with theory of mind research).

Jesse J. Prinz (PhD Chicago), Associate Professor of Philosophy, University of North Carolina at Chapel Hill. He learned neuroscience while teaching in the PNP Program at WUSTL. Author of three books and editor of the forthcoming *Handbook of Philosophy of Psychology* (Oxford University Press), Prinz concentrates his research on categorization, emotion, moral psychology, the nature/nurture debate, and the neural basis of consciousness.

C. Matthew Stewart is a resident at Johns Hopkins University in Otolaryngology, Head and Neck Surgery. He holds a combined MD/PhD from the University of Texas Medical Branch, with a specialization in neuroscience. His research focusses on the interactions of the vestibular and visual systems and how these interactions contribute to recovery of vestibular function after disease, injury, or surgery.

Evan Thompson (PhD Toronto), Canada Research Chair and Associate Professor of Philosophy, York University. While a doctoral student, he studied neuroscience in the laboratory of the late Francisco J. Varela at the Institut des Neurosciences in Paris. This work led to their writing, with Eleanor Rosch, *The Embodied Mind: Cognitive Science and Human Experience* (MIT Press, 1991). Thompson is the author or editor of two other books on vision and many papers and book chapters.

Special thanks to Steven Davis (PhD Pittsburgh), Professor of Philosophy, Carleton University, without whom this volume would not exist.

Introduction

Andrew Brook and Pete Mandik

A small movement dedicated to applying neuroscience to traditional philosophical problems and using philosophical methods to illuminate issues in neuroscience began 20–25 years ago and has been gaining momentum ever since. The central thought behind it is that certain basic questions about human cognition, questions that have been studied in many cases for millennia, will be answered only by a philosophically sophisticated grasp of what contemporary neuroscience is teaching us about how the human brain processes information.

The evidence for this proposition is now overwhelming. The philosophical problem of perception has been transformed by new knowledge about the vision systems in the brain. Our understanding of memory has been deepened by knowing that two quite different systems in the brain are involved in short- and long-term memory. Knowing something about how language is implemented in the brain has transformed our understanding of the structure of language, especially the structure of many breakdowns in language. And so on. On the other hand, a great deal is still unclear about the implications of this new knowledge of the brain. Are cognitive functions localized in the brain in the way assumed by most recent work on brain imaging? Does it even make sense to think of cognitive activity being localized in such a way? Does knowing about the areas active in the brain when we are conscious of something hold any promise for helping with long-standing puzzles about the nature and role of consciousness? And so on.

A group of philosophers and neuroscientists dedicated to informing each other's work has grown up. It is a good time to take stock of where this movement now is and what it is accomplishing. The Cognitive

Science Programme at Carleton University and the McDonnell Centre
for Philosophy and the Neurosciences at Simon Fraser University organized a conference, the McDonnell/Carleton Conference on Philosophy
and the Neurosciences, at Carleton University, Ottawa, Canada, October
17–20, 2002, around this theme. Many of the essays in the current volume
are derived from work presented at that conference, though all of them
go well beyond what was presented there. The aim of the volume is to
achieve a comprehensive 'snapshot' of the current state of the art in the
project to relate philosophy and neuroscience.

One of the special features of the authors in this volume is that, with
one exception, they all have at least PhD-level training or the equivalent
in both neuroscience and philosophy. (The exception is the first editor.)
The chapters are clustered around five themes:

- data and theory in neuroscience
- neural representation and computation
- visuomotor transformation
- color vision
- consciousness

History of Research Connecting Philosophy and Neuroscience

Prior to the 1980s, very little philosophical work drew seriously on scientific work concerning the nervous system or vice versa. Descartes specu-
lated in (1649) that the pineal gland constituted the interface between
the unextended mind and the extended body and did some anatomy
in laboratories (including on live, unanaesthetized animals; in his view,
animals do not have the capacity to feel pain), but he is at most a modest
exception.

Coming to the 20th century, even when the identity of mind with brain
was promoted in the mid-20th century by the identity theorists, also called
state materialists, they drew upon very little actual brain science. Instead,
the philosophy was speculative, even somewhat fanciful. Some examples:
Herbert Feigl (1958/1967) proposed an autocerebroscope whereby people could directly observe their own mental/neural processes. This was
science fiction, not science fact or even realistic scientific speculation.
Much discussion of identity theory involved the question of the identification of pain with C-fibre firings (U. T. Place 1956 and J. J. C. Smart
1959). But it has been known for a very long time that the neural basis
of pain is much more complicated than that (see V. G. Hardcastle 1997
for a recent review).

There were a few exceptions to the general ignorance about neuroscience among philosophers prior to the 1980s. Thomas Nagel (1971) is an example. This paper discusses the implications of experiments with commissurotomy (brain bisection) patients for the unity of consciousness and the person. D. C. Dennett (1978) discusses the question of whether a computer could be built to feel pain according to a thorough and still interesting summary of what was known about pain neurophysiology at the time. Barbara von Eckardt Klein (1975) discussed the identity theory of sensations in terms of then-current work on neural coding by Vernon Mountcastle, Benjamin Libet, and Herbert Jasper. But these exceptions were very much the exception.

The failure of philosophers of the era to draw on actual neuroscientific work concerning psychoneural identities could not be blamed on any lack of relevant work in neuroscience. David Hubel and Torsten Wiesel's (1962) Nobel Prize–winning work on the receptive fields of visual neurons held great promise for the identification of the perception of various visual properties with various neural processes. A decade earlier, Donald Hebb (1949) had tried to explain cognitive phenomena like perception, learning, memory, and emotional disorders in terms of neural mechanisms.

In the 1960s, the term 'neuroscience' emerged as a label for the interdisciplinary study of nervous systems. The Society for Neuroscience was founded in 1970. (It now has 25,000 members.) In the 1970s, the term 'cognitive science' was adopted as the label for interdisciplinary studies of 'cognition' – the mind as a set of functions for processing information. The idea of information processing might not have been much more than a uniting metaphor, but real effort was put into implementing the relevant functions in computational systems (artificial intelligence). Cognitive Science became institutionalized with the creation of the Cognitive Science Society and the journal *Cognitive Science* in the late 1970s. However, it has not grown the way neuroscience has. After 30 years, the Cognitive Science Society has about 1,500 members.

Until the 1980s, there was very little interaction between neuroscience and cognitive science. On the philosophical front, this lack of interaction was principled (if wrong-headed). It was based on a claim, owing to functionalists such as Jerry Fodor (1974) and Hilary Putnam (1967), that since cognition could be multiply realized in many different neural as well as non-neural substrates, nothing essential to cognition could be learned by studying neural (or any other) implementation. It is the cognitive functions that matter, not how they are implemented in this, that, or the other bit of silicon or goopy wetware.

The 1980s witnessed a rebellion against this piece of dogma. Partly this was because of the development of new and much more powerful tools for studying brain activity, fMRI (functional magnetic resonance imaging; the 'f' is usually lowercase for some reason) brain scans in particular. In the sciences, psychologist George Miller and neurobiologist Michael Gazzaniga coined the term 'cognitive neuroscience' for the study of brain implementation of cognitive functioning. Cognitive neuroscience studies cognition in the brain through such techniques as PET (positron emission tomography) and fMRI that allow us to see how behaviour and cognition, as studied by cognitive scientists, are expressed in functions in the brain, as studied by neuroscientists. The idea of relating cognitive processes to neurophysiological processes was not invented in the 1980s, however. For example, in the 1970s, Eric Kandel (1976) proposed explaining simple forms of associative learning in terms of presynaptic mechanisms governing transmitter release. T. V. P. Bliss and T. Lomo (1973) related memory to the cellular mechanisms of long-term potentiation (LTP).

In philosophy, an assault on the functionalist separation of brain and mind was launched with the publication of Patricia (P. S.) Churchland's *Neurophilosophy* in 1986 (a book still in print). Churchland's book has three main aims:

1. to develop an account of intertheoretic reduction as an alternative to the account from logical positivist philosophy of science;
2. to show that consciousness-based objections to psychoneural reduction don't work; and
3. to show that functionalist/multiple realizability objections to psychoneural reduction don't work.

A later neurophilosophical rebellion against multiple realizability was led by W. Bechtel and J. Mundale (1997). Their argument was based on the way in which neuroscientists use psychological criteria in determining what counts as a brain area.

With this sketch of the history of how the philosophy and neuroscience movement emerged, let us now look at particular topic areas in order to lay out some of the relevant history, examine what is going on currently, and connect the area to the contributions in this volume. By and large, the topics of primary interest in the philosophy of neuroscience are those that relate the mind/brain issue to concerns from the philosophy of science and the philosophy of mind. In fact, it is not always easy to distinguish philosophy of mind from philosophy of science in the philosophy and neuroscience movement. For example, the philosophy of mind question 'Are cognitive processes brain processes?' is closely related to the philosophy of science question 'Are psychological theories reducible to neurophysiological theories?' Either way, neurophilosophical interest is mostly concerned with research on the brain that is relevant to the mind (I. Gold and D. Stoljar, 1999, explore the relationship of neuroscience and the cognitive sciences in detail). There are a few exceptions, however. An important philosophical study of areas of neuroscience not directly relevant to cognition is found in P. Machamer et al. (2000), who discuss philosophically individual neurons, how neurons work, and so on.

First we will examine two big background topics: (1) neuroscience and the philosophy of science; and (2) reductionism versus eliminativism in neuroscience and cognitive science. Then we will turn to some of the areas in which philosophy and neuroscience are interacting.

Neuroscience and the Philosophy of Science

Much early philosophy of science held a central place for the notion of law, as in the Deductive-Nomological theory of scientific explanation or the Hypothetico-Deductive theory of scientific theory development or discussions of intertheoretic reduction. While the nomological view of science seems entirely applicable to sciences such as physics, there is a real question as to whether it is appropriate for life sciences such as biology and neuroscience. One challenge is based on the seeming teleological character of biological systems. Mundale and Bechtel (1996) argue that a teleological approach can integrate neuroscience, psychology, and biology.

Another challenge to the hegemony of nomological explanation comes from philosophers of neuroscience, who argue that explanations in terms of laws at the very least need to be supplemented by explanations in terms of mechanisms (Bechtel and R. C. Richardson 1993; Machamer et al. 2000). Here is how their story goes. Nomological explanations, as conceived by the Deductive-Nomological model, involve showing that a description of the target phenomenon is logically deducible from a statement of general law. Advocates of the mechanistic model of explanation claim that adequate explanations of certain target phenomena can be given by describing how the phenomena result from various processes and subprocesses. For example, cellular respiration is explained by appeal to various chemical reactions and the areas in the cell where these reactions take place. Laws are not completely abandoned but they are supplemented (P. Mandik and Bechtel 2002).

A related challenge to logical positivist philosophy of science questions whether scientific theories are best considered as sets of sentences. Paul (P. M.) Churchland (1989), for example, suggests that the vector space model of neural representation should replace the view of representations as sentences (more on vector spaces later in this section). This would completely recast our view of the enterprise of scientific theorizing, hypothesis testing, and explanation. The issue is directly connected to the next one.

Reductionism Versus Eliminativism

There are three general views concerning the relation between the psychological states posited by psychology and the neurophysiological processes studied in the neurosciences:

1. The autonomy thesis: While every psychological state may be (be implemented by, be supervenient on) a brain state, types of psychological states will never be mapped onto types of brain states. Thus, each domain needs to be investigated by distinct means (see Fodor 1974).

Analogy: Every occurrence of red is a shape of some kind, but the color-type, redness, does not map onto any shape-type. Colors can come in all shapes and shapes can be any color (see A. Brook and R. Stainton 2000, chapter 4, for background on the issue under discussion here).

2. Reductionism: Types of psychological states will ultimately be found to be types of neurophysiological states.

The history of science has been in no small part a history of reduction: Chemistry has been shown to be a branch of physics, large parts of biology have been shown to be a branch of chemistry. Reductivists about cognition and psychology generally believe that cognition and psychology, or much of them, will turn out to be a branch of biology.

3. Eliminativism (also called eliminative materialism): Psychological theories are so riddled with error and psychological concepts are so weak when it comes to building a science out of them (for example, phenomena identified using psychological concepts are difficult if not impossible to quantify precisely) that psychological states are best regarded as talking about nothing that actually exists.

Eliminativist arguments are antireductivist in the following way: They argue that there is no way to reduce psychological theories to neural theories or at best no point in doing so.

Philosophers of neuroscience generally fall into either the reductionist or the eliminativist camps. Most are mainly reductionists – most, for example, take the phenomena talked about in the 'cognitive' part of cognitive neuroscience to be both perfectly real and perfectly well described using psychological concepts – but few are dogmatic about the matter. If some psychological concepts turn out to be so confused or vague as to be useless for science, or to carve things up in ways that do not correspond to what neuroscience discovers about what structures and functions in the brain are actually like, then most people in the philosophy and neuroscience movement would cheerfully eliminate rather than try to reduce these concepts. Few are total eliminativists. Even the most radical people in the philosophy and neuroscience movement accept that *some* of the work of cognitive science will turn out to have enduring value.

Some philosophers of neuroscience explicitly advocate a mixture of the two. For instance, Paul and Patricia Churchland seem to hold that 'folk psychology' (our everyday ways of thinking and talking about ourselves as psychological beings) will mostly be eliminated, but many concepts of scientific psychology will be mapped onto, or 'reduced' to, concepts of neuroscience. For example, the Churchlands seem to hold that 'folk concepts' such as belief and desire don't name anything real but that scientific psychological concepts such as representation do (as long as we keep our notion of representation neutral with respect to various theories of what representations are), and that many kinds of representation will ultimately be found to be identical to some particular kind of neural state or process (P. S. Churchland 1986).

We cannot go into the merits of reductivist versus eliminativist claims, but notice that the truth of eliminativism will rest on at least two things:

1. The first is what the current candidates for elimination actually turn out to be like when we understand them better. For example, eliminativists about folk psychology often assume that folk psychology views representations as structured something like sentences and computations over representations as something very similar to logical inference (P. M. Churchland 1981; S. Stich 1983; P. S. Churchland 1986). Now, there are *explicit* theories that representation is like that. Fodor (1975), for example, defends the ideas that all thought is structured in a language – a language of thought.

But it is not clear that any notion of what representations are like is built into the *very folk concept* of representation. The picture of representation and computation held by most neuroscientists is very different from the notion that representations are structured like sentences, as we will see when we get to computation and representation, and so *if* the sententialist idea is built into folk psychology, then folk psychology is probably in trouble. But it is not clear that any such idea is built into folk psychology.

2. The second thing on which the truth of eliminativism will depend is what exactly reduction is like. This is a matter of some controversy (C. Hooker 1981; P. S. Churchland 1986). For example, can reductions be less than smooth, with some bits reduced, some bits eliminated, and still count as reductions? Or what if the theory to be reduced must first undergo some rejigging before it can be reduced? Can we expect theories dealing with units of very different size and complexity (as in representations in cognitive science, neurons in neuroscience) to be reduced to one another at all? And how much revision is tolerable before reduction fails and we have outright elimination and replacement on our hands? J. Bickle (1998) argues for a revisionary account of reduction. R. McCauley (2001) argues that reductions are usually between theories at roughly the same level (intratheoretic), not between theories dealing with radically different basic units (intertheoretic).

These big issues in the philosophy of neuroscience have been hashed and rehashed in the past 25 years. As we have seen, most people in the philosophy and neuroscience movement have arrived at roughly the same position on them. Thus, while they certainly form the background to current work, none of the contributions to this volume takes them up.

On many other topics, we are far from having a settled position. The chapters in this volume focus on these topics. Specifically, they contribute to the issues of

1. localization and modularity
2. role of introspection
3. three specific issues in the area of neural computation and representation:
   - the architecture, syntax, and semantics of neural representation
   - visuomotor transformation
   - color vision

and

4. consciousness.

We have grouped the contributions to the first two topics under the heading 'Data and Theory in Neuroscience'. Otherwise, the book examines these topics in the order just given.

Data and Theory: Localization, Modularity, and Introspection

Localization

A question with a long history in the study of the brain concerns how localized cognitive function is. Early localization theorists included the phrenologists Franz Gall and Johann Spurzheim. Pierre Flourens was a severe contemporary critic of the idea in the early 1800s.

Localizationism reemerged in the study of the linguistic deficits of aphasic patients of J. B. Bouillaud, Ernest Auburtin, Paul Broca, and Carl Wernicke in the mid-1800s. Broca noted a relation between speech production deficits and damage to the left cortical hemisphere, especially in the second and third frontal convolutions. Thus was 'Broca's area' born. It is considered to be a speech production locus in the brain. Less than two decades after Broca's work, Wernicke linked linguistic comprehension deficits with areas in the first and second convolutions in the temporal cortex now called 'Wernicke's area'.

The lesion/deficit method of inferring functional localization raises several questions of its own, especially for functions such as language for which there are no animal models (von Eckardt Klein 1978). Imaging technologies help alleviate some of the problems encountered by lesion/deficit methodology (for instance, the patient doesn't need to die before the data can be collected!). We mentioned two prominent imaging techniques earlier: positron emission tomography, or PET, and functional magnetic resonance imaging, or fMRI. Both have limitations, however. The best spatial resolution they can achieve is around 1 mm. A lot of neurons can reside in a 1 mm by 1 mm space! And there are real limitations on how short a time span they can measure, though these latter limitations vary from area to area and function to function. Especially in fMRI, resolution improves every year, however.

In PET, radionuclides possessing excessive protons are used to label water or sugar molecules that are then injected into the patient's bloodstream. Detectors arranged around the patient's head detect particles emitted in the process of the radioactive decay of the injected nuclides. PET thus allows the identification of areas high in blood flow and glucose utilization, which is believed to be correlated with the level of neural and glial cell activity (a crucial and largely untested, maybe untestable, assumption). PET has been used to obtain evidence of activity in the anterior cingulate cortex correlated with the executive control of attention, for example, and to measure activity in neural areas during linguistic tasks like reading and writing (D. Caplan et al. 1999). For a philosophical treatment of issues concerning PET, see R. Stufflebeam and Bechtel (1997).

fMRI measures the amount of oxygenation or phosphorylation in specific regions of neural tissue. Amounts of cell respiration and cell ATP utilization are taken to indicate the amount of neural activity. fMRI has been used to study the localization of linguistic functions, memory, executive and planning functions, consciousness, memory, and many, many other cognitive functions. Bechtel and Richardson (1993) and Bechtel and Mundale (1997) discuss some of the philosophical issues to do with localization.

In this volume, **Valerie Hardcastle** and **Matthew Stewart** present compelling evidence in 'Localization and the Brain and Other Illusions' that even a system as simple and biologically basic as oculomotor control is the very reverse of localized. To the contrary, it involves contributions from units dispersed widely across the cortex. They also show that a given nucleus can be involved in many different information-processing and control activities. They point out that the brain's plasticity – its capacity to recover function by using new areas when damage to an area affects function – holds the same implication. (They also make the point that these assays into how the brain actually does something undermine the claim that we can study cognitive function without studying the brain.)

Modularity

The question of localization connects to another big question in cognitive neuroscience, namely, modularity. Fodor (1983) advanced a strong modularity thesis concerning cognitive architecture. According to Fodor, a module is defined in terms of the following properties: (1) domain specificity, (2) mandatory operation, (3) limited output to central processing, (4) rapidity, (5) information encapsulation, (6) shallow outputs, (7) fixed neural architecture, (8) characteristic and specific breakdown patterns, and (9) characteristic pace and sequencing of development. He then argues that most of the brain's peripheral systems are modular, sometimes multimodular, while the big central system in which the thinking, remembering, and so on is done is emphatically not.

Fodor's account can be resisted in two ways. One is to argue that he has an overly restricted notion of what a module has to be like. The other is to argue that no matter how characterized, there are precious few if any modules in the brain. This is the conclusion that Hardcastle and Stewart's research supports. Another body of evidence supporting the same conclusion is back projection. For example, temporal cortical areas implicated in high levels of visual processing send back projections to lower-level areas in the primary visual cortex, which in turn send back projections to even lower areas in the lateral geniculate nuclei and ultimately back to the retina. I. Appelbaum (1998) argues for similar phenomena in speech perception: Higher-level lexical processing affects lower-level phonetic processing. In fact, neuroscientific research shows that back projections are to be found everywhere. But where there are back projections, there cannot be encapsulated modules.

Introspection

In a variety of ways, the advent of sophisticated imaging of brain activity has created a new reliance on introspections – self-reports about what is going on in oneself consciously. Introspection has been in bad odour as a research tool for more than 100 years. Researchers claim that introspective claims are unreliable in the sense that they are not regularly replicated in others. Subjects confabulate (make up stories) about what is going on in themselves when needed to make sense of behaviour. Introspection has access only to a tiny fraction of what is going on in oneself cognitively. And so on. Neuroscience researchers were forced back onto introspection because the only access as cognitive phenomena that they have to the things that they want to study in the activity of the brain is the access the subject him- or herself has. Many find it ironic that this most scientific of approaches to human nature has been forced to fall back onto a technique rejected as unscientific 100 years ago!

In this volume, two contributions focus on the role of introspection in neuroscience. **Evan Thompson**, **Antoine Lutz**, and **Diego Cosmelli** argue in 'Neurophenomenology: An Introduction for Neurophilosophers' that first-person reports about real-time subjective experience made after training and practise can play and should play a vital role in neuroscience, especially the neuroscience of consciousness. They call their approach neurophenomenology; their chapter provides a comprehensive overview of the approach and the context out of which it arose, together with two extremely interesting examples of neurophenomenology actually at work experimentally.

By way of contrast, in 'Out of the Mouths of Autistics: Subjective Report and Its Role in Cognitive Theorizing', **Victoria McGeer** urges that first-person utterances other than introspective reports can also be an important source of evidence. She studies autistics and their self-reports. Because of the extensive cognitive deficits found in autistic people, many hold that their first-person utterances are suspect. McGeer argues that behind this belief is an assumption that the job of first-person utterances is to describe what is going on in people's heads. If, instead, we treat self-reports as *expressions* of the underlying cognitive and affective systems, then we can see that there is a great deal to be learned from them. She then draws two implications from this analysis. At the philosophical level, she argues than once we abandon the model of first-person utterance as report, there will no longer be any conflict between the use of subjective report and the requirement that evidence be public. At the empirical level, she urges a radical rethinking of autism in terms of developmental connections between profound sensory abnormalities and the emergence of specific higher-order cognitive deficits.

Neural Computation and Representation

As we said, the topic of neural computation and representation is huge. Just the issue of neural representation alone is huge. Contributions to the latter topic can be thought of as falling into three groups, though the boundaries between them are far from crisp.

The neurophilosophical questions concerning computation and representation nearly all assume a definition of computation in terms of representation transformation. Thus, most questions concerning computation and representation are really questions concerning representation. There are three general kinds of question: questions to do with architecture, questions to do with syntax, and questions to do with semantics. The question of architecture is the question of how a neural system having syntax and semantics might be structured. The question of syntax is the question of what the formats or permissible formats of the representations in such a system might be and how representations interact with one another on the basis of their forms alone. The question of semantics is the question of how it is that such representations come to represent – how they come to have content, meaning.

Architecture of Neural Representation
In this volume, two papers are devoted to neural architecture, one on the issue in general, one on how time might be represented neurally.

In 'Moving Beyond Metaphors: Understanding the Mind for What It Is', **Chris Eliasmith** suggests that past approaches to understanding the mind, including symbolism, connectionism, and dynamicism, fundamentally rely on metaphors for their underlying theory of mind. He presents a new position that is not metaphorical and synthesizes the strengths of these past approaches. The new view unifies representational and dynamical descriptions of the mind, and so he calls it R&D Theory. In R&D Theory, representation is rigorously defined by encoding and decoding relations. The variables identified at higher levels can be considered state variables in control theoretical descriptions of neural dynamics. Given the generality of control theory and representation so defined, this approach is sufficiently powerful to unify descriptions of cognitive systems from the neural to the psychological levels. R&D Theory thus makes up for the absence of a proper dynamical characterization of cognitive systems characteristic of sententialism and shows how, contrary to dynamicist arguments (T. van Gelder, 1998), one can have both representation and dynamics in cognitive science.

Rick Grush, in 'Brain Time and Phenomenological Time', focusses on the structures in a neural system that represent time. His target is not objective time, actual persistence, but rather the subjective time of behaviour: the temporal representation that is analogous to egocentric space (in contrast to objective or allocentric space). There are two parts to his theory. The first concerns the neural construction of states that are temporally indexed (see also Grush 2002); mechanisms of emulation can be augmented to maintain a temporally 'thick' representation of the body and environment. The second part of the story concerns how these temporally indexed states acquire specifically temporal phenomenal content. Building on Grush (1998 and 2001) on spatial representation, the author urges that these temporally indexed states are imbued with 'egocentric' temporal content for an organism, to the extent that they cue and guide temporally relevant aspects of sensorimotor skills.

Neural Syntax

The standard way of interpreting synaptic events and neural activity patterns as representations is to see them as constituting points and trajectories in vector spaces. The computations that operate on these representations will then be seen as vector transformations (P. M. Churchland 1989). This is thus the view adopted in much neural network modelling (connectionism, parallel distributed processing). The system is construed as having network nodes (neurons) as its basic elements, and representations are states of activations in sets of one or more neurons. (Bechtel and A. A. Abrahamsen 2002; Andy Clark 1993).

Recently, work in neural modelling has started to become even more fine-grained. This new work does not treat the neuron as the basic computational unit, but instead utilizes compartmental modelling whereby activity in and interactions between patches of the neuron's membrane may be modelled (J. Bower and D. Beeman 1995). Thus, not only are networks of neurons viewed as performing vector transformations but so too are individual neurons.

Neural syntax consists of the study of the information-processing relationships among neural units, whatever one takes the relevant unit to be. Any worked-out story about the architecture of neural representation will hold implications for neural syntax, for what kind of relationships neural representations will have to other neural representations such that they can be combined and transformed computationally. Eliasmith and Grush see their stories as holding such implications.

Neural Semantics

Cognitive science and cognitive neuroscience are guided by the vision of information-processing systems. A crucial component of this vision is that states of the system carry information about or represent aspects of the external world (see A. Newell 1980). Thus, a central role is posited for intentionality, a representation being about something, mirroring Franz Brentano's (1874) insistence on its importance (he called it "the mark of the mental", only slightly an exaggeration) a century before.

How do neural states come to have contents? There are two broad answers to this question that have been popular in philosophy. They both appear in the philosophy of neuroscience: the functional role approach and the informational approach. Proponents of functional role semantics propose that the content of a representation, what it is about, is determined by the functional/causal relations it enters into with other representations (N. Block 1986). For informational approaches, a representation has content, is about something, in virtue of certain kinds of causal interactions with what it represents (F. Dretske 1981, 1988). In philosophy of neuroscience, Paul Churchland has subscribed to a functional roles semantics at least since 1979. His account is further fleshed out in terms of state-space semantics (P. M. Churchland 1989, 1995). However, certain aspects of Churchland's 1979 account of intentionality also mirror informational approaches.

The neurobiological paradigm for informational semantics is the feature detector – for example, the device in a frog that allows it to detect flies. J. Y. Lettvin et al. (1959) identified cells in the frog retina that responded maximally to small shapes moving across the visual field. Establishing that something has the function of detecting something is difficult. Mere covariation is often insufficient. Hubel and Wiesel (1962) identified receptive fields of neurons in striate cortex that are sensitive to edges. Did they discover edge detectors? S. R. Lehky and T. Sejnowski (1988) challenge the idea that they did, showing that neurons with similar receptive fields emerge in connectionist models of shape-from-shading networks. (See P. S. Churchland and Sejnowski 1992 for a review.) K. Akins (1996) offers a different challenge to informational semantics and the feature detection view of sensory function through a careful analysis of thermoperception. She argues that such systems are not representational at all.

As was true of neural syntax, any worked-out story about the architecture of neural representation will hold implications for neural semantics, for the question of how neural representations can come to have content or meaning, be about states of affairs beyond themselves. Eliasmith and Grush see their stories as holding implications for this issue, too.

In addition to these chapters, in 'The Puzzle of Temporal Experience', **Sean Kelly** tackles a very specific kind of representational content, the representation and conscious experience of temporality, and the neuroscience of same. He starts from Kant's famous distinction between a succession of independent representations and a representation of a single, unified, temporally extended object, the distinction between a succession of representations and a representation of succession. He shows how neither Specious Present Theory nor Kantian/Husserlian Retention Theory gives us a satisfying account of our experience of the passage of time. Since these are the only going accounts, how we manage to represent the passage of time is deeply puzzling. He then shows that there is a conceptual/phenomenological distinction to be made between the pure visual experience of motion and the visual experience of an object as moving, and that neuroscientists have tended to conflate the two. In response, he proposes experiments that could confirm a double dissociation between the two and recommends that we look for a certain kind of short-term visual storage that could serve as the basis for the experience of a single, unified, temporally extended object moving through space. This short-term visual storage should have a certain phenomenology associated with it, but in isolation will be unlike any kind of memory, including visual iconic memory, that we are familiar with. Since representation of objects as persisting through time is a completely general feature of representation, Kelly's paper identifies something that we have to understand if we are to understand neural representation. He concludes the chapter with a revealing overview of the current state of neuroscience on the issue.

Visuomotor Transformation

Grush's chapter and, less directly, Eliasmith's also connect to the first of the two more specific topics to do with neural representation examined in the book, visuomotor transformation, that is to say, the use of visual information to guide motor control.

In 'Grasping and Perceiving Objects', **Pierre Jacob** starts from the A. D. Milner and M. A. Goodale (1995) hypothesis that we have two complementary visual systems, vision-for-perception and vision-for-action, based on a double dissociation between two kinds of disorder found in brain-lesioned human patients: visual form agnosia and optic ataxia. Milner and Goodale claim that this functional distinction mirrors the anatomical distinction between the ventral pathway and the dorsal pathway in the visual system of primates. Using psychophysical experiments in normal human beings based on visual illusions, he makes some new discoveries about how visuomotor representation and conscious visual perception relate. He shows that these findings have major implications for theses advanced by others.

The chapter by **Pete Mandik**, 'Action-Oriented Representation', relates to Jacob's. Focussing on the claim that spatial perception and motor output are interdependent, Mandik asks how best to characterize this interdependence. There are two broad approaches. One favours the positing of mental representations mediating between perception and action; the other opposes the idea. He favours the former proposal, urging that sensorimotor interdependence is best accounted for by a novel theory of representational content whereby the most primitive forms of representation are those that have the function of commanding actions. Other representations, including sensory representations of spatial properties, depend on these motor command representations. His argument draws heavily on both neurophysiology and computer simulations and hold striking implications for both antirepresentationalism and competing representational accounts.

Color Vision

The second of our two more specific topics having to do with neural representation is color vision.

In 'Chimerical Colors', **Paul Churchland** presents a stunning example of neurophilosophy at work. He shows that by exploiting shifts in experienced color due to tiredness and habituation, experiences of color can be brought about where the colors do not exist in nature and, what is even more striking, could not exist in nature according to an extremely long-held and well-confirmed color theory. They are impossible. (Indeed, some of them cannot even be represented by a color sample, which makes the argument for these cases a bit difficult to present!) Churchland's chapter is also a nice example of neurophenomenology at work. (Whether he would accept this description of his method is, of course, another question.)

Focussing on perceived color similarity, in 'Opponent Processing, Linear Models, and the Veridicality of Color Perception', **Zoltán Jakab** argues against views of color experience that hold that the representational content of a color experience is exhausted by the color property in an object for which the experience stands. To the contrary, he argues, color experience arises from processing that distorts the stimulus features that are its canonical causes in numerous ways, thereby largely constructing our world of perceived color. Perceived color similarity is a systematic misrepresentation of the corresponding stimuli. From an evolutionary perspective, that distortion either is indifferent for the organism's fitness or may even serve its interests. However, in another crucial respect, color vision is veridical (or at least highly reliable): Where it indicates a salient surface difference, there indeed exists a physical surface difference of value to the organism.

An earlier version of Jakab's chapter won the prize for best paper by a graduate student at the 2002 Philosophy and Neuroscience Conference, where many of the chapters in this volume originated.

Consciousness

Most of the philosophical interest in consciousness started from the question of whether consciousness could possibly be a physical process, even a brain process. A common view in philosophy of neuroscience is that if there is anything appropriately given the name 'consciousness' (on this topic, however, there are few genuine eliminativists), it must be physical and, furthermore, explicable in terms of neurophysiology – no explanatory autonomy allowed.

Against this view, Nagel (1974) argued that because conscious experience is subjective, that is, directly accessible by only the person who has it, we are barred from ever understanding it fully, including whether and if so how it could be physical. For example, even if we knew all there is to know about bat brains, we would not know what it is like to be a bat because bat conscious experience would be so different from human conscious experience. Later, F. Jackson (1986), C. McGinn (1991), D. Chalmers (1996), and others extended this line of thought with new arguments and more sharply delineated conclusions.

Like many neurophilosophers, **Jesse Prinz** simply sidesteps this debate. Using recent neurobiology and cognitive psychology in 'A Neurofunctional Theory of Consciousness', Prinz argues that consciousness arises when mechanisms of attention allow intermediate-level perceptual systems to access working memory. He then supports this view by appeal to multiple other modalities, which suggests that consciousness has a uniform material basis. He then draws out the implications of his analysis for the traditional mind-body problem. Both leading current approaches, functionalism and identity theory or radical reductivism, fail to appreciate the extent, he argues, to which the solution to the mind-body problem may rely on multiple levels of analysis, with constitutive contributions at relatively abstract psychological levels and levels that are often dismissed as merely implementational.

In a similar spirit, Akins (1993a, 1993b) and others urge that, for example, an investigation of bat neurophysiology can in fact tell us a lot about what bat 'subjectivity' would be like. It would be boring and myopic. Mandik (2001) actually tries to say what subjectivity would be like neurophysiologically. It would consist in, or at least be built on, what neuroscientists call egocentric representations. Paul Churchland (1995) had tried the same thing earlier, explaining consciousness in terms of recurrent connections between thalamic nuclei (particularly 'diffusely projecting' nuclei like the intralaminar nuclei) and cortex. He showed how thalamocortical recurrency might account for certain features of consciousness, such as the effects of short-term memory on consciousness, and dreaming during REM. However, his proposal was a bit short on detail.

Many sceptics about the very possibility of a complete neuroscience of consciousness are unmoved by these analyses. They suspect that the researcher either has changed the topic to something that can be understood neuroscientifically or at best is coming up with an account merely of neural correlates of consciousness (NCCs), not consciousness itself (for an introduction to the issues here, see Thompson et al., this volume, section 8). One common way of arguing that consciousness cannot be anything neural or even physical is via the notion of what philosophers call *qualia:* the introspectible aspects of conscious experiences, what it is like to be conscious of something. Antiphysicalists argue that there could be beings who are behaviourally, cognitively, and even physically exactly like us, yet either have radically different conscious experience or no conscious experience at all. For example, they might see green where we see red (inverted spectrum) but, because of their training and so on, they use color words, react to colored objects, and even process information about color exactly as we do. Or, they have no conscious experience or color or anything else – they are zombies – yet use words for conscious experiences, react to experienced objects, and even process information about represented things exactly as we do.

One way to argue that representations can have functionality as representations without consciousness is to appeal to cases of blindsight and inattentional blindness. Due to damage to the visual cortex, blindsight patients have a scotoma, a 'blind spot', in part of their visual field. Ask them what they are seeing there and they will say 'Nothing'. However, if you ask them instead to *guess* what is there, they guess with far better than chance accuracy. If you ask them to reach out to touch whatever might be there, they reach out with their hands turned in the right way and fingers and thumb at the right distance apart to grasp anything that happens to be there. And so on (see L. Weiskrantz, 1986).

Inattentional blindness and related phenomena come in many different forms. In one form, a subject fixates (concentrates) on a point and is asked to note some feature of an object introduced on or within a few degrees of fixation. After a few trials, a second object is introduced, in the same region but usually not in exactly the same place. Subjects are not told that a second object will appear. When the appearance of the two objects is followed by 1.5 seconds of masking, at least one-quarter of the subjects and sometimes almost all subjects have no awareness of having seen the second object. (For more on this fascinating group of phenomena, see A. Mack, http://psyche.cs.monash.edu.au/v7/psyche-7-16-mack.htm or Mack and Rock 1998.)

There is a sense in which the inattentionally blind are not conscious of what they missed: They did not notice and cannot report on the item(s). However, their access to the missed items is extensive, much more extensive than the access that blindsight patients have to items represented in their scotoma. When the second object is a word, for example, subjects clearly encode it and process its meaning. Evidence? When asked shortly after to do, for example, a stem completion task (i.e., to complete a word of which they are given the first syllable or so), they complete the word in line with the word they claim not to have seen much more frequently than controls do. Thus, subjects' access to words that they miss involves the processing of semantic information. If so, their access to the missed words is much like our access to items in our world when we are conscious of them but not conscious of being thus conscious. Thus, far from inattentional blindness suggesting that representations can have full functionality without consciousness, the phenomenon tends to pull in the opposite direction. It is at least fully compatible with the idea that sufficient representational complexity *just is* a form of consciousness.

So, what should we say of the claim that at least some element of some kind of consciousness is not neural or even physical at all? Philosophers who care about neuroscience tend to have three kinds of reaction to this claim:

1. They just ignore the claim (most cognitive scientists),

or

2. They throw science at it and attempt implicitly or explicitly to produce the kind of account that is supposed to be impossible (Dennett 1978; C. L. Hardin 1988; Austen Clark 1993; Akins 1993a, 1993b, 1996; Hardcastle 1997),

or

3. They try to show that the claim is wrong (or incoherent, or in some other way epistemically troubled) (Dennett 1991, 1995; M. Tye 1993; Brook and P. Raymont, forthcoming).

In 'Making Consciousness Safe for Neuroscience', **Andrew Brook** urges that neither (1) nor (2) is the best course of action. Ignoring the antiphysicalist claim or throwing science at it will leave many – and not just dyed-in-the-wool antiphysicalists – feeling that the real thing, consciousness itself, has been left out, that the researcher has covertly changed the subject and is talking about something else, not consciousness. This is how many react, for example, to suggestions that consciousness is some form of synchronized firing of neurons. 'Surely,' they react, 'you could have the synchronized firing without consciousness. If so, consciousness is not synchronized firing of neurons. Maybe this firing pattern is a *neural correlate* of consciousness, but it is not *what consciousness is*. Ignoring this reaction, Brook argues, is a bad idea – it is not going to fade away on its own.

Moreover, neuroscience is not going to be relevant. The most effective appeal to neuroscience in this context is probably the kind of appeal mounted in Thompson et al., this volume. Because neurophenomenology puts first-person conscious experience front and centre, it does not even have the appearance of leaving consciousness out, changing the topic. However, even such consciousness-centred work can still be accused of studying mere correlates, of not telling us anything about the nature of *consciousness*. According to Brook, what we need to do instead is to tackle head-on the urge to split consciousness off from cognition and the brain and the antiphysicalist arguments that aim to support the urge, to show that attempts to split consciousness off from cognition and the brain do not succeed.

The aim of this volume is to provide a representative, fairly comprehensive snapshot of the work currently going on in the philosophy and neuroscience movement.

References

Akins, K. (1993a) What Is It Like to Be Boring and Myopic? In B. Dahlbom, ed. *Dennett and His Critics*. New York: Basil Blackwell.

Akins, K. (1993b) A Bat Without Qualities. In M. Davies and G. Humphreys, eds. *Consciousness: Psychological and Philosophical Essays*. New York: Basil Blackwell.

Akins, K. (1996) Of Sensory Systems and the 'Aboutness' of Mental States. *Journal of Philosophy* 93, 337–372.

Appelbaum, I. (1998) Fodor, Modularity, and Speech Perception. *Philosophical Psychology* 11, 317–330.

Bechtel, W., and Abrahamsen, A. A. (2002) *Connectionism and the Mind: Parallel Processing, Dynamics, and Evolution in Networks*. Oxford: Blackwell.

Bechtel, W., and Mundale, J. (1997) Multiple Realizability Revisited. *Proceedings of the Australian Cognitive Science Society*.

Bechtel, W., and Richardson, R. C. (1993) *Discovering Complexity: Decomposition and Localization as Scientific Research Strategies*. Princeton, NJ: Princeton University Press.

Bickle, J. (1998) *Psychoneural Reduction: The New Wave*. Cambridge, MA: MIT Press.

Biro, J. (1991) Consciousness and Subjectivity in Consciousness. In E. Villaneuva, ed. *Philosophical Issues*. Atascadero, CA: Ridgeview.

Bliss, T. V. P., and Lomo, T. (1973) Long-Lasting Potentiation of Synaptic Transmission in the Dentate Area of the Anaesthetized Rabbit Following Stimulation of the Perforant Path. *Journal of Physiology (London)* 232 (2): 331–356.

Block, N. (1986) Advertisement for a Semantics for Psychology. In P. A. French, ed. *Midwest Studies in Philosophy* X: 615–678.

Bower, J., and Beeman, D. (1995) *The Book of GENESIS*. New York: Springer-Verlag.

Brentano, F. (1874) *Psychology from an Empirical Standpoint*. A. C. Pancurello, D. B. Tyrrell, and L. L. McAlister, trans. New York: Humanities.

Brook, A., and Raymont, P. (forthcoming) *A Unified Theory of Consciousness*. Cambridge, MA: MIT Press.

Brook, A., and Stainton, R. (2000) *Knowledge and Mind*. Cambridge, MA: MIT Press/A Bradford Book.

Caplan, D., Carr, T., Gould, J., and Martin, R. (1999) Language and Communication. In Zigmond et al.

Chalmers, D. (1996) *The Conscious Mind*. Oxford: Oxford University Press.

Churchland, P. M. (1979) *Scientific Realism and the Plasticity of Mind*. Cambridge: Cambridge University Press.

Churchland, P. M. (1981) Eliminative Materialism and the Propositional Attitudes. *Journal of Philosophy* 78, 67–90.

Churchland, P. M. (1989) *A Neurocomputational Perspective: The Nature of Mind and the Structure of Science*. Cambridge, MA: MIT Press.

Churchland, P. M. (1995) *The Engine of Reason, The Seat of the Soul*. Cambridge, MA: MIT Press.

Churchland, P. S. (1986) *Neurophilosophy*. Cambridge, MA: MIT Press.

Churchland, P. S., and Sejnowski, T. (1992) *The Computational Brain*. Cambridge, MA: MIT Press.

Clark, Andy (1993) *Associative Engines*. Cambridge, MA: MIT Press.

Clark, Austen (1993) *Sensory Qualities*. Cambridge: Cambridge University Press.

Dennett, D. C. (1978) Why You Can't Make a Computer That Feels Pain. In his *Brainstorms*. Montgomery, VT: Bradford Books.

Dennett, D. C. (1991) *Consciousness Explained*. New York: Little Brown.

Dennett, D. C. (1995) The Path Not Taken. *Behavioral and Brain Sciences* 18 (2): 252–253.

Descartes, R. (1649) *Les passions de l'âme*, Amsterdam: Lodewijk Elsevier, and Paris: Henry le Gras. In Adam, C. and Tannery, P. (1964–1974) *Œuvres de Descartes*. Paris: J. Vrin, vol. XI.

Dretske, F. (1981) *Knowledge and the Flow of Information*. Cambridge, MA: MIT Press.

Dretske, F. (1988) *Explaining Behavior*. Cambridge, MA: MIT Press.

Feigl, H. (1958/1967) *The "Mental" and the "Physical": The Essay and a Postscript*. Minneapolis: University of Minnesota Press.

Fodor, J. A. (1974) Special Sciences (or: The Disunity of Science as a Working Hypothesis). *Synthese* 28, 97–115.

Fodor, J. A. (1975) *The Language of Thought*. New York: Crowell.

Fodor, J. A. (1983) *The Modularity of Mind*. Cambridge, MA: MIT Press/Bradford Books.

Gold, I., and Stoljar, D. (1999) A Neuron Doctrine in the Philosophy of Neuroscience. *Behavioral and Brain Sciences* 22 (5), 809–830.

Grush, R. (1998) Skill and Spatial Content. *Electronic Journal of Analytic Philosophy* 6 (6), http://ejap.louisiana.edu/EJAP/1998/grusharticle98.html.

Grush, R. (2001) The Semantic Challenge to Computational Neuroscience. In Peter Machamer, Rick Grush, and Peter McLaughlin, eds. *Theory and Method in the Neurosciences*. Pittsburgh, PA: University of Pittsburgh Press.

Grush, R. (2002) Cognitive Science. In Peter Machamer and Michael Silberstein, eds. *The Blackwell Guide to the Philosophy of Science*. London: Blackwell.

Hardcastle, V. G. (1997) When a Pain Is Not. *Journal of Philosophy* 94 (8): 381–406.

Hardin, C. L. (1988) *Colour for Philosophers*. Indianapolis: Hackett.

Hebb, D. (1949) *The Organization of Behavior*. New York: Wiley.

Hooker, C. (1981) Towards a General Theory of Reduction. Part I: Historical and Scientific Setting. Part II: Identity in Reduction. Part III: Cross-Categorial Reduction. *Dialogue* 20: 38–59.

Hubel, D., and Wiesel, T. (1962) Receptive Fields, Binocular Interaction and Functional Architecture In the Cat's Visual Cortex. *Journal of Physiology (London)* 160, 106–154.

Jackson, F. (1986) What Mary Didn't Know. *Journal of Philosophy* 83 (5), 291–295.

Kandel, E. (1976) *Cellular Basis of Behavior*. San Francisco: W. H. Freeman.

Lehky, S. R., and Sejnowski, T. (1988) Network Model of Shape-from-Shading: Neural Function Arises from Both Receptive and Projective Fields. *Nature* 333, 452–454.

Lettvin, J. Y., Maturana, H. R., McCulloch, W. S., and Pitts, W. H. (1959) What the Frog's Eye Tells the Frog's Brain. *Proceedings of the IRF*.

Machamer, P., Darden, L., and Craver, C. (2000) Thinking About Mechanisms. *Philosophy of Science* 67, 1–25.

Mack, A., and Irvin Rock. Inattentional Blindness. *PSYCHE* 5(3), May 1999, http://psyche.cs.monash.edu.au/v5/psyche-5-03-mack.html.

Mack, A., and Rock, I. (1998) *Inattentional Blindness*. Cambridge, MA: MIT Press.

Mandik, P. (2001) Mental Representation and the Subjectivity of Consciousness. *Philosophical Psychology* 14 (2): 179–202.

Mandik, P., and Bechtel, W. (2002) Philosophy of Science. In Lynn Nadel, ed. *The Encyclopaedia of Cognitive Science*. London: Macmillan.

McCauley, R. (2001) Explanatory Pluralism and the Co-evolution of Theories of Science. In W. Bechtel, P. Mandik, J. Mundale, and R. S. Stufflebeam, eds. *Philosophy and the Neurosciences: A Reader*. Oxford: Basil Blackwell.

McGinn, C. (1991) *The Problem of Consciousness: Essays Towards a Resolution*. Oxford: Basil Blackwell.

Milner, A. D., and Goodale, M. A. (1995) *The Visual Brain in Action*. Oxford: Oxford University Press.

Mundale, J., and Bechtel, W. (1996) Integrating Neuroscience, Psychology, and Evolutionary Biology Through a Teleological Conception of Function. *Minds and Machines* 6, 481–505.

Nagel, T. (1971) Brain Bisection and the Unity of Consciousness. *Synthèse* 22, 396–413.

Nagel, T. (1974) What Is It Like to Be a Bat? *Philosophical Review* LXXXIII, 4 (October): 435–450.

Newell, A. (1980) Physical Symbol Systems. *Cognitive Science* 4, 135–183.

Place, U. T. (1956) Is Consciousness a Brain Process? *The British Journal of Psychology* XVII (1), 44–50.

Putnam, H. (1967) Psychological Predicates. In W. H. Capitan and D. D. Merrill, eds. *Art, Mind and Religion.* Pittsburgh, PA: University of Pittsburgh Press.

Smart, J. J. C. (1959) Sensations and Brain Processes. *Philosophical Review* 68, 141–156.

Stich, S. (1983) *From Folk Psychology to Cognitive Science.* Cambridge, MA: MIT Press.

Stufflebeam, R., and Bechtel, W. (1997) PET: Exploring the Myth and the Method. *Philosophy of Science (Supplement)* 64 (4): S95–S106.

Tye, M. (1993) Blindsight, the Absent Qualia Hypothesis, and the Mystery of Consciousness. In C. Hookway, ed. *Philosophy and the Cognitive Sciences.* Cambridge: Cambridge University Press.

van Gelder, T. (1998) Mind as Motion: Explorations in the Dynamics of Cognition. *Journal of Consciousness Studies* 5 (3): 381–383.

von Eckardt Klein, B. (1975) Some Consequences of Knowing Everything (Essential) There Is to Know About One's Mental States. *Review of Metaphysics* 29, 3–18.

von Eckardt Klein, B. (1978) Inferring Functional Localization from Neurological Evidence. In E. Walker, ed. *Explorations in the Biology of Language.* Cambridge, MA: MIT Press.

Weiskrantz, L. (1986) *Blindsight: A Case Study and Implications.* Oxford: Clarendon Press.

Zigmond, M., Bloom, F., Landis, S., Roberts, J., and Squire, L., eds. (1999) *Fundamental Neuroscience.* San Diego, CA: Academic Press.

DATA AND THEORY IN NEUROSCIENCE

1

Localization in the Brain and Other Illusions

Valerie Gray Hardcastle and C. Matthew Stewart

We are all probably too aware of the latest push in neuroscience to localize brain and cognitive function. Headlines scream with claims of scientists locating the morality center of the brain or the home for violence and aggression. It seems that we have taken phrenology inside the head and are now feeling our way around brain bumps in search of the exact location of any and all thought.

We trace this trend to three independent moves in academia. First, the recent and very popular emphasis on evolutionary psychology and its concomitant emphasis on modularity fuel the need to find the modules of thought that Mother Nature supposedly crafted during the Pleistocene. Second, there has been a renewed interest in philosophy of science and elsewhere in mechanistic explanations. We are told at every turn that neuroscience explanations are explanations of mechanism. How else to isolate mechanisms except through localization studies? Third and finally, the fairly recent emergence of cognitive neuroscience and fMRI studies as the methodological exemplars for cognitive science push us to see the brain as a static machine that only instantiates what psychologists tell us it does. Further, almost no one does just straight cognitive psychology any more – you have to connect your psychological experiments to some sort of imaging study these days. This approach, with its infamous subtraction method, can only lead to increased putative localization of putative cognitive functions.

By now, just about everyone interested should be familiar with the methodological problems inherent in fMRI or other imaging techniques as a way to find the place where some process or other occurs in the brain. Indeed, we have contributed to that whining ourselves (Hardcastle and Stewart 2002). In this chapter, we are going to set those concerns aside and instead turn our attention to a related but larger theoretical problem concerning localization as an explanatory goal. We shall argue by example that brain plasticity and concomitant multifunctionality belie any serious hope of localizing functions to specific channels or areas or even modalities.

We should also by now already be familiar with how cognitive neuroscientific explanations are supposed to work, at least in a cartoon fashion. Psychologists and other cognitive theoreticians are supposed to issue learned statements detailing the steps involved in some cognitive process or other. These steps have been isolated through countless variations on some reaction-time or error-based protocol. Cognitive neuroscientists then take these coarse-grained statements and use them to design imaging studies that essentially repeat the countless variations of the reaction time or error-based protocols. With these imaging studies, we can in principle isolate the regions in the brain that contribute, underlie, instantiate, or are otherwise related to the cognitive steps. Without doubt, the imaging studies will uncover more steps, or different steps, or new steps, which will feed back into the cognitive studies as we all refine what we believe happens in the head when we think. Ideally, once we get a good story about the causal chain that leads to whatever outputs we are interested in, we can then try to connect it to single cell and lesion studies in other animals that would further isolate and refine our localized functional modules. It is this last step in the explanatory process that we want to focus upon – what it is that single cell and lesion recordings might actually be telling us about the viability of localization as a methodological principal.

In other words, we want to take seriously Patricia Churchland's notion of reductive coevolution (Churchland 1986) and see what lesion studies can tell us about the principles behind cognitive function. We should mention that we believe that Churchland was wildly optimistic in the mid-1980s when she spoke about reductive coevolution as the way the mind/brain sciences actually operate. Instead, we think she should have offered this vision as a normative view of what we should be doing in cognitive science. For what we actually find today are localization studies that antecedently take cognitive psychology as theoretically primary, and therefore, the engine that drives much experimentation in cognitive neuroscience. However, theoretical information must flow in both directions, and one ignores what neuroscience has to say in favor of psychology's message at one's peril.

1 The Eyes Have It

Given all the work one must do to get good data out of single cell recordings or controlled lesions studies (cf. Hardcastle and Stewart 2003), it is important to choose what you are going to study very carefully. You want a system that is simple enough in structure so that you have a fairly good idea what you are going to be recording from, yet you want its functional contributions complex enough that with your recordings, you will be able to learn something about brain function in general. In other words, you want to study something known, but not too well known.

The oculomotor system is a good model system to study because it is quite simple structurally, but quite interesting functionally (cf. Robinson 1981). We can think of the eyeball as rotating around a fixed point, so that there is only one "joint" in the motor control system. Only two muscles rotate the eye in any one plane. Ocular muscles are straight parallel fibers such that the force of each fiber is applied directly to the eyeball; they are also reciprocally innervated and usually do not actively co-contract. Tendons wrap around the globe so that the moment arm of the muscles does not depend on eye position. Since the eyeball is not used to apply forces to external loads as with most other muscles, much of the circuitry required by other muscle systems (the stretch reflex, for example) to deal with a variety of changes in loads is absent or rudimentary.

In addition, the system's function is conceptually easy to understand. We know what the oculomotor system does and why. It is comprised of three major subsystems in afoveate, lateral-eyed animals: the saccadic system, the vestibulo-ocular reflex system, and the optokinetic system. The saccadic system reorients eyes quickly in space. The next two systems prevent images from moving on the retina when an animal's head (or body) turns. The vestibulo-ocular reflex (VOR) and optokinetic responses (OKRs) work together to optimize image stabilization on the retina during object and subject motion. At the low to very low frequency range of head motion, VOR performance contributes less and less to eye movements. During normal (physiologic) locomotion, the VOR dominates eye movement generation due to shorter latencies for the two pathways (for these animals, about 20 msec for VOR and 60–80 msec for OKR).

Foveate animals, like humans, have two additional oculomotor systems. We also have a vergence system that puts images of targets located at various distances on the fovea simultaneously for binocular vision, and (especially in primates) a smooth pursuit system to track moving targets with smooth eye movements and to keep their images relatively still on the fovea.

We currently know enough about each subsystem network and consequent stereotyped motor behavior that we can specify their functions mathematically. This frees neuroscientists to investigate *how* the networks are actually doing what we know them to be doing. For most other motor control systems, we don't know precisely what function the system is supposed to be performing, and so any models we develop are much more speculative.

2 Vestibular Compensation

We will focus on the VOR. This reflex subsystem is one system we have that provides oculomotor command signals to stabilize retinal images during head movements. For example, if you stare at this word – TARGET – while shaking your head back and forth, your eyes move equal and opposite to the angular motion of your head so that the image of "TARGET" is stabilized on your retina, even though both the globe of your eye and your head are moving.

The profound scientific interest in the vestibular system in general, and in hemilabyrinthectomy (HL) – which is a lesion of the labyrinthine structure in the ear – in particular, is driven by the unexpected degree of change after perturbations. In a normal animal or person, very slight disturbances of the vestibular system lead to overwhelming, incapacitating dysfunction. In an HL animal or patient, the severity of the vestibular dysfunction overwhelms all homeostatic mechanisms for posture, gaze, and control of locomotion.

These effects were first noticed in 1824 when Pierre Flourens published a treatise describing the effects of manipulating and destroying different parts of the vestibular semicircular canals in pigeons. He described the gross motor symptoms of surviving animals, even asking the commissioners of the Royal Academy of Science to observe certain behaviors prior to sacrifice (Flourens 1824). From his point of view, the severity of responses was completely unexpected given the microscopic size of the canal and the miniscule deflections performed by probing the components within the vestibule, and, we think, that reaction continues today.

Equally remarkable, though, is the degree of recovery in HL subjects after the immediate postoperative period. After being given time to recover, many animals and humans appear to have normal locomotion.

Patients with HL surgery can even operate motor vehicles, a task that forces active and passive head movements in a visual environment characterized by huge differences in motion between the fixed roadway, other moving vehicles traveling at a variety of speeds, and the inside of the patient's vehicle, which is also moving. (It is important to note, however, that many species of animals do have permanent observable motor changes, such as head tilt, following HL surgery. Sensitive experiments can also find such changes in humans (Hain et al. 1987; Takahashi et al. 1990).)

The challenge is to account for the recovery of dynamic function after HL. What are the mechanisms behind such profound recoveries? Since labyrinthine structures do not regenerate, and peripheral neurons continue to fire abnormally, whatever the brain is doing to recover has to be a central effect (Schaefer and Meyer 1973). Single neuron recordings from a variety of animals indicate that the vestibular nucleus (VN) on the same side of the brain as the lesion recovers a partial degree of normal function as the brain learns to compensate for its injury.

Moreover, scientists also believe that whatever the mechanism is, it is likely to be a general procedure the brain uses for recovery, for we find similar resting rate recoveries of the sort we see with the ipsilateral vestibular nuclei following denervation in the lateral cuneate nucleus, the trigeminal nucleus, and the dorsal horn, among other areas (Loeser et al. 1968; Kjerulf et al. 1973). Neuroscientists argue that the phenomenon of vestibular compensation should serve as a model for studying brain plasticity in general. For example, F. A. Miles and S. G. Lisberger conclude that "the VOR is a particularly well-defined example of a plastic system and promises to be a most useful model for studying the cellular mechanisms underlying memory and learning in the central nervous system" (1981, p. 296). In their review of vestibular compensation, P. F. Smith and I. S. Curthoys concur that "the recovery of resting activity in the ipsi VN following UL is an expression of a general CNS process which functions to offset the long-term changes in tonic synaptic input which would otherwise be caused by denervation" (1989, p. 174; see also Galiana et al. 1984).

3 Models of Compensation

M. E. Goldberger (1980) gives a general framework for modeling motor recovery after injury (see also Xerri et al. 1988; Xerri and Zennou 1988; Berthoz 1988). There are three categories of processes in this framework: *sensory substitution,* in which different sensory receptors trigger the same behavior as occurred before the operation or injury; *functional substitution,* in which the precise neural pathways differ, but the same neural subsystem is used as before; and *behavioral substitution,* in which the organism devises an entirely new behavior that was not originally part of its normal repertoire. Which framework is correct turns out to have major implications for how to understand both brain plasticity and attempts to localize brain function.

A process of sensory substitution for HL recovery would mean that the motor outputs in question remain the same after the surgery, but a different set of sensory signals or systems evokes and controls them. In this case, the intact vestibular system would be considered intra sensory, with its signals including both somatosensory afferent information and special sensory afferent information given by some other system.

Functional substitution processes involve altering the neural elements within a given system. With respect to HL, what counts as "the system" is difficult to define. For example, foveating on a moving target by the moving eyes of a moving subject uses at least smooth pursuit, vestibular-ocular reflexes, vestibular-colic reflexes, cortical and subcortical motion processing, and cerebellar activity related to locotion, eye movement, and vestibular input and output. It becomes problematic, to say the least, to characterize these elements as a single system. At the same time, it is clear that vestibular, somatosensory, and visual inputs are all required for normal functioning, and so most descriptions of "sensory" substitution are actually a sort of functional substitution, if we understand what a system is in a suitably broad way. Authors end up blurring the distinction between functional and sensory substitution by using such terms as "multi-modal substitution" (Xerri et al. 1988; Zennou-Azogui et al. 1994). We will use the term "sensory/function substitution" to reflect their fundamental connection, if not actual identity.

Behavioral substitution entails that a different motor system approximate the original motor output in an attempt to restore a behavior. For example, short-term adaptations to reversing prisms leads to saccadic eye movements instead of our normal ocular reflexes (Jones et al. 1988). However, this is an example of a normal subject with an intact visual system adapting to novel conditions. It doesn't tell us whether other modalities help HL patients recover vestibular function.

Some experiments have shown that, remarkably enough, *vision* is of central and critical importance after HL surgery in foveate animals, such as the cat (Courjon et al. 1977; Putkonen et al. 1977; Zennou-Azogui et al. 1994) and primates (Lacour and Xerri 1980; Lacour et al. 1981; Fetter et al. 1988), and recently in the afoveate chicken (Goode et al. 2001). The eyes literally help the ears do their work. These experiments, which combined visual and vestibular lesions, tell us that the visual system plays some role in either behavioral or sensory/functional substitution in VOR recovery.

This conclusion makes sense if the same neurons in the vestibular nuclei are used in recovery, for they must be getting orientation information from somewhere other than the (now ablated) semicircular canals. Since recovery occurs too quickly for the growth of new connections to explain how our brains are compensating for their loss, then some other sensory system must already be feeding into the vestibular system. One possibility is the visual system. Perhaps as animals try to orient toward targets, error signals from the retina help the vestibular system compute head location. If this be the case, then disrupting visual processing in the brain stem should also disrupt vestibular compensation.

Matthew Stewart and his colleagues have lesioned the first relay in horizontal field motion processing in the brain stem, the pretectal nucleus of the optic tract (NOT), in monkeys trained to fixate on visual targets under a variety of conditions (Stewart et al. 1999; Stewart 2002). They performed bilateral, ibotenic acid lesions of the NOT in two rhesus monkeys that subsequently received the HL surgery. These combined lesions studies, when set against the results of a rhesus monkey receiving HL alone, would test the hypothesis that the horizontal visual field motion information provided by the NOT is *necessary* for the development of vestibular compensation. In short, they found that, using HVOR as an indicator of vestibular compensation, the combined NOT-HL lesion did in fact lead to a deficit in vestibular compensation when compared to the HL surgery alone.

Both sensory/functional substitution and behavioral substitution remain viable explanations for HL recovery. As you can see in the Figure 1.1 schematic, there is a direct anatomical connection from the NOT to the VNC. At the same time, examination of postrotatory gain recovery shows that there is a difference in processing of vestibular information from the intact labyrinth. In other words, both inter- and intrasensory signals underwrite vestibular compensation, which supports both substitution models.

Let us consider the proposed mechanisms for behavioral substitution (Berthoz 1988), namely, that vision merely provides a static error signal about the position of the head in space, allowing for central connections

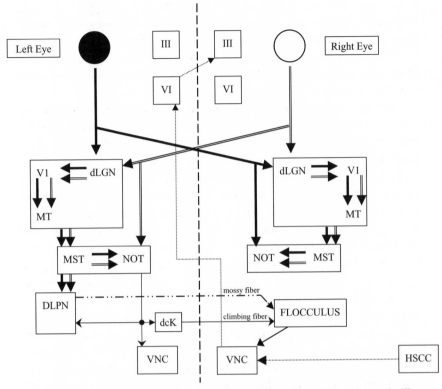

FIGURE 1.1. Basic model of visual-vestibular interaction. This schematic illus-trates the relevant anatomic connections and interactions between the visual and vestibular systems relating to horizontal background motion, horizontal angular acceleration, and horizontal eye movements evoked by retinal (OKR and OKN) or vestibular (VOR) stimulation. The left side of the figure illustrates the projections of the direct and indirect OKN pathways, and three projections of the left nucleus of the optic tract (NOT): to the dorsal cap of KOOY (dcK), whose output provides climbing fiber input into the flocculus; to the dorsolateral pontine nucleus (DLPN), whose output includes mossy fiber input into the flocculus; and to the vestibular nuclear complex (VNC), especially the medial vestibular nucleus and nucleus prepositus hypoglossi. The right side of the figure illustrates the projections of the brainstem VOR pathways: primary afferent input for the horizontal semicircular canal (HSCC); VNC output to the contralateral abducens nucleus (VI); and an interneuron projection from VI to the oculomotor nucleus (III). The direct pathway includes a set of retinal afferent connections: the lateral genicular nucleus, dorsolaterally (dLGN) to striate cortex (v1) to the middle temporal area (MT). This region of cerebral cortex has efferent connections from the middle superior temporal area (MST) to the NOT. The NOT also receives retinal afferent input, predominantly from the contralateral eye, and input from the contralateral NOT, not shown. The direct pathway

to gaze motor outputs. Goode et al. (2001) have shown that chicks reared in stroboscopic light develop abnormal HVOR gain responses. That is, rearing a chick in stroboscopic light causes visual-vestibular deficits. Moreover, chicks recovering from ototoxic doses of streptomycin in stroboscopic light continue to present with behavioral difficulties, even though they would receive head orientation information under these conditions.

In addition, the behavioral substitution hypothesis predicts that there should be no differences in the monkeys' recovery in Stewart's study. All three animals should have shown equal recovery if "vision could be used only to provide a spatial reference (visual straight ahead, or visual vertical) to the gaze system but may not directly be instrumental for compensation" (Berthoz 1988), since all three animals have access to such visual information. The behavioral substitution hypothesis is now untenable. We conclude that vestibular compensation is a product of sensory/functional substitution.

4 Brain Plasticity and Localization

To put matters more informally, these results show that both vestibular and visual information feed into the same nucleus, and this nucleus can use the inputs from either modality to calculate head location. Moreover and more importantly, the visual inputs remain "hidden" until a lesion or other sort of brain damage reveals them. If we simply looked at what a normal brain was doing under normal conditions, we would not see any extravestibular contributions. If neuroscientists are right and vestibular recovery gives us a general model for brain plasticity, and if these results we report are to be believed, then we would expect to find such intrasensory connections everywhere (or at least in lots of places).

Indeed, suspecting that our brain systems overlap dovetails with studies of monkey cortex. When the median nerve of the hand is severed in adult owl or squirrel monkeys, areas 3b and 1 in somatosensory cortex (areas

FIGURE 1.1. *(continued)* includes MST projections to DLPN. The indirect pathway modifies eye movements by the retinal-NOT-flocculus-VNC-VI-III pathway. The indirect pathway may be activated under experimental conditions by sustained OKN stimulation, a rare event with normal subjects in the natural environment. The direct pathway, responsible for the rapid aspects of OKN, is likely in generating a retinal-visual world mismatch through sensing natural motions of the head or the environment and then evoking OKR. *Source:* Adapted from Mustari et al. 1994; Tusa et al. 2001.

which normally respond to medial nerve stimulation) begin responding to inputs from other nerves in the hand (Merzenich et al. 1983; see also Clark et al. 1988). M. M. Merzenich argues that given the rapidity of the response, silencing the radial nerve inputs "unmasks" secondary inputs from other afferent nerves. At first, areas 3b and 1 only crudely represented the dorsal surfaces of the hand. Over time, they transformed into highly topographic representations. A month after surgery, the areas that used to respond to the median nerve now completely responded to the alternative inputs in standard hypercolumn form. Apparently, using mechanisms still unknown, the brain capitalizes on previous secondary connections to compensate for lost inputs.

So what's our point here? Single cell and lesion studies of brain compensation paint a very different general picture of brain structure and function than what we get with the plethora of fMRI studies floating about. First, brain computational practices are plastic, but only plastic up to a point. Brains try to work with the connections they already have, but many of these connections do not reveal themselves until insult or injury occurs. Hence, the common practice in cognitive neuroscience of taking snapshots of brain functioning cannot reveal the true complexity of what is going on computationally. At best, it will only reveal dominant functionality, and then, because of the gross resolution of most imagining techniques, only in a highly simplified form.

Second, searching for *the* function of any area is a fool's errand. The same area could be doing different things, depending on what else is happening in the rest of the brain. Or, perhaps more accurately, the brain might emphasize or privilege one process over another in the same area, depending on circumstance. So, what was visual is now vestibular. It doesn't make sense to ask, as many cognitive neuroscientists are prone to do, What does this area *really* do? For the answer will always be that it depends. Note that we can't even ask what the modality of a particular area is, for that too will depend on circumstances.

Third, one must be wary of reading cognitive function straightforwardly off lesion experiments. When you or Mother Nature ablates an area, it isn't the case that you now have the same system with one of its components simply removed, a Tonka-toy model of the brain. The brain always struggles to compensate for its losses, and change is always occurring. Uncovering exactly how the brain cognizes is going to be much more difficult and much more subtle than we ever anticipated.

Cognitive psychology has a lot to offer the mind/brain sciences, to be sure. However, we don't take its theoretical dominance in cognitive neuroscience always to be a good thing. Neuroscience has more to offer than sorting out the instantiation of cognitive function. If we pay attention, it can give us new ways of thinking about how the mind/brain operates; it can give us a new theoretical paradigm, if you will.

And if we don't pay attention to what neuroscience is trying to tell us, then we will continue to buy into psychology's assumptions that investigating the brain won't tell us anything fundamentally new about how we think, that the brain is essentially an unchanging machine that merely underlies our cognitive capacities, and that our research project is to continue to refine our stepwise computational model of thought. All of these assumptions, we think, are wrong. Neuroscience is more than brain phrenology, and cognitive psychology is less than theoretical neuroscience.

Acknowledgments

This chapter was first presented at the Philosophy and Neuroscience Conference hosted by Carleton University. We are very grateful for the many thoughtful comments given by the audience. This research was supported in part by a grant from the McDonnell Project in Philosophy and Neuroscience.

References

Berthoz, A. 1988. The role of gaze in compensation of vestibular dysfunction: The gaze substitution hypothesis. *Progress in Brain Research* 7: 411–420.

Churchland, P. S. 1986. *Neurophilosophy.* Cambridge, MA: MIT Press.

Clark, S. A., Allard, T., Jenkins, W. M., and Merzenich, M. M. 1988. Receptive fields in the body-surface map in adult cortex defined by temporally correlated inputs. *Nature* 332: 444–445.

Courjon, J. H., Jeannerod, M., Ossuzio, I., and Schmid, R. 1977. The role of vision in compensation of vestibulo-ocular reflex after hemilabyrinthectomy in the cat. *Experimental Brain Research* 8: 235–248.

Fetter, M., Zee, D. S., and Proctor, L. R. 1988. Effect of lack of vision and of occipital lobectomy upon recovery from unilateral labyrinthectomy in rhesus monkey. *Journal of Neurophysiology* 59: 394–407.

Flourens, P. 1824. Recherches experimentales sur les propriétés et functions du système nerveux dans les animaux vertebras. Paris: Balliere.

Galiana, H. L., Flohr, H., and Melvill Jones, G. 1984. A re-evaluation of inter-vestibular nuclear coupling: Its role in vestibular compensation. *Journal of Neurophysiology* 51: 258–275.

Goldberger, M. E. 1980. Motor recovery after lesions. *Trends in Neuroscience* 3: 288–291.

Goode, C. T., Maney, D. L., Rubel, E. W., and Fuchs, A. F. 2001. Visual influences on the development and recovery of the vestibulo-ocular reflex in the chicken. *Journal of Neurophysiology* 85: 1119–1128.

Hain, T. C., Fetter, M., and Zee, D. S. 1987. Head-shaking nystagmus in patients with unilateral peripheral vestibular lesions. *American Journal of Otolaryngology* 8: 36–47.

Hardcastle, V. G., and Stewart, C. M. 2002. What do brain data really show? *Philosophy of Science* 69: 72–82.

Hardcastle, V. G., and Stewart, C. M. 2003. The art of single cell recordings. *Biology and Philosophy* 18: 195–208.

Jones, M., Guitton, D., and Berthoz, A. 1988. Changing patterns of eye-head coordination during 6 hours of optically reversed vision. *Experimental Brain Research* 69: 531–544.

Kjerulf, T. D., O'Neal, J. T., Calvin, W. H., Loeser, J. D., and Westrum, L. E. 1973. Deafferentation effects in lateral cuneate nucleus of the cat: Correlation of structural changes with firing patterns changes. *Experimental Neurology* 39: 86–102.

Lacour, M., and Xerri, C. 1980. Compensation of postural reactions to free-fall in the vestibular neurectomized monkey: Role of the visual motion cues. *Experimental Brain Research* 40: 103–110.

Lacour, M., Vidal, P. P., and Xerri, C. 1981. Visual influences in vestibulospinal reflexes during vertical linear motion in normal and hemilabyrinthectomized monkeys. *Experimental Brain Research* 43: 383–394.

Loeser, J. D., Ward, Jr., A. A., and White, Jr., L. E. 1968. Chronic deafferentation of human spinal cord neurons. *Journal of Neurosurgery* 29: 48–50.

Merzenich, M. M., Kaas, J. H., Sur, M., Nelson, R. J., and Felleman, D. J. 1983. Progression of change following median nerve section in the cortical representation of the hand in areas 3b and 1 in adult owl and squirrel monkeys. *Neuroscience* 10 (3): 639–665.

Miles, F. A., and Lisberger, S. G. 1981. Plasticity in the vestibulo-ocular reflex: A new hypothesis. *Annual Review of Neuroscience* 4: 273–299.

Mustari, M. J., Fuchs, A. F., Kaneko, C. R. S., and Robinson, F. R. 1994. Anatomical connections of the primate pretectal nucleus of the optic tract. *Journal of Comparative Neuroanatomy* 349: 111–128.

Putkonen, P. T. S., Courjon, J. H., and Jeannerod, M. 1977. Compensation of postural effects of hemilabyrinthectomy in the cat. A sensory substitution process? *Experimental Brain Research* 28: 249–257.

Robinson, D. A. 1981. The use of control systems analysis in the neurophysiology of eye movements. *Annual Review of Neuroscience* 4: 463–503.

Schaefer, K.-P., and Meyer, D. L. 1973. Compensatory mechanisms following labyrinthine lesions in the guinea pig: A simple model of learning. In H. P. Zippel (ed.), *Memory and Transfer of Information*. New York: Plenum Press.

Smith, P. F., and Curthoys, I. S. 1989. Mechanisms of recovery following a unilateral labyrinthectomy: A review. *Brain Research Reviews* 14: 155–180.

Stewart, C. M. 2002. *Visual-Vestibular Interactions During Compensation*. Doctoral dissertation. University of Texas Medical Branch at Galveston.

Stewart, C. M., Perachio, A. A., Mustari, M. J., and Allen, T. C. 1999. Effects of the pretectal nucleus of the optic tract and hemilabyrinthectomy on vestibulo-ocular reflex compensation in rhesus monkey. *Society for Neuroscience Abstracts*. Maimi: Society for Neuroscience.

Takehashi, S., Fetter, M., Koenig, E., and Dichgans, J. 1990. The clinical signifi-cance of head-shaking nystagmus in the dizzy patient. *Acta Otolaryngology* 109: 8–14.

Tusa, R. J., Mustari, M. J., Burrows, A. F., and Fuchs, A. F. 2001. Gaze-stabilizing deficits and latent nystagmus in monkeys with brief, early-onset visual deprivation: Eye movement recordings. *Journal of Neurophysiology* 86: 651–661.

Xerri, C., and Zennou, Y. 1988. Sensory, functional, and behavioral substitu-tion processes in vestibular compensation. In M. Lacour, M. Toupet, and P. Denise (eds.), *Vestibular Compensation, Facts, Theories, and Clinical Perspectives.* Paris: Elsevier.

Xerri, C., Lacour, M., and Borel, L. 1988. Multimodel sensory substitution process in vestibular compensation. In H. Flohr (ed.), *Post-lesion Neuronal Plasticity.* Berlin: Springer-Verlag.

Zennou-Azogui, Y., Xerri, C., and Harlay, F. 1994. Visual sensory substitution in vestibular compensation: Neuronal substrates in the alert cat. *Experimental Brain Research* 98: 457–473.

2

Neurophenomenology: An Introduction for Neurophilosophers

Evan Thompson, Antoine Lutz, and Diego Cosmelli

1 Introduction

One of the major challenges facing neuroscience today is to provide an explanatory framework that accounts for both the subjectivity and neurobiology of consciousness. Although neuroscientists have supplied neural models of various aspects of consciousness, and have uncovered evidence about the neural correlates of consciousness (or NCCs),[1] there nonetheless remains an 'explanatory gap' in our understanding of how to relate neurobiological and phenomenological features of consciousness. This explanatory gap is conceptual, epistemological, and methodological:

- An adequate conceptual framework is still needed to account for phenomena that (i) have a first-person, subjective-experiential or phenomenal character; (ii) are (usually) reportable and describable (in humans); and (iii) are neurobiologically realized.[2]
- The conscious subject plays an unavoidable epistemological role in characterizing the explanandum of consciousness through first-person descriptive reports. The experimentalist is then able to link first-person data and third-person data. Yet the generation of first-person data raises difficult epistemological issues about the relation of second-order awareness or meta-awareness to first-order

---

[1] For recent neural models of consciousness, see Tononi and Edelman (1998), Freeman (1999a, 1999b), Dehaene and Naccache (2001), Parvizi and Damasio (2001), Engel and Singer (2001), Crick and Koch (2003). For NCC evidence, see Metzinger (2000) and Rees, Kreiman, and Koch (2002).

[2] For a recent overview of this aspect of the explanatory gap see Roy, Petitot, Pachoud, and Varela (1999).

experience (e.g., whether second-order attention to first-order experience inevitably affects the intentional content and/or phenomenal character of first-order experience).[3]

• The need for first-person data also raises methodological issues (e.g., whether subjects should be naïve or phenomenologically trained).

Neurophenomenology is a neuroscientific research program whose aim is to make progress on these issues associated with the explanatory gap.[4] In this chapter we give an overview of the neurophenomenological approach to the study of consciousness.

Whereas 'neurophilosophy' is rooted in analytic philosophy (Churchland 1986, 2002), neurophenomenology, as its name suggests, has strong roots in the tradition of phenomenological philosophy. In recent years, a new current of phenomenological philosophy has emerged in Europe and North America, one that goes back to the source of phenomenology in Edmund Husserl's philosophy, but is influenced by cognitive science and analytic philosophy of mind, and aims to contribute to these fields (see Petitot, Varela, Pachoud, and Roy 1999; Thompson 2004; and the new journal *Phenomenology and the Cognitive Sciences*). Neurophenomenology belongs to this current. Our aim in this chapter is to communicate the main ideas of neurophenomenology in the form of a review for a neurophilosophical audience, who may not be as familiar with phenomenology and its potential in relation to the neuroscientific study of consciousness.[5]

Neurophenomenology also grows out of the enactive approach in cognitive science, which has strong ties to phenomenology (Varela, Thompson, and Rosch 1991). We discuss this cognitive science background of neurophenomenology in section 2. In section 3 we discuss the motivation for neurophenomenology in current affective-cognitive neuroscience. Section 4 is devoted to explaining phenomenological conceptions of consciousness. Section 5 gives an overview of neurodynamical approaches to the study of large-scale brain activity. Section 6 is concerned

---

3 See Hurlbert and Heavey (2001), Jack and Shallice (2001), Jack and Roepstorff (2002), Lambie and Marcel (2002), and Schooler (2002).
4 Neurophenomenology was introduced by F. J. Varela in the mid-1990s (Varela 1996, 1997b, 1999). For subsequent discussions, see Bitbol (2002), Le Van Quyen and Petitmengin (2002), Lutz (2002), Rudrauf, Lutz, Cosmelli, Lachaux, and Le Van Quyen (2003), Varela and Thompson (2003), Lutz and Thompson (2003), and Hanna and Thompson (2004).
5 For an introduction to phenomenology written for a cognitive science audience see Thompson and Zahavi (in press).

with first-person methods of investigating experience. Section 7 reviews two experimental, neurophenomenological studies of ongoing conscious states and large-scale brain activity. Section 8 focuses on research on the neural correlates of consciousness, as seen from a neurophenomenological perspective. Section 9 concludes the chapter by returning to the issue of the explanatory gap.

2 The Enactive Approach

Neurophenomenology is an offshoot of the enactive approach in cognitive science (Varela, Thompson, and Rosch 1991). The word 'enactive' was coined by F. J. Varela in order to describe and unify under one heading several related ideas. The first idea is that organisms are autonomous agents that actively generate and maintain their identities, and thereby define their own cognitive domains. The second idea is that the nervous system is an autonomous system: It actively generates and maintains its own coherent patterns of activity, according to its operation as an organizationally closed sensorimotor network of interacting neurons. The third idea is that cognitive structures emerge from such recurrent sensorimotor couplings of body, nervous system, and environment.[6] Like phenomenology, the enactive approach thus emphasizes that the organism defines its own point of view on the world. As we will see, this autonomy perspective has important implications for how we conceptualize brain activity and its relationship to mental activity, including consciousness.

Let us start with the notion of an autonomous system. 'Autonomous' literally means self-governing or conforming to its own law. In the theory of complex (nonlinear dynamic) systems, autonomous is used to describe a system that is a self-generating and self-maintaining whole, such as a cell or multicellular organism (Varela 1979; Varela and Bourgine 1991). In general, to specify any system, one needs to describe its organization – the

---

[6] Varela thought of the name 'the enactive approach' in the summer of 1986 in Paris when he and Thompson began writing *The Embodied Mind* (Varela, Thompson, and Rosch 1991). Earlier he had been using 'the hermeneutic approach' to emphasize the affiliation of his ideas to the philosophical school of hermeneutics – an affiliation also emphasized by other theorists of embodied cognition at the time (see Varela, Thompson, and Rosch 1991, pp. 149–150). The first and second ideas summarized here were presented in Varela's 1979 book, *Principles of Biological Autonomy*. They were developed with Humberto Maturana, and grew out of Maturana's earlier work on the biology of cognition (Maturana 1969, 1970; Maturana and Varela 1980, 1987). The third idea was presented by Varela, Thompson, and Rosch (1991) and by Thompson, Palacios, and Varela (1992), and was elaborated by Varela in a number of subsequent papers (e.g., Varela 1991, 1997a).

set of relations that defines it as a system. In a complex dynamic system, the relevant relations are ones between dynamic processes. An autonomous system is a network of processes, in which (i) the processes recursively depend on each other for their generation and their realization as a network; and (ii) the processes constitute the system as a unity in whatever domain they exist (Varela 1979, p. 55). The paradigmatic example is a living cell. The constituent processes in this case are chemical; their recursive interdependence takes the form of a self-producing, metabolic network that also produces its own membrane; and this network constitutes the system as a unity in the biochemical domain. This kind of self-production in the biochemical domain is known as *autopoiesis* (Maturana and Varela 1980). Autopoiesis is the fundamental and paradigmatic case of biological autonomy.

According to Varela, an autonomous system can be precisely defined as a system that has *organizational closure* and *operational closure* (Varela 1979, pp. 55–60). The term 'closure' does not mean that the system is materially and energetically closed to the outside world (which, of course, is impossible). On the contrary, autonomous systems are thermodynamically far from equilibrium systems, which incessantly exchange matter and energy with their surroundings. 'Organizational closure' describes the self-referential (circular and recursive) network of relations that defines the system as a unity. At any given instant or moment, this self-referential network must be maintained; otherwise the system is no longer autonomous and no longer viable in whatever domain it exists. 'Operational closure' describes the recursive, re-entrant, and recurrent dynamics of the system. The system changes state on the basis of its self-organizing dynamics (in coupling with an environment), and the product of its activity is always further self-organized activity within the system (unless its operational closure is disrupted and it disintegrates).[7] Biological examples abound – single cells, microbial communities, nervous systems, immune systems, multicellular organisms, ecosystems, and so on. Such systems need to be seen as sources of their own activity, and as specifying their own informational or cognitive domains, not as

---

7 'Closure' is used here in its algebraic sense: An operation K exhibits closure in a domain D if every result of its operation yields results within D. Thus, the operation of a system has operational closure if the results of its activity remain within the system itself. This notion of operational closure has nothing to do with the idea of a materially closed system – a system that is closed to interactions with what lies beyond it. Autonomous systems are and must be coupled to their surroundings; the issue is the nature of the dynamics that defines them and on the basis of which they interact with what lies beyond them.

transducers or functions for converting input instructions into output products. In other words, the autonomous nature of these systems needs to be recognized.

In the case of the brain and nervous system, this autonomous systems perspective implies that the endogenous, self-organizing dynamics of neural activity should be the reference point for relating brain processes to mental processes. The overall organization of the brain reflects a principle of reciprocity: if area A connects to area B, then there are reciprocal connections from B to A (Varela 1995; Varela, Lachaux, Rodriguez, and Martinerie 2001). Moreover, if B receives most of its incoming influence from A, then it sends the larger proportion of its outgoing activity back to A, and only a smaller proportion onward (Freeman 2000, p. 224).

Nevertheless, traditional neuroscience has tried to map brain organization onto a hierarchical, input-output processing model, in which the sensory end is taken as the starting point. Perception is described as proceeding through a series of feedforward or bottom-up processing stages, and top-down influences are equated with back-projections or feedback from higher to lower areas. Walter Freeman (1999a, 1999b, 2000) aptly describes this view as the 'passivist-cognitivist view' of the brain.

From an enactive viewpoint, things look rather different. Brain processes are recursive, re-entrant, and self-activating, and do not start or stop anywhere. Instead of treating perception as a later stage of sensation and taking the sensory receptors as the starting point for analysis, the enactive approach treats perception, emotion, and cognition as dependent aspects of intentional action, and takes the brain's self-generated, endogenous activity as the starting point for analysis. This activity arises far from the sensors – in the frontal lobes, limbic areas, or temporal and associative cortices – and reflects the organism's states of expectancy, preparation, emotional tone, attention, and so on, states necessarily active at the same time as the sensory inflow, and that shape that inflow in a meaningful way (Varela, Lachaux, Rodriguez, and Martinerie 2001; Engel, Fries, and Singer 2001). Freeman describes this view as the 'activist-pragmatist' view of the brain. Whereas a passivist-cognitivist view would describe such internal states as acting in a top-down manner on sensory processing, from an activist-pragmatist or enactive view, 'top down' and 'bottom up' are simply heuristic terms for what in reality is a large-scale network that integrates incoming and endogenous activities on the basis of its own internally established reference points (Varela, Lachaux, Rodriguez, and Martinerie 2001). Hence, from an enactive viewpoint, we need to look to this large-scale dynamic network in order to understand how cognition, intentional action, and consciousness emerge through self-organizing neural activity.

Yet understanding such large-scale dynamic activity poses considerable challenges for the experimentalist, because it is highly labile and variable from trial to trial and cannot be fully controlled from the outside. Similarly, mental activity is variable from moment to moment, in the form of fluctuations in emotion, quality of attention, motivation, and so forth. Thus, a large part of the subject's internal activity at both psychological and experiential levels is externally uncontrollable and therefore unknown. The working hypothesis of neurophenomenology is that disciplined, first-person accounts of the phenomenology of mental processes can provide additional, valid information about these externally uncontrollable aspects of mental activity, and that this information can be used to detect significant patterns of dynamic activity at the neural level.

3 Neurophenomenology and Affective-Cognitive Neuroscience

The enactive approach in its neurophenomenological development converges with the growing recognition in affective-cognitive neuroscience that much more detailed and refined first-person descriptions of conscious experience are needed in order to characterize the explanandum of consciousness and relate it to the complexity of brain activity (Jack and Roepstorff 2002, 2003; Lutz and Thompson 2003). Neurophenomenology stresses the importance of gathering first-person data from phenomenologically trained subjects as a heuristic strategy for describing and quantifying the physiological processes relevant to consciousness. The general approach, at a methodological level, is (i) to obtain richer first-person data through disciplined phenomenological explorations of experience, and (ii) to use these original first-person data to uncover new third-person data about the physiological processes crucial for consciousness. Thus, one central aim of neurophenomenology is to generate new data by incorporating refined and rigorous phenomenological explorations into the experimental protocols of neuroscientific research on consciousness.

The term 'neurophenomenology' pays homage to phenomenological traditions in both Western philosophy (Petitot, Varela, Pachoud, and Roy 1999; Moran 2000) and Asian philosophy (Gupta 1998; Wallace 1998; Williams 1998). Phenomenology in this broad sense can be understood as the project of providing a disciplined characterization of the phenomenal invariants of lived experience in all of its multifarious forms. By 'lived experience' we mean experiences as they are lived and verbally articulated in the first-person, whether they be lived experiences of perception, action, memory, mental imagery, emotion, attention, empathy, self-consciousness, contemplative states, dreaming, and so on. By 'phenomenal invariants' we mean categorical and structural features of experience that are phenomenologically describable both across and within the various forms of lived experience. By 'disciplined characterization' we mean a phenomenological mapping of experience based on the use of both first-person methods for increasing one's sensitivity to one's experience and second-person methods for facilitating this process and eliciting descriptive accounts of experience (Varela and Shear 1999a; Petitmengin 2001; Depraz, Varela, and Vermersch 2003). The importance of this sort of phenomenological investigation of experience for cognitive science was already proposed and extensively discussed by Varela, Thompson, and Rosch (1991) as an integral part of the enactive approach. It was then subsequently elaborated by Varela (1996, 1997b, 1999) into the neuroscientific research program of neurophenomenology.

The use of first-person and second-person phenomenological methods to obtain original and refined first-person data is central to neurophenomenology. It seems true both that people vary in their abilities as observers and reporters of their own experiences and that these abilities can be enhanced through various methods. First-person methods are disciplined practices that subjects can use to increase their sensitivity to their own experiences at various time-scales (Varela and Shear 1999a; Depraz, Varela, and Vermersch 2003). These practices involve the systematic training of attention and self-regulation of emotion. Such practices exist in phenomenology, psychotherapy, and contemplative meditative traditions. Using these methods, subjects may be able to gain access to aspects of their experience, such as transient affective state and quality of attention, that otherwise would remain unnoticed and unavailable for verbal report. The experimentalist, on the other hand, using phenomenological accounts produced with these methods, may be able to gain access to physiological processes that otherwise would remain opaque, such as the variability in brain response as recorded in neuroimaging experiments (see section 7). Thus, at a methodological level, the neurophenomenological rationale for using first-person methods is to generate new data – both first-person and third-person – for the science of consciousness.

The working hypothesis of neurophenomenology in an experimental context is that phenomenologically precise first-person data produced by employing first-person/second-person methods provide strong constraints on the analysis and interpretation of the physiological processes relevant to consciousness. In addition, third-person (biobehavioural) data produced in this manner might eventually constrain first-person data, so that the relationship between the two would become one of dynamic reciprocal constraints (Varela 1996; Lutz 2002). Reciprocal constraints means not only (i) that the subject is actively involved in generating and describing specific phenomenal invariants of experience, and (ii) that the neuroscientist is guided by these first-person data in the analysis and interpretation of physiological data, but also (iii) that the (phenomenologically enriched) neuroscientific analyses provoke revisions and refinements of the phenomenological accounts, as well as facilitate the subject's becoming aware of previously inaccessible or phenomenally unavailable aspects of his or her mental life. Preliminary examples of this third step can be found in neurophenomenological studies of epilepsy (Le Van Quyen and Petitmengin 2002) and pain (Price, Barrell, and Rainville 2002).

To establish such reciprocal constraints, both an appropriate candidate for the physiological basis of consciousness and an adequate theoretical framework to characterize it are needed. Neurophenomenology is guided by the theoretical proposal (discussed in section 5) that the best current candidate for the neurophysiological basis of consciousness is a flexible repertoire of dynamic large-scale neural assemblies that transiently link multiple brain regions and areas. This theoretical proposal is shared by a number of researchers, though specific models vary in their details (Varela 1995; Tononi and Edelman 1998; Freeman 1999a, 1999b; Engel and Singer 2001). In this approach, the framework of dynamic systems theory is essential for characterizing the neural processes relevant to consciousness (see Le Van Quyen 2003). Neurophenomenology is thus based on the synergistic use of three fields of knowledge:

1. (NPh1) First-person data from the careful examination of experience with specific first-person methods.
2. (NPh2) Formal models and analytic tools from dynamic systems theory, grounded on an enactive approach to cognition.
3. (NPh3) Neurophysiological data from measurements of large-scale, integrative processes in the brain.

The remaining sections of this chapter cover topics that fall within this threefold schema.

4 Concepts of Consciousness

A number of different concepts of consciousness can be distinguished in current neuroscientific and philosophical discussions of consciousness:

- *Creature consciousness:* consciousness of an organism as a whole insofar as it is awake and sentient (Rosenthal 1997).
- *Background consciousness* versus *state consciousness:* overall states of consciousness, such as being awake, being asleep, dreaming, being under hypnosis, and so on (Hobson 1999), versus specific conscious mental states individuated by their contents (Rosenthal 1997; Chalmers 2000). (The coarsest-grained state of background consciousness is sometimes taken to be creature consciousness (Chalmers 2000).)
- *Transitive consciousness* versus *intransitive consciousness:* object-directed consciousness (consciousness-of), versus non-object-directed consciousness (Rosenthal 1997).
- *Access consciousness:* mental states whose contents are accessible to thought and verbal report (Block 2001). According to one important theory, mental contents are access conscious when they are 'globally available' in the brain as contents of a 'global neuronal workspace' (Dehaene and Naccache 2001; Baars 2002).
- *Phenomenal consciousness:* mental states that have a subjective-experiential character (there is something 'it is like' for the subject to be in such a state) (Nagel 1979; Block 2001).
- *Introspective consciousness:* meta-awareness or second-order awareness of a first-order conscious state. The second-order conscious state is usually understood as a form of access consciousness, and the first-order state as a phenomenally conscious state (Jack and Shallice 2001; Jack and Roepstorff 2002; Lambie and Marcel 2002).
- *Prereflective self-consciousness:* primitive self-consciousness; self-referential awareness of subjective experience that does not require active reflection or introspection (Wider 1997; Williams 1998; Gupta 1998; Zahavi 1999; Kriegel 2003).

The relationships of these concepts to one another are unclear and currently the subject of debate. A great deal of debate has centered on the relation between access consciousness and phenomenal consciousness.[8] Although a notional distinction can be drawn between

---

[8] See Block (1997) and replies by Dennett (1997), Chalmers (1997), Church (1997), and Burge (1997).

awareness in the sense of cognitive access and awareness in the sense of subjective experience, it is not clear whether this distinction corresponds to a difference in kind of awareness or merely a difference in degree of awareness. Put another way, it is not clear whether 'access consciousness' and 'phenomenal consciousness' are two separate concepts, or that if they are, they pick out two different properties, rather than one and the same property. Some theorists argue that it is conceptually possible for there to be phenomenally conscious contents that are inaccessible to thought and verbal report (Block 1997, 2001). Others argue that this notion of subjective experience without access is incoherent, and that all consciousness is access consciousness (Dennett 1997, 2001).

How does this debate look from a broadly phenomenological perspective? Before addressing this question, we need to review two concepts central to phenomenology, the concepts of *intentionality* and *pre-reflective self-consciousness*.

Intentionality

According to phenomenology, conscious states and processes are intentional in a broad sense: They aim toward or intend something beyond themselves. Phenomenologists distinguish different types of intentionality. In a narrow sense, intentionality is defined as object-directedness. In a broader sense, intentionality is defined as openness toward the world or what is 'other' ('alterity'). In either case, the emphasis is on denying that the mind is self-enclosed (Zahavi 2003, 2004).

Object-directed experiences are ones in which we are conscious *of* something in a more or less determinate sense. When we see, we see something; when we remember, we remember something; when we hope or fear, we hope for or fear something, and so on. These kinds of *transitive consciousness* are characterized by the intending of an object (which need not exist). 'Object' in its etymological sense means something that stands before one. Something standing before one lies beyond, over against, or outside of one. Object-directed experiences can thus be understood as experiences in which one is conscious as of something distinct from oneself as a present subject, whether this be a past event remembered, a thing perceived in the settings around one, a future event feared or hoped for, and so on.

Phenomenologists call this act-object relation the *correlational structure of intentionality*. 'Correlational' does not mean the constant conjunction of two terms that could be imagined to exist apart, but the invariant structure

of intentional act/intentional object. Object-directed intentional experiences necessarily comprise these two inseparable poles. In Husserlian
phenomenological language, these two poles of experience are known
as the *noema* (the object as experienced) and the *noesis* (the mental act
that intends the object).

Many kinds of experience, however, are not object-directed in this
sense. Such experiences include bodily feelings of pain, moods such as
undirected anxiety, depression, and elation, and 'absorbed skilful coping'
(Dreyfus 1991) in everyday life. These types of experience either are not
or need not be 'about' any intentional object. They are not directed
toward a transcendent object, in the sense of something experienced as
standing over against oneself as a distinct subject. Put another way, they
do not have a subject-object structure.

Philosophers who limit intentionality to object-directedness or transitive consciousness deny that experiences like these are intentional (e.g.,
Searle 1983). Nevertheless, they do qualify as intentional in the broader
phenomenological sense of being open to what is other or having a world-
involving phenomenal character. Thus, bodily feelings such as pain are
not self-enclosed without openness to the world. On the contrary, they
present things in a certain affective light or atmosphere, and thereby
deeply influence how we perceive and respond to things. A classic example is Jean-Paul Sartre's discussion of feeling eyestrain and fatigue as a
result of reading late into the night (1956, pp. 332–333).[9] The feeling
manifests first not as an intentional object of some higher-order state of
transitive consciousness, but as a trembling of the eyes and a blurriness
of the words on the page. One's body and immediate environment dis-
close themselves in a certain manner through the feeling. In the case
of moods, although they are not object-directed in the manner of intentional emotions, such as a feeling of sympathy for a loved one or a
feeling of envy for a rival, they are nonetheless hardly self-enclosed without reference to the world. On the contrary, as Martin Heidegger (1996)
analyzes at length in *Being and Time,* moods reveal our embeddedness in
the world, and (as he sees it) make possible more circumscribed forms of
directedness in everyday life. In absorbed skilful activities, such as driving,
dancing, or writing, one's experience is not that of relating to a distinct
intentional object but of being engaged and immersed in a fluid activity, which takes on a subject-object structure only during moments of

---

9 Our use of this example is due to Gallagher (1986) and Zahavi (2004).

breakdown or disruption (see Dreyfus and Dreyfus 1982; Dreyfus 1991, 2002).[10]

These cases of intransitive or non-object-directed experience illustrate the difference between a thematic, explicit, or focal, object-directed mode of consciousness, and a nonreflective, implicit, tacit sensibility. This intransitive sensibility constitutes our primary openness to the world. Edmund Husserl and M. Merleau-Ponty use the term 'operative intentionality' to describe this kind of intentionality, because it is constantly operative in a prereflective and involuntary way, without being engaged in any explicit cognitive project (Husserl 2001, p. 206; Merleau-Ponty 1962, p. xviii).

Two additional distinctions are important in the context of distinguishing object-directed and transitive intentionality from intransitive and operative intentionality.[11] The first is between *activity* and *passivity*. Activity means taking a cognitive position in acts of attending, judging, valuing, wishing, and so on. Passivity means being involuntarily affected, moved, and directed. Whenever one is active, one is also passive, because to be active is to react to something that has affected one. Every kind of active position taking presupposes a prior (and ongoing) passivity. The second distinction is between *affection* and *receptivity*. Affection belongs to the sphere of passivity and means being affectively influenced or perturbed. Receptivity means responding to an involuntary affection (affective influence) by noticing or turning toward it. Every receptive action presupposes a prior affection. As Dan Zahavi explains: 'Receptivity is taken to be the first, lowest, and most primitive type of intentional activity, and consists in responding to or paying attention to that which is affecting us passively. Thus, even receptivity understood as a mere "I notice" presupposes a prior affection' (Zahavi 1999, p. 116). In other words, receptivity is the lowest active level of attention, at the fold, as it were, between passivity and activity (a differentiation that can be made only dynamically and relatively, not statically and absolutely). To be affected by something is not yet to be presented with an intentional object, but to be drawn to turn one's attention toward whatever exerts the affection.

---

[10] A more difficult case is certain contemplative or meditative states, in which it is reported that awareness is experienced as without an object (Foreman 1990; Austin 1998). How to describe these states in relation to the phenomenological notion of intentionality is a complicated issue that we cannot pursue here.

[11] These distinctions emerge in Husserl's lectures during the 1920s on the phenomenology of 'passive synthesis' in experience. See Husserl (2001).

Thus, whatever comes into relief in experience through attention must have already been affecting one and must have some kind of 'affective force' or 'affective allure' in relation to one's attention and motivations. Attentiveness at any level is motivated in virtue of something's affective allure, an idea familiar to psychologists as motivated attention (Derryberry and Tucker 1994). Depending on the nature and force of the allure, as well as one's motivations, one may yield to the allure passively or involuntarily, voluntarily turn one's attention toward it, or have one's attention captured or repulsed by it. Allure implies a dynamic gestalt or figure-ground structure: Something becomes noticeable, at whatever level of attentiveness, due to the strength of its allure, emerging into affective prominence, salience, or relief, while other things become less noticeable due to comparative weakness of their allure (Husserl 2001, p. 211). Whatever exercises affective allure without our turning to it attentively (even at a level of bare noticing) is said to be 'pregiven', and whatever succeeds in gaining attention is said to be 'given'. Thus the given – the mode or way in which something appears to one – has to be understood dynamically as emergent in relation to the pregiven.[12] This dynamic interplay of passivity and activity, affection and receptivity, expresses a constantly operative and affectively 'saturated intentionality' (Steinbock 1999). Object-directed and transitive intentional experiences emerge only against the background of this precognitive, operative intentionality (Merleau-Ponty 1962, p. xviii).

---

[12] Does Husserl's phenomenology imply belief in an uninterrupted and theory-neutral 'given' in experience, the so-called philosophical myth of the given? This is a difficult and complicated question. There is not one but several different notions of the given in philosophy, and Husserl's thought developed considerably over the course of his life, such that he held different views at different times regarding what might be meant by the given. Suffice it to say that it is mistaken to label Husserl as a philosopher of the given in the sense originally targeted by Wilfrid Sellars (1956) or by critics of the notion of theory-neutral observation, such as Thomas Kuhn (1970), for at least these four reasons: First, the given in the phenomenological sense is not nonintentional sense-data but the world as it appears to us. Second, the phenomenality of the world is not understandable apart from the constitutive relation subjectivity and consciousness bear to it. Third, as discussed here, whatever counts as 'given' has to be seen as dynamically emergent in relation to what is 'pregiven', and the transition from pregiven to given depends on the subject's motivations, interests, and attentional capacities. Finally, the given comprises not simply phenomenal presence in a narrow sense (what is facing me right now) but also absence and possibility (the sides of the bottle I cannot see but that are available for me to see through movement). For recent discussions of some of these issues, see Botero (1999), Steinbock (1999), and Roy (2004).

Prereflective Self-Consciousness

As mentioned, the term 'phenomenal consciousness' refers to mental states that have a subjective and experiential character. In T. Nagel's words, for a mental state to be conscious is for there to be something it is like for the subject to be in that state (Nagel 1979). Various notions are used to describe the properties characteristic of phenomenal consciousness – qualia, sensational properties, phenomenal properties, the subjective character of experience – and there is considerable debate about the relation between these properties and other properties of mental states, such as their representational content or their being cognitively accessible to thought and verbal report (access consciousness). From a phenomenological perspective, Nagel's original term, 'the subjective character of experience', is the most fitting because it calls attention to the subjectivity of conscious mental states. Every phenomenally conscious state – be it a perception, an emotion, a recollection, or (for Husserl and William James) a conscious thought[13] – has a certain subjective character, corresponding to what it is like to live through or undergo that state.

The subjective character of experience, however, needs to be analyzed further.[14] Phenomenologists are careful to distinguish between 'phenomenality' and 'subjectivity'. Phenomenality corresponds not to *qualitative properties of consciousness* (*qualia*), but to the *qualitative appearance of the world for consciousness*. As Merleau-Ponty puts it, qualia are not elements *of* consciousness but qualities *for* consciousness (1962, p. 5). Subjectivity, on the other hand, corresponds to the *first-personal givenness* of experience. First-personal givenness can be explained by taking perceptual experience as an example. The *object* of my perceptual experience – say, the wine bottle on the table – is intersubjectively accessible, in the sense that it can be given to others in the same way it is given to me (from the vantage point of my seat at the table, in this lighting, as half full, and so on). My *perceptual experience* of the bottle, on the other hand, is given directly only to me. We can both perceive one and the same wine bottle, but we each have our own distinct perceptions of it, and we cannot share these perceptions. In the same sense, we cannot share each other's thoughts,

---

[13] See Zahavi (2004) for discussion of Husserl's view that conscious thoughts, including abstract beliefs, have a phenomenal quality or subjective experiential character.

[14] The next two paragraphs and note 15 draw from Thompson and Zahavi (in press). See also Zahavi (2004).

emotions, memories, pains, and so on. A more precise way to formulate this point is to say that you have no access to the first-personal givenness of my experience, and I have no access to the first-personal givenness of yours.[15]

This notion of first-personal givenness is closely related to the concept of prereflective self-consciousness. When one is directly and noninferentially aware of one's occurrent thoughts, perceptions, and feelings, they are characterized by a first-personal givenness that immediately reveals them as one's own. This first-personal givenness is not something the experiences could lack while still being experiences. In other words, it is a necessary or essential feature of their being experiences at all. It is their first-personal givenness that makes the experiences subjective. To put it differently, first-personal givenness entails a built-in self-reference, a primitive, experiential self-referentiality. When I am aware of an occurrent thought, perception, or feeling from the first-person perspective,

---

[15] One might object that this formulation misleadingly suggests that experiences can be given in more than one way, that is, from a perspective other than the first-person perspective. The objection is that either an experience is given from a first-person perspective or it is not given at all (and hence is not an experience). The phenomenological reply, however, is that although any experience must always be given from a first-person perspective (otherwise it is not an experience), it can also be given from a *second-person perspective* in empathy. To deny this claim commits one to the view that we never experience the thoughts and feelings of another person in any sense but can only infer their existence from physical behaviour. This view involves a highly problematic dichotomy between inner experience and outer behaviour. In face-to-face encounters, we are confronted neither with a mere physical body nor with a hidden psyche, but with a unified intentional whole. An account of subjectivity and intersubjectivity must start from a correct understanding of the relation between mind and body. Experiences are not internal in the sense of being hidden inside the head (or anywhere else); they express and manifest themselves in bodily gestures and actions. When one sees the face of another person, one sees it as friendly or angry, happy or sad, and so on – it is the very face that expresses these emotions. Moreover, bodily behaviour in general is meaningful and intentionally related to the world, and intentional relations cannot be reduced to inside–outside relations. There simply is no clear sense in which behaviour is external and perceptually available, and experience internal and perceptually inaccessible. These considerations imply that a proper account of subjectivity and intersubjectivity must be based on an understanding that the body of the other is fundamentally different from a mere material object, and that our perception of the other's bodily presence is accordingly unlike our perception of material objects. The other is given in its bodily presence as a lived body according to a distinctive form of intentional experience phenomenologists call empathy. In empathy, the experiences of another person are given from a second-person perspective and have second-person givenness. Empathy allows us to experience the feelings, desires, and beliefs of others in a more or less direct manner. It constitutes a unique form of intentionality, alongside other forms, such as perception, recollection, and imagination.

the experience is given immediately and noninferentially as mine. I do not first scrutinize the experience and then identify it as mine. Accordingly, self-awareness cannot be equated with reflective (explicit, thematic, introspective) self-awareness, as some philosophers and cognitive scientists have claimed. Not only is reflective self-awareness only one type of self-awareness, but it also presupposes a prereflective (implicit, tacit) self-awareness. Self-awareness does not happen only when one realizes reflectively or introspectively that one is perceiving, feeling, or thinking something. Rather, it is legitimate to speak of a primitive but basic type of self-awareness whenever one is acquainted with an experience from a first-person perspective. For an experience to be given in a first-personal mode of presentation to me is for it to be given (at least tacitly) as my experience. For an experience to be given as my experience is for me to be aware of it (at least tacitly) as mine. For me to be aware of it as mine is for me to be self-aware. To be aware of oneself is thus not to apprehend a self apart from experience but to be acquainted with an experience in its first-personal mode of presentation, as it were, from 'within'. The subject or self at issue here is not something standing opposed to, apart from, or beyond experience, nor is it some homuncular entity within experience; it is rather a feature or function of the givenness of experience. In short, the first-personal givenness of experience constitutes the most basic form of selfhood (see Zahavi and Parnas 1998; Zahavi 1999).

In summary, any object-directed, transitive conscious experience, in addition to being of or about its intentional object, is prereflectively and intransitively manifest to itself, in virtue of its first-personal givenness. Such self-manifesting awareness is a primitive form of self-consciousness in the sense that it (i) does not require any subsequent act of reflection or introspection, but occurs simultaneously with awareness of the object; (ii) does not consist in forming a belief or making a judgment; and (iii) is passive in the sense of being spontaneous and involuntary.

This conception of prereflective self-consciousness can be linked to the correlational structure of intentionality, discussed earlier (the act/object or noesis/noema structure). Experience involves not simply awareness of its object (noema) but also tacit awareness of itself as process (noesis). In consciously seeing an object, one is at the same time tacitly and prereflectively aware of one's seeing. In visualizing a mental image, one is at the same time tacitly and prereflectively aware of one's visualizing. This tacit self-awareness has often been explicated by phenomenologists (most notably Merleau-Ponty) as involving a form of nonobjective bodily self-awareness, an awareness of one's lived body or embodied subjectivity, correlative to experience of the intentional object (Merleau-Ponty 1962; Wider 1997; Zahavi 2002). The roots of such prereflective bodily self-awareness sink to the involuntary and precognitive level of operative intentionality, the dynamic interplay of affection and receptivity, discussed earlier.

Phenomenology thus converges with certain proposals about 'primary affective consciousness' or 'core consciousness' coming from affective neuroscience (Panksepp 1998b; Damasio 1999). As J. Parvizi and A. Damasio state, neuroscience needs to explain both 'how the brain engenders the mental patterns we experience as the images of an object' (the noema), and 'how, in parallel . . . the brain also creates a sense of self in the act of knowing . . . how each of us has a sense of "me" . . . how we sense that the images in our minds are shaped in our particular perspective and belong to our individual organism' (Parvizi and Damasio 2001, pp. 136–137). In phenomenological terms, this second issue concerns the noetic aspect of 'ipseity' or the intransitive sense of I-ness in experience. As a number of neuroscientists have emphasized, this core of self-awareness is grounded on bodily processes of life regulation, emotion, and affect, such that cognition and intentional action are emotive (Panksepp 1998a, 1998b; Damasio 1999; Watt 1999; Freeman 2000), a theme central also to phenomenology (Merleau-Ponty 1962; Jonas 1966; Husserl 2001).

Phenomenal Consciousness and Access Consciousness Revisited

Let us return now to the distinction between phenomenal consciousness and access consciousness. How does this distinction look from the phenomenological perspective just sketched?

It may be helpful to anchor this discussion in an example. N. Block (1997, pp. 386–387) gives the example of being engaged in an intense conversation while a power drill rattles away outside the window. Engrossed in the conversation, one does not notice the noise, but then eventually and all of sudden one does notice it. Block's proposal is that insofar as one is aware of the noise all along, one is phenomenally conscious of it but not access conscious of it. When one notices the noise, one becomes access conscious of it (and perhaps also realizes that one has been hearing it all along), so that one now has both phenomenal consciousness and access consciousness of the noise.

J. K. O'Regan and A. Noë (2001, p. 964) dispute this description, claiming that one does not hear the drill until one notices it. One's auditory system may respond selectively to the noise, but one makes no use of the information provided thereby, nor is one poised to make any use of that information, until one notices the drill. Hence, there is no ground for thinking that we have a case of phenomenal consciousness without access consciousness. In the absence of access, there is no phenomenal consciousness.

From a phenomenological point of view, both descriptions seem somewhat flat. First of all, the phenomenological difference between not noticing and then noticing a noise is treated statically, as if it were a discrete state transition, with no extended temporal dynamics. Secondly, there is no differentiation, either at a static conceptual level or within the temporal dynamics of the experience, between prereflective and reflective aspects. One may notice a noise in a prereflective or unreflective sense, in which case one inattentively experiences the noise without positing it as a distinct object of awareness.[16] One may also notice a noise in the sense of turning one's attention to it or having one's attention be captured by it, in which case it does become posited as a distinct object of awareness. Finally, at a prereflective level, there is no differentiation between moments of comparatively weak and strong affective force on the part of the noise as the experience unfolds.[17]

---

[16] O'Regan and Noë might deny this claim because they argue that all consciousness is transitive consciousness, and all transitive consciousness is fully attentional: to be conscious of *X* is to put all one's sensorimotor and attentional resources onto *X*, that is, to be 'actively probing' *X*. Besides neglecting intransitive and prereflective self-consciousness, this account leaves out the moment of passive affection prerequisite to active probing, and thereby leaves attention unmotivated: What could motivate attention to the sound other than its affecting one inattentively? For further discussion, see Thompson (forthcoming). See also Ellis (2001).

[17] Compare Husserl's description of this sort of case: '[A] soft noise becoming louder and louder takes on a growing affectivity in this materially relevant transformation; the vivacity of it in consciousness increases. This means that it exercises a growing pull on the ego. The ego finally turns toward it. However, examining this more precisely, the modal transformation of affection has already occurred prior to the turning toward. Along with a certain strength that is at work under the given affective circumstances, the pull proceeding from the noise has so genuinely struck the ego that it has come into relief for the ego, even if only in the antechamber of the ego. The ego already detects it now in its particularity even though it does not yet pay attention to it by grasping it in an attentive manner. This "already detecting" means that in the ego a positive tendency is awakened to turn toward the object, its "interest" is aroused – it becomes an acutely active interest in and through the turning toward in which this positive tendency, which goes from the ego-pole toward the noise, is fulfilled in the striving-toward. Now we understand the essential modal transformation that has occurred here. First an increasing affection; but the affective pull is not yet, from the standpoint of the ego, a counter-pull, not yet a responsive tendency toward the allure issuing from the object, a tendency that for its part can assume the new mode of an attentively grasping tendency. There are

These considerations suggest that Block is right that hearing the sound before noticing it is a case of phenomenal consciousness. Yet there seems no good reason to believe that it is a case of phenomenal consciousness without access consciousness (which is the burden of Block's argument). One does consciously hear the sound before noticing it, if 'noticing' means turning one's attention to it. The sound is experienced prereflectively or unreflectively. One lives through the state of being affected by the sound without thematizing the sound or one's affectedness by it. This unreflective consciousness counts as phenomenal consciousness, because the phenomenality (appearance) of the sound has a subjective character, and the sound's affective influence is given first-personally. Hence, it does not seem right to say that one has no experience of the sound at all until one notices it (at least if 'notice' means focal attention to the sound). Nevertheless, no compelling reason has been given to believe that this experience is not also a case of access consciousness. After all, one is poised to make use of one's inattentive and unreflective hearing of the sound. Thus, the content of prereflective experience is at least access*ible*, even if it is not accessed explicitly. Jennifer Church makes a similar point:

> [T]he access*ibility* (i.e., the access *potential*) of the hearing experience is evident from the fact that I do eventually access it. Further, it seems that I *would* have accessed it sooner had it been a matter of greater importance – and thus, in a still stronger sense, it was accessible all along. Finally, it is not even clear that it was not *actually* accessed all along insofar as it rationally guided my behaviour in causing me to speak louder, or move closer, and so forth. (Church 1997, p. 426)

On the other hand, if we imagine that one is not cognitively poised in any way to rely on the sound, then we would need a reason to believe that one is nonetheless phenomenally conscious of it, rather than simply discriminating (differentially responding to) it nonconsciously, but no reason is forthcoming simply from this example.

The terminology of 'access consciousness' and 'phenomenal consciousness' is foreign to traditional phenomenology, and so it would be problematic to use this distinction to explicate phenomenological ideas. On the one hand, phenomenologists emphasize that most of experience

further distinctions that can be made here, but they do not concern us at this time' (Husserl 2001, p. 215). This description is explicitly temporal and dynamic; it displays phenomenal consciousness as characterized by continuous graded transformations of accessibility or access potential; and it roots modal transformations of consciousness in the dynamics of affect and movement tendencies (emotion).

is lived through unreflectively and inattentively, with only a small portion being thematically or attentively given. This view differentiates phenomenology from views that identify the contents of consciousness strictly with the contents of focal attention. On the other hand, so far as we can see, phenomenological analyses provide no reason to think that at this unreflective, implicit level, it is conceptually possible for there to be experiences that are not accessed by or accessible to other mental states and processes guiding behaviour.

In summary, according to phenomenology, lived experience comprises prereflective, precognitive, and affectively valenced mental states. These states are subjectively lived through, and thus have an experiential or phenomenal character, but their contents are not thematized. These states are also necessarily states of prereflective self-awareness (they have first-personal givenness); otherwise they do not qualify as conscious at all. Although not explicitly accessed in focal attention, reflection, introspection, and verbal report, they are accessible in principle: They are the kind of states that can become available to attention, reflection, introspection, and verbal report. In order to make active use of this potential access, however, first-person/second-person methods of phenomenological attentiveness and explication may be required. Thus, whereas many theorists discuss access consciousness and phenomenal consciousness in largely static terms, a phenomenological approach would reorient the theoretical framework by emphasizing the dynamics of the whole intentional structure of consciousness (the noetic-noematic structure), in particular, the structural and temporal dynamics of *becoming aware* at prereflective and reflective levels. Of particular concern to neurophenomenology is the process whereby implicit, unthematized, and intransitively lived-through aspects of experience can become thematized and verbally described, and thereby made available in the form of intersubjective, first-person data for neuroscientific research on consciousness. The rationale for introducing first-person methods into affective-cognitive neuroscience is to facilitate this process (see section 6).

5 Neurodynamics and Large-Scale Integration

As we mentioned in section 2, from an enactive perspective, the endogenous, self-organizing dynamics of large-scale neural activity is the appropriate level for characterizing neural processes in relation to mental activity. Neuroscience leaves little doubt that specific cognitive acts (such as visual recognition of a face) require the transient integration of widely distributed and continually interacting brain areas. It is also widely accepted that the neural activity crucial for consciousness involves the transient and continual orchestration of scattered mosaics of functionally specialized brain regions, rather than any single brain process or structure. Hence, a common theoretical proposal is that each moment of conscious awareness involves the transient selection of a distributed neural population that is both integrated or coherent, and differentiated or flexible, and whose members are connected by reciprocal and transient dynamic links.[18] A prelude to understanding the neural processes crucial for consciousness is to identify the mechanisms and dynamic principles of this large-scale activity. This problem is known as the large-scale integration problem (Varela, Lachaux. Rodriguez, and Martinerie 2001). Large-scale brain processes typically display endogenous, self-organizing behaviours, which are highly variable both from trial to trial and across subjects, and cannot be fully controlled by the experimentalist. Hence, affective-cognitive neuroscience faces at least a twofold challenge: (i) to find an adequate conceptual framework to understand brain complexity, and (ii) to relate brain complexity to conscious experience in an epistemologically and methodologically rigorous way.

Brain Complexity

For the first challenge, neurophenomenology follows the strategy, now shared by many researchers, of using the framework of complex dynamic systems theory. (This strategy corresponds to NPh2 in section 3.) According to the dynamical framework, the key variable for understanding large-scale integration is not so much the activity of the system's components, but rather the dynamic nature of the links between them. The neural counterpart of subjective experience is thus best studied not at the level of specialized circuits or classes of neurons (Crick and Koch 1998), but through a collective neural variable that describes the emergence and change of patterns of large-scale integration (Varela, Lachaux, Rodriguez, and Martinerie 2001). One recent approach to defining this collective variable of large-scale integration is to measure transient patterns of phase synchronization and desynchronization between oscillating neural populations at multiple frequency bands (Varela, Lachaux, Rodriguez, and Martinerie 2001; Engel, Fries, and Singer 2001). A

---

[18] See Varela (1995), Tononi and Edelman (1998), Freeman (1999a, 1999b), Dehaene and Naccache (2001), Engel and Singer (2001), Thompson and Varela (2001), Crick and Koch (2003).

number of researchers have hypothesized that such dynamic large-scale integration is involved in or required for certain generic features of consciousness, such as unity (Varela and Thompson 2003), integration and differentiation (Tononi and Edelman 1998), transitoriness and temporal flow (Varela 1995, 1999), and awareness of intentional action (Freeman 1999b).

We can explain the reasoning behind this approach in the form of two working hypotheses (Varela 1995; see also David, Cosmelli, Lachaux, Baillet, Garnero, and Martinerie 2003; Le Van Quyen 2003):

Hypothesis I: For every (type of) cognitive act, there is a singular and specific (type of) large-scale neural assembly that underlies its emergence and operation.

A neural assembly can be defined as a distributed subset of neurons with strong reciprocal connections. In the context of large-scale integration, a neural assembly comprises not only clusters of strongly interacting pyramidal cells and interneurons at a local scale but also more remote areas actively connected by excitatory pathways. On the one hand, there are reciprocal connections within the same cortical area, or between areas at the same level of the network. On the other hand, there are reciprocal connections that link different levels of the network in different brain regions to the same assembly (Varela, Lachaux, Rodriguez, and Martinerie 2001). Because of these strong interconnections (recall the principle of reciprocity mentioned in section 2), a large-scale neural assembly can be activated or ignited from any of its smaller subsets, whether sensorimotor or internal. These assemblies have a transient, dynamic existence that spans the time required to accomplish an elementary cognitive act (such as visual recognition of a face) and for neural activity to propagate through the assembly. Various empirical, theoretical, and phenomenological considerations suggest that the time-scale of such neurocognitive activity – whether it be a perception/action state, passing thought or memory, or emotional appraisal – is in the range of fractions of a second.[19] During these successive time intervals, there is competition between different neural assemblies: When a neural assembly is ignited from one or more of its smaller subsets, it either reaches a distributed coherence or is swamped by the competing activations of overlapping neural assemblies. If the assembly holds after its activation,

---

[19] See Varela, Thompson, and Rosch (1991, pp. 72–79), Dennett and Kinsbourne (1992), Varela (1995, 1999), Lewis (2000), Crick and Koch (2003), VanRullen and Koch (2003).

one can assume it has a transitory efficient function. The holding time is bound by two simultaneous constraints: (1) It must be larger than the time for spike transmission either directly or through a small number of synapses (from a few milliseconds to several tens of milliseconds); and (2) it must be smaller than the time it takes for a cognitive act to be completed, which is on the order of several hundreds of milliseconds. Thus, the relevant neuronal processes are distributed not only in space but also over periods of time that cannot be compressed beyond a certain limit (a fraction of a second).

Hypothesis I is a strong one, for it predicts that only one dominant or major neural assembly is present during a cognitive act, and that physiological correlates associated with the assembly should be repeatedly detected for different token realizations of the same act (for instance, in an oddball discrimination task or go/no-go task in the laboratory). The formation of the dominant assembly involves the selection (incorporation and rejection) of multiple neuronal activities distributed over both cortical and subcortical areas. The total flexible repertoire of such dynamic neural assemblies provides the 'dynamic core' of cognition (Tononi and Edelman 1998; Le Van Quyen 2003).

Given this first hypothesis, the issue arises of how large-scale dynamic integration is actually accomplished in the brain so as to produce a flow of coherent and adaptive cognitive acts. The basic intuition is that a specific neural assembly arises through a kind of 'temporal glue'. The most well studied candidate for this temporal integrative mechanism is neural phase synchronization. Neuronal groups exhibit a wide range of oscillations, from theta to gamma ranges (spanning 4–80 hertz or cycles a second), and can enter into precise synchrony or phase-locking over a limited period of time (a fraction of a second).[20] A growing body of evidence suggests that synchronization on a millisecond time-scale serves as a mechanism of brain integration (Varela, Lachaux, Rodriguez, and

---

[20] Synchrony in this context refers to the relation between the temporal structures of the signals regardless of signal amplitude. Two signals are said to be synchronous if their rhythms coincide. In signals with a dominant oscillatory mode, synchronization means the adjustment of the rhythmicity of two oscillators in a *phase-locking*:

$$n\phi_1(t) - m\phi_2(t) = const$$

where $\phi_1 t$, $\phi_2 t$ are the instantaneous phases of the oscillators, and n, m are integers indicating the ratios of possible frequency locking. It is usually assumed for simplicity that $n = m = 1$, but evidence for 1:2 and 1:3 phase synchrony also exists (see Lachaux, Rodriguez, Martinerie, and Varela 1999; Varela, Lachaux, Rodriguez, and Martinerie 2001; Le Van Quyen 2003).

Martinerie 2001; Engel, Fries, and Singer 2001). This idea can be stated in the form of a second hypothesis (Varela 1995):

Hypothesis II: A specific neural assembly is selected through the fast, transient phase-locking of activated neurons belonging to subthreshold competing neural assemblies.

Neural phase synchrony is a multiscale phenomenon occurring in local, regional, and long-range networks. It is useful to distinguish between two main scales, short-range and long-range. Short-range integration occurs over a local network (e.g., columns in primary visual cortex), distributed over an area of approximately 1 centimeter, through monosynaptic connections with conduction delays of 4 to 6 milliseconds. Most electrophysiological studies in animals have dealt with short-range synchronies or synchronies between adjacent areas corresponding to a single sensory modality. These local synchronies have usually been interpreted as a mechanism of 'perceptual binding' – the selection and integration of perceptual features in a given sensory modality (e.g., visual Gestalt features) (Singer 1999). Large-scale integration concerns neural assemblies that are farther apart in the brain, and connected through polysynaptic pathways with transmission delays greater than 8 to 10 milliseconds (for a review, see Varela, Lachaux, Rodriguez, and Martinerie 2001). In this case, phase synchrony cannot be based on the local cellular architecture but must instead reside in distant connections (cortico-cortical fibers or thalamocortical reciprocal pathways). These pathways correspond to the large-scale connections that link different levels of the network in different brain regions to the same assembly.

Long-distance phase synchronization is hypothesized to be a mechanism for the transient formation of a coherent macroassembly that selects and binds multimodal networks (such as assemblies between occipital and frontal lobes, or across hemispheres, which are separated by dozens of milliseconds in transmission time). Phase synchrony measures have predictive power with respect to subsequent neural, perceptual, and behavioural events (Engel, Fries, and Singer 2001). Animal and human studies demonstrate that specific changes in synchrony occur during arousal, sensorimotor integration, attentional selection, perception, and working memory, all of which are crucial for consciousness (for reviews, see Engel, Fries, and Singer 2001; Varela, Lachaux, Rodriguez and Martinerie, 2001). It has also been hypothesized that whether a local process participates directly in a given conscious state depends on whether it participates in a coherent, synchronous global assembly (Dehaene and

Naccache 2001; Engel and Singer 2001). Evidence for these hypotheses comes from studies at coarser levels of resolution than the microscale of single neuron activity, namely, the intermediate or mesoscale of local field potentials (the summated dendritic current of local neural groups), and the macroscale of scalp recordings in EEG (electroencephalography) and MEG (magnetoencephalography). These studies provide direct evidence for long-range synchronizations between widely separated brain regions during cognitive tasks.

Long-distance phase synchronies occur in a broad range of frequencies. Fast rhythms (above 15 hertz) in gamma and beta frequencies meet the requirement for fast neural integration, and thus are thought to play a role in conscious processes on the time scale of fractions of a second (Varela 1995; Tononi and Edelman 1998). Yet neural synchrony must also be understood in the context of the slower alpha and theta bands (4–12 hertz), which play an important role in attention and working memory (Sarnthein, Petsche, Rappelsberger, Shaw, and von Stein 1998; von Stein, Chiang, and König 2000; von Stein and Sarnthein 2000; Fries, Reynolds, Rorie, and Desimone 2001), sensorimotor integration (O'Keefe and Burgess 1999; Kahana, Seelig, and Madsen 2001), and probably emotion (Lewis in press). This evidence supports the general notion that phase synchronization subserves not simply the binding of sensory attributes but also the overall integration of all dimensions of a cognitive act, including associative memory, affective tone and emotional appraisal, and motor planning (Damasio 1990; Varela 1995; Varela, Lachaux, Rodriguez, and Martinerie 2001). More complex nonlinear forms of cross-band synchronization, so-called generalized synchrony (Schiff, So, Chang, Burke, and Sauer 1996), are thus also expected, and may indeed prove more relevant in the long run to understanding large-scale integration than strict phase synchronization (Friston 2000a, 2000b; Varela, Lachaux, Rodriguez, and Martinerie 2001; Le Van Quyen 2003).

It should be emphasized that large-scale integration must involve not only the establishment of dynamic links but also their active uncoupling or dismantling, in order to make way for the next cognitive moment. Thus, not only phase synchronization but also desynchronization or phase-scattering is likely to play an important role in large-scale integration (Varela 1995; Rodriguez, George, Lachaux, Martinerie, Renault, and Martinerie Varela 1999; Varela, Lachaux, Rodriguez, and Martinerie 2001, p. 236). A dynamic of synchronization and desynchronization, combined with retention or holdover of elements from the previous assembly at the beginning of each new one, might be the neural basis for the apparent unity and moment-to-moment transitoriness in the temporal flow of experience (Varela 1999).

In summary, Hypotheses I and II taken together imply that a succession of cognitive events is correlated to a succession of bursts of phase synchronization and desynchronization between either remote or close neural signals in large-scale neural assemblies. The neurophenomenological approach to brain complexity involves testing these working hypotheses, as well as developing new dynamic systems techniques for investigating more complex forms of cross-band relations (Lachaux, Chavez, and Lutz 2003; David, Cosmelli, and Friston 2004).

Relating Brain Complexity to Experience: Varela's 'Core Hypothesis'
The second challenge mentioned earlier is to relate brain complexity to conscious experience in a rigorous way. Let us address this point first at a theoretical level and then at a methodological level.

At a theoretical level, a working hypothesis is needed to specify explicitly what the link between conscious experience and large-scale integration via synchrony is supposed to be. An early hypothesis about this relation was proposed by Varela (1995). His idea was that the large-scale neural assembly that dominates over competing subthreshold assemblies constitutes a transient self-referential pole for the interpretation of current neural activity, and that conscious mental states are, in effect, neural interpretations of neural events. Similar proposals have since been put forward by Tononi and Edelman (1998), and recently by Crick and Koch (2003). Varela's version takes the form of rephrasing Hypotheses I and II in the following 'core hypothesis':

Core Hypothesis: 'Mental-cognitive states are interpretations of current neural activity, carried out in reference to a transient, coherency-generating process generated by that nervous system.'

We cite here Varela's original formulation (Varela 1995, pp. 90–91). A clearer rendering of what he means would be: *A mental-cognitive state is a neural interpretation of current neural activity, carried out by and in reference to a transient and coherent large-scale assembly, generated by the nervous system of an embodied and situated agent.*

The mental-cognitive states of concern in this discussion are conscious mental states. A conscious mental state is one having a subjective or first-personal character. Its content involves an appearance of something to someone (it is phenomenal and subjective). The issue is how to relate phenomenal content and subjectivity to neural processes on the microtemporal scale of a fraction of a second. A. R. Damasio has noted that many neurobiological proposals about consciousness focus on the problem of the specific phenomenal (e.g., sensory) contents of consciousness, while neglecting the problem of subjectivity or selfhood, 'the problem of how, in parallel with engendering mental patterns for an object, the brain also engenders a sense of self in the act of knowing' (Damasio 1999, p. 9). Varela's core hypothesis tries to keep both aspects together. On the one hand, the specific contents of current neural events derive from the organism's sensorimotor coupling and from the endogenous activity of its nervous system. On the other hand, these distributed and local neural events are always being evaluated from the point of view of the large-scale assembly dominant at the moment. This assembly acts as a global and dynamic self-referential pole – a sort of transient virtual self – for the selection of local activity.[21]

To say that a mental-cognitive state is a neural 'interpretation' of current neural activity means that distributed and local neural events are never taken at face value but are always 'seen' or 'evaluated' from the point of view of the assembly most dominant at the time. For example, a specific mental-cognitive act, such as visual recognition of a face, will be lived differently depending on factors such as arousal, motivation, attention, and associative memories unique to the individual. The neural events specific to the recognition will not be taken at face value but will be shaped and modified by the dominant large-scale assembly integrating frontal and limbic activities. Such endogenous large-scale activity is always present and forms a global background for evaluating local events generated exogenously and endogenously (see also Freeman 2000). On the other hand, these local events modulate and contribute to the emergence of a dominant assembly from competing subthreshold assemblies. This circular causality sustains a kind of 'neural hermeneutics' or 'neural hermeneutical circle', in which a preexisting dominant assembly (a background preunderstanding) shapes the meaning of novel events (motivates an interpretation), while being modified by those events (the interpretation leads to a new understanding). For instance, as we will see when looking at some experimental work, the visual recognition of an image is experienced differently depending on the subjective context leading up to the moment of perception, and these differences in

---

[21] Of course, this transient pole of integration must be embedded in the brainstem and body-regulation processes constitutive of the proto-self (dynamic neural patterns that map the body state) and core consciousness (primitive self-consciousness) (see Parvizi and Damasio 2001).

antecedent subjective context and subsequent perception are reflected in corresponding differences in the local and long-distance patterns of phase synchronization and desynchronization in large-scale assemblies (Lutz, Lachaux, Martinerie, and Varela 2002).

According to the core hypothesis, the large-scale neurodynamics of mental states should exhibit three generic characteristics (Le Van Quyen 2003): (1) *metastability:* the coherence-generating process constantly gives rise to new large-scale patterns without settling down or becoming trapped in any one of them (Kelso 1995; Friston 2000a, 2000b); (2) *rapid integration:* these patterns are generated quickly, on a time period of 100 to 300 milliseconds; and (3) a *global self-reference pole:* the coherence-generating process provides a global self-referential pole for the selection (incorporation and rejection) of ongoing local neuronal activities. The inevitable multiplicity of concurrent potential assemblies is evaluated in relation to this dynamic reference pole, until one assembly is transiently stabilized and expressed behaviourally, before bifurcation to another. A map of conceptual and mathematical frameworks to analyze these spatiotemporal large-scale brain phenomena has recently been proposed by M. Le Van Quyen (2003).

The core hypothesis is not meant to be an a priori internalist identity thesis of either a type-type (mental properties = neural properties) or token-token (mental events = neural events) form. It is meant rather as an empirical hypothesis about how to distinguish in the brain the neural events underlying or directly contributing to a distinct mental state (on a fraction-of-a-second time scale) from other neural events going on in the nervous system. Nor is there any suggestion that the former sort of distinctly neurocognitive events is *sufficient* for mental states. According to the hypothesis, every mental state (or cognitive act) requires the formation of a specific large-scale neural assembly. It does not follow, however, that the internal neural characteristics of such assemblies are sufficient for their correlative mental states. On the contrary, the somatic and dynamic sensorimotor context of neural activity is also crucial (Thompson and Varela 2001; Hurley and Noë 2003; Noë and Thompson 2004a, 2004b). According to the enactive approach, mental states depend crucially on the manner in which neural processes are embedded in the somatic and environmental context of the organism's life, and hence, it is doubtful that there is such a thing as a minimal internal neural correlate, even a complex dynamical one, whose intrinsic properties are sufficient for conscious experience.

Finally, the transient, coherence-generating process referred to in the core hypothesis is hypothesized to be none other than large-scale integration via phase synchronization (a mechanism that will probably have to be expanded to include generalized synchrony). Thus, among the various ways to define the state variable of large-scale coherence, one recent approach is to use as a 'dynamical neural signature' the description and quantification of transient patterns of local and long-distance phase synchronies occurring between oscillating neural populations at multiple frequency bands (Rodriguez, George, Lachaux, Martinerie, Renault, and Varela 1999; Lutz, Lachaux, Martinerie, and Varela 2002; Cosmelli, David, Lachaux, Martinerie, Garnero, Renault, and Varela 2004). (This proposal corresponds to a specific working hypothesis under the heading of NPh3 in section 3.) Current neurophenomenology thus assumes that local and long-distance phase synchrony patterns provide a plausible neural signature of subjective experience.

Relating Brain Complexity to Conscious Experience: Methodology
At a methodological level, neurophenomenology's new and original proposal is to test the core hypothesis by incorporating careful phenomenological investigations of experience into neurodynamical studies in an explicit and rigorous way. The aim is to integrate the phenomenal structure of subjective experience into the real-time characterization of large-scale neural activity. The response to the second challenge of relating brain complexity to conscious experience is, accordingly, to create experimental situations in which the subject is actively involved in identifying and describing phenomenal categories and structural invariants of experience that can be used to identify and describe dynamical neural signatures of experience and structural invariants of brain activity. As we will see (section 7), a rigorous relationship between brain complexity and subjective experience is thereby established, because original phenomenal categories and structural invariants are explicitly used to detect original neurodynamical patterns. Such joint collection and analysis of first-person and third-person data instantiates methodologically the neurophenomenological hypothesis that neuroscience and phenomenology can be related to each other through reciprocal constraints (Varela 1996). The long-term aim is to produce phenomenological accounts of real-time subjective experience that are sufficiently precise and complete to be both expressed in formal and predictive dynamical terms and shown to be realized in specific neurodynamical properties of brain activity. Such threefold, phenomenological, formal, and neurobiological descriptions of consciousness could provide a robust and predictive way to link reciprocally the experiential and neurophysiological domains.

6 First-Person Methods

For this approach to move forward, better methods for obtaining phenomenological data are needed. The role of first-person methods is to meet this need. First-person methods are disciplined practices that subjects can use to increase their sensitivity to their experience from moment to moment (Varela and Shear 1999b). They involve systematic training of attention and emotional self-regulation. Such methods exist in phenomenology (Moustakas 1994; Depraz 1999), psychology (Price and Barrell 1980; Price, Barrell, and Rainville 2002), psychotherapy (Gendlin 1981; Epstein 1996), and contemplative meditative traditions (Varela, Thompson, and Rosch 1991; Wallace 1999). Some are routinely used in clinical and health programs (Kabat-Zinn 1990), and physiological correlates and effects of some of these practices have been investigated (Austin 1998; Davidson, Kabat-Zinn, Schumacher, Rosenkranz, Muller, Santorelli, Urbanowski, Harrington, Bonus, and Sheridan 2003). The relevance of these practices to neurophenomenology derives from the capacity for sustained awareness of experience they systematically cultivate. This capacity enables tacit, preverbal, and prereflective aspects of subjective experience – which otherwise would remain unreflectively lived through – to become subjectively accessible and describable, and thus available for intersubjective and objective (biobehavioural) characterization.

First-person methods vary depending on the phenomenological, psychological, or contemplative framework. The following schema, taken from Varela and Shear (1999a, p. 11), calls attention to certain generic features of first-person methods:

- *Basic attitude: suspension* (of inattentive immersion in the content of experience), *redirection* (of awareness to ongoing process of experience), and *receptive openness* (to newly arising experience). This stance or posture must be adopted in one's own first-person singular case.
- *Intuitive filling-in:* stabilizing and sustaining the basic attitude so that awareness of the experiencing process becomes fuller in content. This process requires practice and involves the second-person perspective of a trainer or coach.
- *Expression and intersubjective validation:* this step of verbal description and intersubjective evaluation is required in order for there to be any valid first-person (phenomenological) data. It implicates and is always situated in a first-person plural perspective.

In phenomenology, the basic attitude of suspension, redirection, and receptive openness is known as the 'epoché'. This term originally comes from Greek scepticism, where it means to refrain from judgment, but Husserl adopted it as a term for the suspension, neutralization, or bracketing of our naïve realistic inclinations about the relationship between the world and our experience. As a procedure carried out in the first-person by the phenomenologist, the epoché is a practiced mental gesture of redirecting one's attention from the intentional object of experience to one's experiencing of the object, such that the object is now taken strictly as the objective correlate of the experiencing act or process. The basic principle of phenomenological analysis is that analysis of the experiencing process, rather than the specific objects of experience, discloses the invariant structures and factors within and across different types of experience.

The three phases of suspension, redirection, and receptive openness form a dynamic cycle. The first phase induces a transient suspension of beliefs or habitual thoughts about what is experienced. The aim is to bracket explanatory belief-constructs in order to adopt an open and unprejudiced descriptive attitude. This attitude is an important prerequisite for gaining access to experience as it is lived prereflectively. The second phase of redirection proceeds on this basis: Given an attitude of suspension, one's attention can be redirected from its habitual immersion in the object toward the lived character of the experiencing process. The epoché thus mobilizes and intensifies the tacit self-awareness of experience by inducing an explicit attitude of bare attention to the experiencing process. 'Bare attention' means noticing, witnessing, or being present to what is happening in one's experience, without explanation or judgment. This requires acceptance of whatever is happening, or in other words, a receptive openness. Thus, an attitude of receptivity or letting-go is encouraged, in order to broaden the field of experience to new horizons. Many aspects of experience are not noticed immediately but require multiple instances and variations in order to emerge. The repetition of the same task in an experimental setting, for instance, enables new contrasts to arise, and validates emerging phenomenal categories and structural invariants. Training and practice are therefore a necessary component to cultivate all three phases and to enable the emergence and stabilization of phenomenal invariants.

Downstream from this threefold cycle is the phase of verbalization or expression. The communication of phenomenal invariants provides the crucial step whereby this sort of first-person knowledge can be intersubjectively shared and calibrated, and eventually related to objective data.

This account of the procedural steps of the epoché represents an attempt to fill a lacuna of phenomenology, which has emphasized theoretical analysis and description, to the neglect of the actual pragmatics of the epoché (Depraz 1999). By contrast, the pragmatics of 'mindfulness' in the Buddhist tradition is far more developed (Varela, Thompson, and Rosch 1991). This is one reason that the description of the structural dynamics of becoming aware, as well as attempts to develop a more pragmatic phenomenology, have drawn from Buddhist traditions of mental discipline (Varela, Thompson, and Rosch 1991; Depraz, Varela, and Vermersch. 2000, 2003). One can also point to a recent convergence of theories and research involving introspection (Vermersch 1999), the study of expertise and intuitive experience (Petitmengin-Peugeot 1999; Petitmengin 2001), phenomenology (Depraz 1999), and contemplative mental self-cultivation (Wallace 1999). This convergence has also motivated and shaped the description of the generic features of first-person methods (see Depraz, Varela, and Vermersch 2000, 2003).

The use of first-person methods in affective-cognitive neuroscience clearly raises important methodological issues. One needs to guard against the risk of the experimentalist either biasing the phenomenological categorization or uncritically accepting it. D. C. Dennett (1991) introduced his method of 'heterophenomenology' (phenomenology from a neutral third-person perspective) in part as a way of guarding against these risks. His warnings are well taken. Neurophenomenology asserts that first-person methods are necessary to gather refined first-person data, but not that subjects are infallible about their own mental lives, nor that the experimentalist cannot maintain an attitude of critical neutrality. First-person methods do not confer infallibility upon subjects who use them, but they do enable subjects to thematize important but otherwise tacit aspects of their experience. At the time of this writing, Dennett has not addressed the issue of the scope and limits of first-person methods from his heterophenomenological viewpoint, and so it is not clear where he stands on this issue. A full exchange on this issue would require discussion of the different background epistemological and metaphysical differences between phenomenology and heterophenomenology concerning intentionality and consciousness. There is not space for such a discussion here.[22] We will therefore restrict ourselves to a comment

---

[22] For discussion of some aspects of this issue, see Thompson, Noë, and Pessoa (1999). A forthcoming special issue of *Phenomenology and the Cognitive Sciences* devoted to heterophenomenology will contain a paper by Thompson addressing the relations among heterophenomenology, phenomenology, and neurophenomenology.

about heterophenomenology as a method for obtaining first-person reports. Our view is that to the extent that heterophenomenology rejects first-person methods, it is too limited a method for the science of consciousness because it is unable to generate refined first-person data. On the other hand, to the extent that heterophenomenology acknowledges the usefulness of first-person methods, then it is hard to see how it could avoid becoming in its practice a form of phenomenology, such that the supposed opposition between 'hetero' and 'auto' phenomenology would no longer apply.

Another issue concerns the way that attention to experience can modify or affect the intentional content and phenomenal character of experience (Lambie and Marcel 2002). First-person methods rely on a sheer witnessing or noticing of the process of experiencing, a bare attention without judgment. Their effect on experience is, accordingly, different from introspection (in the introspectionist sense), which uses active attention to the objects of experience as its basis (thus, the introspectionists used active attention to a sensory stimulus as the basis for introspecting a sensation or image). Bare attention is not intrusive but maintains a light touch. First-person methods intensify or heighten awareness by mobilizing this kind of nonjudgemental attention.

This last point raises a related issue – the modification of experience by phenomenological training. If first-person methods work to transform awareness from an unstable and inattentive mode to a more stable and attentive one, then it follows that experience is being trained and reshaped. One might therefore object that one mode of experience is replacing another, and hence, the new mode of experience cannot be used to provide insight into the earlier mode of untrained experience. Although at a superficial level there may seem to be an incompatibility between gaining insight into something and transforming it, this way of looking at things is misguided. There need not be any inconsistency between altering or transforming experience (in the way envisaged) and gaining insight into experience through such transformation. If there were, then one would have to conclude that no process of cognitive or emotional development could provide insight into experience before the period of such development. Such a view is extreme and unreasonable. The problem with the objection is its assumption that experience is a static given, rather than dynamic, plastic, and developmental. Indeed, it is hard to see how the objection could even be formulated without presupposing that experience is a fixed, predelineated domain, related only externally to the process of becoming aware, such that this process would have to supervene from outside, instead of being motivated and called forth from within experience itself. First-person methods are not supposed to be a way of accessing such a (mythical) domain; they are supposed to be a way of enhancing and stabilizing the tacit self-awareness already intrinsic to experience, thereby 'awakening' experience to itself.

It is also to be expected that the stabilization of new phenomenal invariants in experience, resulting from using first-person methods, will be associated with specific short-term or long-term changes in brain activity. It has been shown, for instance, that category formation during learning is accompanied by changes in the ongoing dynamics of the cortical stimulus representation (Ohl, Scheich, and Freeman 2001). But the fact that phenomenological training can modify experience and brain dynamics is not a limitation but an advantage. Anyone who has acquired a new cognitive skill (such as stereoscopic fusion, wine tasting, or a second language) can attest that experience is not fixed but is dynamic and plastic. First-person methods help to stabilize phenomenal aspects of this plasticity so that they can be translated into descriptive first-person reports. As C. Frith writes in a recent comment on introspection and brain imaging: 'A major programme for 21st century science will be to discover how an experience can be translated into a report, thus enabling our experiences to be shared' (Frith 2002, p. 374). First-person methods help 'tune' experience, so that such translation and intersubjective corroboration can be made more precise and rigorous. The issue of the generality of data from trained subjects remains open but seems less critical at this stage of our knowledge than the need to obtain new data about the phenomenological and physiological processes constitutive of the first-person perspective.

Frith (2002), following A. I. Jack and A. Roepstorff (2002), also comments that 'sharing experiences requires the adoption of a second-person perspective in which a common frame of reference can be negotiated' (p. 374). First-person methods help to establish such a reference frame by incorporating the mediating 'second-person' position of a trainer or coach. Neurophenomenology thus acknowledges the intersubjective perspective involved in the science of consciousness (Thompson 2001; Depraz and Cosmelli 2004). Subjects needs to be motivated to cooperate with the experimentalist and empathetically to understand their motivations; and reciprocally the experimentalist needs to facilitate subjects' finding their own phenomenal invariants. Without this reciprocal, empathetically grounded exchange, there is no refined first-person data to be had.

7 Experimental Studies

In this section, we review two experimental studies of consciousness and large-scale cortical dynamics. Each study used a visual perception protocol (depth perception and binocular rivalry, respectively). In the first study, individual subjects gave detailed trial-by-trial accounts of their experience, and these descriptions were used to define stable experiential categories or phenomenal invariants for each individual subject. These phenomenal invariants were then used to detect and interpret neural imaging data (Lutz, Lachaux, Martinerie, and Varela 2002). In the second study, a strict structural invariant of experience for all subjects was defined and then used as a heuristic to reveal a corresponding structural invariant of cortical dynamics (David, Cosmelli, Hasboun, Garnero 2003; Cosmelli 2004; Cosmelli, David, Lachaux, Martinerie, Garnero, Renault, and Varela 2004). Both studies were explicitly motivated by the methodology and hypotheses of the neurophenomenological approach.

The Subjective Context of Perception

When awake and alert subjects are stimulated during an experiment, their brains are not idle or in a state of suspension but engaged in cognitive activity. The brain response derives from the interaction between this ongoing activity and the afferent stimulation that affects it. Yet because this ongoing activity has not been carefully studied, most of the brain response is not understood: Successive exposure to the same stimulus elicits highly variable responses, and this variability is treated as unintelligible noise (and may be discarded by techniques that average across trials and/or subjects). The source of this variability is thought to reside mainly in fluctuations of the subjective cognitive context, as defined by the subject's attentional state, spontaneous thought processes, strategy to carry out the task, and so on. Although it is common to control, at least indirectly, for some of these subjective factors (such as attention, vigilance, or motivation), the ongoing subjective mental activity has not yet been analyzed systematically.

One strategy would be to describe in more detail this ongoing activity by obtaining verbal reports from human subjects. These reports should reveal subtle changes in the subject's experience, whether from trial to trial, or across individuals. This type of qualitative first-person data is usually omitted from brain-imaging studies, yet if methodological precautions are taken in gathering such data, they could be used to shed light on cognition via a joint analysis with quantitative measures of neural activity. Following this approach, a pilot neurophenomenological study (Lutz, Lachaux, Martinerie, and Varela 2002) investigated variations in subjective experience for one limited aspect of visual perception, namely, the emergence of an illusory 3D figure during the perceptual fusion of 2D random-dot images with binocular disparities.

The task began with subjects fixating for seven seconds on a dot pattern containing no depth cues. At the end of this 'preparation period', the pattern was changed to a slightly different one with binocular disparities. Subjects then had to press a button as soon as the 3D shape had completely emerged. Throughout the trial, dense-array EEG signals were recorded, and immediately after the button-press subjects gave a brief verbal report of their experience. In these reports, they labeled their experience using phenomenal categories or invariants that they themselves had found and stabilized during a prior training session. The recording session thus involved the simultaneous collection of first-person data (phenomenological reports) and third-person data (electrophysiological recordings and behavioural measures of button-pressing reaction time).

In the training session, subjects intensively practiced performing the task in order to improve their perceptual discrimination and to enable them to explore carefully variations in their subjective experience during repeated exposure to the task. Subjects were instructed to direct their attention to their own immediate mental processes during the task and to the felt-quality of the emergence of the 3D image.

This redirection of awareness to the lived quality of experience corresponds to the epoché described in the previous section. Its aim is to intensify the tacit self-awareness of experience by inducing a more explicit awareness of the experiencing process correlated to a given experiential content. More simply put, the aim is to induce awareness not simply of the 'what' or object-pole of experience (the 3D percept) but also of the necessarily correlated 'how' or act-pole of experience (the performance of perceptual fusion and its lived or subjective character). As described earlier, this practice of becoming aware involves the three interlocking phases of suspension, redirection, and receptive openness.

In this pilot study, these phases were either self-induced by subjects familiar with them or facilitated by the experimenter through open questions (Petitmengin-Peugeot 1999). For example: Experimenter, 'What did you feel before and after the image appeared?' Subject, 'I had a growing sense of expectation but not for a specific object; however, when the figure appeared, I had a feeling of confirmation, no surprise at all'. Or, 'It was as if the image appeared in the periphery of my attention, but then my attention was suddenly swallowed up by the shape'.

Subjects were repeatedly exposed to the stimuli, and trial by trial they described their experience through verbal accounts, which were recorded on tape. In dialogue with the experimenters, they defined their own stable experiential categories or phenomenal invariants to describe the main elements of the subjective context in which they perceived the 3D shapes. The descriptive verbal reports from a total of four subjects were classified according to the common factor of *degree of preparation* felt by the subject and *quality of perception*. This factor was used to cluster the trials into three main categories, described as follows: Steady Readiness, Fragmented Readiness, and Unreadiness. Subcategories (describing the unfolding of the visual perception, for instance) were also found in individual subjects. These were not investigated in the pilot study.

- *Steady Readiness.* In most trials, subjects reported that they were 'ready', 'present', 'here', or 'well-prepared' when the image appeared on the screen, and that they responded 'immediately' and 'decidedly'. Perception was usually experienced with a feeling of 'continuity', 'confirmation', or 'satisfaction'. These trials were grouped into a cluster SR, characterized by the subjects being in a state of 'steady readiness'.
- *Fragmented Readiness.* In other trials, subjects reported that they had made a voluntary effort to be ready, but were prepared either less 'sharply' (due to a momentary 'tiredness') or less 'focally' (due to small 'distractions', 'inner speech', or 'discursive thoughts'). The emergence of the 3D image was experienced with a small feeling of surprise or 'discontinuity.' These trials formed a second cluster corresponding to a state of 'fragmented readiness'.
- *Unreadiness (Spontaneous Unreadiness, Self-Induced Unreadiness).* In the remaining trials, subjects reported that they were unprepared and saw the 3D image only because their eyes were correctly positioned. They were surprised by it and reported that they were 'interrupted' by the image in the middle of a thought (memories, projects, fantasies, etc.). This state of distraction occurred spontaneously for subjects $S1$ and $S4$, whereas $S2$ and $S3$ triggered it either by fantasizing or by thinking about plans ($S3$), or by visualizing a mental image ($S2$). To separate passive and active distraction, these trials were divided into

two different clusters, 'spontaneous unreadiness' for S1 and S4, and 'self-induced unreadiness' for S2 and S3.[23]

These phenomenal invariants found in the training session were used to divide the individual trials of the recording session into corresponding phenomenological clusters. The EEG signals were analyzed to determine the transient patterns of local and long-distance phase synchronies between electrodes, and separate dynamical analyses of the signals were conducted for each cluster. The phenomenological clusters were thus used as a heuristic to detect and interpret neural activity. The hypothesis was that distinct phenomenological clusters would be characterized by distinct dynamical neural signatures before stimulation (reflecting state of preparation), and that these signatures would then differentially condition the neural and behavioural responses to the stimulus. To test this hypothesis, the behavioural data and the EEG data were analyzed separately for each cluster.

The overall result was that original dynamical categories of neural activity were detected, and hence, the opacity in brain responses (due to their intrinsic variability) was reduced. For an example, we can consider the contrast between the two clusters of Steady Readiness and Spontaneous Unreadiness for one of the subjects. In the first cluster, the subject reported being prepared for the presentation of the stimulus, with a feeling of continuity when the stimulation occurred and an impression of fusion between himself and the percept. In the second cluster, the subject reported being unprepared, distracted, and having a strong feeling of discontinuity in the flux of his mental states when the stimulus was presented. He described a clear impression of differentiation between himself and the percept. These distinct features of subjective experience were correlated with distinct dynamical neural signatures (in which phase-synchrony and amplitude were rigorously separated in the dynamical analysis). During steady preparation, a frontal phase-synchronous ensemble emerged early between frontal electrodes and was maintained on average throughout the trial, correlating with the subject's impression of continuity.

The average reaction time for this group of trials was short (300 milliseconds on average). The energy in the gamma band (30–70 Hz)

---

[23] An intermediate cluster between the first and second clusters was defined for subject S3. This state was described as one of open attention without active preparation. It was unique to this subject who found that this state contrasted sharply with that of prepared steady readiness.

increased during the preparation period leading up to the time of stimulus presentation. This energy shift toward the gamma band occurred in all subjects and was specific to the 'prepared' clusters. The energy in the gamma band was always higher in anterior regions during the prestimulus period for subjects in the 'prepared' clusters than for subjects in the 'unprepared' clusters, whereas the energy in the slower bands was lower. These results suggest that the deployment of attention during the preparation strategy was characterized by an enhancement of the fast rhythms in combination with an attenuation of the slow rhythms. On the other hand, in the unprepared cluster, no stable phase-synchronous ensemble can be distinguished on average during the prestimulus period. When stimulation occurred, a complex pattern of weak synchronization and massive desynchronization or phase-scattering between frontal and posterior electrodes was revealed. A subsequent frontal synchronous ensemble slowly appeared while the phase-scattering remained present for some time. In this cluster, the reaction time was longer (600 milliseconds on average). The complex pattern of synchronization and phase-scattering could correspond to a strong reorganization of the brain dynamics in an unprepared situation, delaying the constitution of a unified cognitive moment and an adapted response. This discontinuity in the brain dynamics was strongly correlated with a subjective impression of discontinuity.

Apart from these patterns common to all subjects, it was also found that the precise topography, frequency, and time course of the synchrony patterns during the preparation period varied widely across subjects. These variations should not be treated as 'noise', however, because they reflect distinct dynamical neural signatures that remained stable in individual subjects throughout several recording sessions over a number of days.

In summary, this study demonstrated that (i) first-person data about the subjective context of perception can be related to stable phase synchrony patterns measured in EEG recordings before the stimulus; (ii) the states of preparation and perception, as reported by the subjects, modulated both the behavioural responses and the dynamic neural responses after the stimulation; and (iii) although the precise shape of these synchrony patterns varied among subjects, they were stable in individual subjects throughout several recording sessions, and therefore seem to constitute a consistent signature of a subject's cognitive strategy or aptitude to perform the perceptual task. More generally, by using first-person methods to generate new first-person data about the structure of subjective experience, and using these data to render intelligible some of the opacity of the brain response, this pilot study illustrates the validity and fruitfulness of the neurophenomenological approach.

Waves of Consciousness

As James and Husserl discussed at great length, conscious experience appears from the first-person perspective as an unceasing yet continually changing flow of moments of awareness. Is it possible to take account of this phenomenal temporality while studying the brain mechanisms underlying consciousness? This question motivated the second study, in which binocular rivalry and magnetoencephalography (MEG) were used to investigate the temporal dynamics of brain activity during conscious perception.

When two different visual patterns are presented simultaneously, one to each eye, the patterns are seen as alternating back and forth at irregular intervals, a phenomenon known as binocular rivalry. In binocular rivalry, as in other forms of multistable perception, the perception changes while the visual stimulation remains the same. Binocular rivalry has thus been used as a probe for finding the neural processes associated specifically with the content of a moment of conscious perception.[24] Yet binocular rivalry also offers an ideal experimental condition to assess the dynamics of cortical activity during ongoing conscious perception, because the perception of a given stimulus fluctuates spontaneously and unpredictably in time.

The rivalrous stimuli used in this experiment were an image of a face and an image of a series of expanding checkerboard rings. The subject was instructed to indicate when one or the other image became dominant by pressing one or the other of two buttons. The expanding rings spanned from 0 to 4 degrees of visual eccentricity five times per second, producing a concomitant evoked cortical response at 5 hertz. This spanning velocity was thus used as a 'frequency tag' to mark neural activity and thereby guide the localization of the cortical network specifically evoked by this pattern and the modulation of this network throughout successive dominance and suppression periods.

Prior to the recording session, subjects familiarized themselves extensively with the rivalry experience for these stimuli. They gave detailed phenomenological accounts, guided by the second-person method of open questions asked by the experimenter (Petitmengin-Peugeot 1999). From these descriptions, the following structural invariant of experience was

---

[24] See Blake (2001) and Blake and Logothetis (2002) for reviews.

found to be strictly valid for all subjects: During binocular rivalry, periods of dominance are *recurrent* through time, while the transitions between periods of dominance are *highly variable* in the way they arise. Although highly general, this invariant is nonetheless a significant constraint on the detection and interpretation of the underlying neural activity. In particular, one can hypothesize that in binocular rivalry, cortical activity during conscious perception may exhibit a corresponding structure of recurrent dynamic patterns with highly variable transitions.[25]

Testing this hypothesis requires an approach that can map recurrent brain sources with no restrictions on the temporal sequence of their activation, and that can explicitly take into account the variable dynamics of the perceptual transitions without averaging, which could destroy potentially relevant information. Hence, novel source localization techniques (for inferring underlying cortical activity from MEG data) and statistical analyses (for studying recurrence and variability without averaging) were developed and tested (David, Garnero, Cosmelli, and Varela 2002; David, Cosmelli, Hasboun, and Garnero 2003).

Phase synchrony analysis was performed on the brain signals reconstructed from the MEG data using these source localization and statistical techniques. The hypothesis was that during conscious perception of the 5 hertz rings, phase synchrony in this frequency band would rise and fall in concert with perceptual dominance. A consistent correlation between the time course of the synchronization of the cortical network and alternations in conscious perception was found. Thus, overall evoked cortical synchrony correlated with conscious perception. To analyze the spatiotemporal dynamics of the specific cortical networks engaged during conscious perception, the ongoing pattern of synchronous brain activity throughout a series of perceptual transitions was followed. A dynamic

---

[25] This hypothesis is reminiscent of Köhler's principle of isomorphism. In *Gestalt Psychology*, Köhler states: 'The principle of isomorphism demands that in a given case the organization of experience and the underlying physiological facts have the same structure' (1947, p. 301). But there is an important difference. Kohler presents isomorphism as a principle that demands a structural correspondence between experience and physiology; hence, his view seems to require isomorphism as an a priori condition of any successful explanation. Pessoa, Thompson, and Noë (1998) call this view *analytic isomorphism*. Our conjecture, on the other hand, is stated as a working hypothesis, rather than a principle. The conjecture serves as a heuristic constraint that guides the search for dynamic patterns of neural activity pertinent to consciousness. Isomorphism serves as a working tool for tracking brain dynamics at the level of large-scale integration, not as an explanatory principle. For further discussion of issues about isomorphism see Pessoa, Thompson, and Noë (1998) and Noë and Thompson (2004a).

buildup of the synchronous network was apparent throughout the perceptual alternations. At the beginning of each transition, very few synchronous pairs were evident. As perceptual dominance developed, the occipital pole showed an increase in local synchrony involving primary visual cortex and more dorsal occipital areas. Long-range coactivation was then established between occipital regions and more frontal areas, including mainly medial frontal regions. This pattern of occipito-frontal distributed network was maintained for several seconds and coincided with full perceptual dominance of the tagged expanding checkerboard rings. Then, as suppression began, long-range synchronous activity fell apart, leaving coactive areas in the occipital pole, and in some cases, inferotemporal regions up to the temporal pole. During full suppression, very few coactive regions were left and brain patterns returned to the pretransition situation. This study thus presented for the first time ongoing patterns of cortical synchronous activation that correlate with the spontaneous stream of conscious perception.

8 Neurophenomenology and the Neural Correlates of Consciousness

A currently popular style of investigation is to search for the neural correlates of consciousness (or NCCs) (Metzinger 2000; Rees, Kreiman, and Koch 2002). The term 'correlate' is potentially misleading, however, for the goal is to discover not mere correlates but neural events that are sufficient for the contents of consciousness (Kanwisher 2001), as can be seen from this definition of an NCC: 'An NCC (for content) is a minimal neural representational system N such that representation of a content in N is sufficient, under conditions C, for representation of that content in consciousness' (Chalmers 2000, p. 31).

As this definition makes plain, the content-NCC approach is based on the assumption that the contents of consciousness match the contents of neural representational systems. Alva Noë and Evan Thompson (2004a) call this assumption the *matching content doctrine*. According to the matching content doctrine, for every conscious experience E, there is a neural representational system N, such that (i) N is the minimal neural representational system whose activation is sufficient for E, and (ii) there is a match between the content of E and the content of N.

The matching content doctrine goes hand in hand with a particular research strategy, which Searle (2000) has called the *building block model* of consciousness. The building block approach proceeds by trying to isolate neural correlates of the contents of consciousness for specific types of individual sensory experiences, such as the visual experience of a perceptually dominant stimulus in binocular rivalry. Although this NCC approach is "initially neutral on issues of causality" (Rees, Kreiman, and Koch 2002, p. 261), the ultimate aim is to determine not mere correlates of individual (types of) conscious states but the causally necessary and sufficient conditions of these states, including a theory or model of the neurophysiological mechanisms involved. The conjecture is that if we could determine the content NCCs for a particular sort of conscious experience, such as the visual experience of color or faces, or perhaps visual experience in general, this finding might generalize to other sorts of conscious experiences (Crick and Koch 1998). To put it another way, if we could determine for color vision, or vision in general, what makes a given content a phenomenally conscious content, then we might be able to determine what makes a content conscious for any modality.

There are a number of problems with both the matching content doctrine and the building block model, and hence, with this NCC approach overall (Noë and Thompson 2004a, 2004b). First, although neural correlates of various types of conscious states have been found experimentally, these correlates do not provide any case of a match between the content of a neural system and the content of a conscious state. For example, there is no content match between a population of cells whose content is specified in terms of the receptive field properties of single neurons and the perceptual content of seeing the image of a face in binocular rivalry. Second, it is doubtful that neural systems, at least as standardly conceived in much of the NCC literature, could match conscious states in content, as a result of the logical and conceptual differences between these two sorts of content (Noë and Thompson 2004a). Indeed, it is not unreasonable to think that the very notion of a content match between the level of neurons and the level of consciousness is a category mistake (Tononi and Edelman 1998; Searle 2004). Third, it is problematic to assume that consciousness is made up of various building blocks corresponding to constituent individual experiences, which are then somehow bound together to constitute the unity (or apparent unity) of consciousness. As J. Searle points out: 'Given that a subject is conscious, his consciousness will be modified by having a visual experience, but it does not follow that the consciousness is made up of various building blocks of which the visual experience is just one' (Searle 2000, p. 572). Finally, to remain 'initially neutral on issues of causality' is a significant limitation, given the need to investigate the hypotheses that (i) each conscious state (on a fraction-of-a-second time-scale) involves the transient selection of a large-scale dynamic neural assembly, and that (ii) the global behaviour of such an assembly modulates and constrains local neural activity, and thereby causally influences mental processes and intentional action (Freeman 1999b; Thompson and Varela 2001; Le Van Quyen and Petitmengin 2002). More generally, in postponing issues of causality, the NCC approach avoids what Daniel Dennett calls 'the Hard Question – And Then What Happens?' (Dennett 1991, p. 255; 2001, p. 225). In other words, the NCC approach runs the risk of treating consciousness as simply the end of the line, instead of asking not only what activity is causally necessary and sufficient for consciousness but also what causal difference that activity makes for other brain processes and the organism overall.

In presenting and criticizing the building block model, Searle (2000) also singles out a different approach to the brain basis of consciousness, which he calls the *unified field model*. Searle argues that the neural substrates of individual conscious states should not be thought of as sufficient for the occurrence of those states, for those states themselves presuppose the background consciousness of the subject. Any given conscious state is a modulation of a preexisting conscious field. According to this unified field approach, an individual experience or conscious state (such as visual recognition of a face) is not a constituent (in the building block sense) of some aggregate conscious state, but is rather a modification within the field of a basal or background consciousness: 'Conscious experiences come in unified fields. In order to have a visual experience, a subject has to be conscious already, and the experience is a modification of the field' (Searle 2000, p. 572).[26] The unified field approach accordingly focuses on (i) the neural basis of the whole unified field at the level of large-scale activity, rather than neural correlates for the contents of particular conscious states; (ii) the neurophysiological processes that characterize the dynamic modulation of the conscious field in time; and (iii) the differences in brain activity across basal or background states, such as dreamless sleep, dreaming, wakefulness, and so on.

Neurophenomenology is consistent with the unified field model. As Searle notes, according to the unified field approach, 'what we have to look for is some massive activity of the brain capable of producing a

---

[26] The idea of consciousness as a unified field of subjective experience has of course long been central to phenomenology (e.g., Gurwitsch 1964).

unified holistic conscious experience' (p. 574). One of the working hypotheses of neurophenomenology is that this massive activity is large-scale integration via phase synchrony (or more complex forms of generalized synchrony). This global self-organization is *metastable* (new patterns constantly arise without the system settling down into any one of them), *rapid* (occurring on a time-scale of 100–300 msec), and constitutes a *fundamental pole of integration* for the selection and modulation of local neuronal activity. It needs to be emphasized that this 'core hypothesis' (Varela 1995) is not meant as a hypothesis about what makes a particular brain process a neural correlate of the content of a particular conscious state. Rather, it is meant to be a hypothesis about what differentiates the neural activity that directly contributes to the emergence of a coherent (unified) conscious state from other neural activity going on in the nervous system.

We need to distinguish, however, between a momentary state of unified consciousness and basal consciousness in the sense of being awake and alert. This is the distinction between state consciousness and background consciousness (see section 4). Searle's proposal is that states of consciousness arise as modifications of the field of basal consciousness, but he does not clearly differentiate between the two notions of state consciousness and background consciousness. The distinction is important, however, not only conceptually and phenomenologically but also neurophysiologically. On the one hand, there is good reason to believe, as Searle notes following G. Tononi and G. Edelman (1998) and R. Llinas, U. Ribary, D. Contreras, and C. Pedroarena (1998), that the coherent large-scale activity that characterizes a transient unified conscious state occurs in thalamocortical networks. Strong cortico-cortical and cortico-thalamic projections are capable of generating a coherent neural process on a fast time-scale (100–300 msec) through ongoing reentrant, recursive, and highly parallel interactions among widely distributed brain areas. On the other hand, basal consciousness, in the sense of being awake and alert, depends fundamentally on brainstem structures and processes (Parvizi and Damasio 2001). Furthermore, basal consciousness includes a primary affective awareness or core consciousness of self (Panksepp 1998b; Damasio 1999). Hence, it seems unsatisfactory to say that consciousness, in the sense of a single, unified conscious field, 'is in large part localized in the thalamocortical system and that various other systems feed information to the thalamocortical system that produces modifications corresponding to the various sensory modalities' (Searle 2000, p. 574). Although the thalamocortical system is clearly crucial for transient conscious states on a millisecond time-scale, midbrain and brainstem structures are crucial for affective core consciousness or sentience. Thus, virtually the entire neuraxis seems essential for consciousness in the widest sense of the term (Watt 1999).

The experimental neurophenomenological studies reviewed here are also more consistent with a unified field model than a building block model. Consider first the binocular rivalry study. Binocular rivalry has been one of the main experimental paradigms for the content NCC approach. In the NCC literature, binocular rivalry is usually presented as revealing a competition between two separate and distinct states that alternately become conscious. The research task is to discover the neural activity that represents what the subject sees (the content NCC) for each distinct conscious state (Crick and Koch 1998). Each conscious state is treated, in effect, as a building block of a composite rivalry experience. Noë and Thompson (2004a, pp. 24–25) have argued, however, that careful phenomenological examination suggests the experience of rivalry is not composite in the building block sense: It is not a composite experience having the normal (nonrivalrous) experience of one image and the normal (nonrivalrous) experience of the other image as constituent repeating elements. Rather, it seems better described as one bistable experience. In other words, the bistable experience of seeing a face/expanding checkerboard ring (for example) is not equivalent to the normal experience of seeing a face, plus the normal experience of seeing an expanding checkerboard ring, plus the two alternating with the appropriate temporal dynamics. As a bistable perception, it is a unique sort of experience, which accordingly must be assessed on its own terms. J. A. Kelso (1995) takes precisely this approach to the perception of multistable figures such as the Necker cube, describing this kind of perception as a metastable state. The neurophenomenological study of binocular rivalry develops this approach further. It takes a dynamic structural approach, rather than a building block or content NCC approach. The temporal dynamics of rivalry is considered to be constitutive of the experience (rather than an added element in building block fashion). By taking into account detailed phenomenological accounts of rivalry, a dynamic pattern of recurrence with highly variable transitions is revealed to be a structural invariant of the rivalry experience, a pattern irreducible to an alternation of two stable and clearly differentiated perceptual experiences (Cosmelli 2004). Using this structural invariant as a guide to detect and interpret neural activity, the neurophenomenological approach is able to show that ongoing patterns of cortical synchronization and desynchronization correlate with fluctuations of conscious perception.

The neurophenomenological study of depth perception provides a particularly nice case for the unified field model. This study examines 'modifications of the already existing field of qualitative subjectivity' (Searle 2000, p. 563), in the form of modifications of perceptual experience that depend on the subjective experience of readiness or unreadiness. To map the neurodynamics of ongoing conscious states, this study makes use of both proximal and distant baselines, corresponding respectively to the 1 second period immediately before stimulation, and the 7 second preparation period leading up to stimulation. Using the proximal baseline enhances the contrast between the synchronous process immediately preceding the arrival of the stimulation and those processes triggered by the stimulation, whereas using the distant baseline reveals a resemblance in synchrony patterns between prestimulus activity and the response induced by the stimulation. From a unified field perspective, these two baselines correspond to different time slices of the conscious field. The distant baseline takes in more, as it were, of the preexisting field of subjective experience, and thus reveals patterns of dynamic activity that are not seen using the proximal baseline. It also enables one to show that the preexisting conscious field is not only modified by, but also differentially conditions the emergence of, the 3D image. Described phenomenologically, the subjective context of readiness/unreadiness leading up to perception is a noetic factor of the field of experience, which differentially conditions the noematic factor of the quality of the depth percept. This neurophenomenological study thus focuses on the dynamic noetic/noematic structure of the 'field of consciousness' (Gurwitsch 1964).

Given the results of this study, it seems appropriate to redefine the temporal interval of interest for the neural correlate of a conscious act. The correlate of depth perception obviously occurs between the appearance of the stimulus and the motor response. Yet this moment of consciousness extends from a previous one, and finds its place within a temporal horizon of retention and anticipation that cannot be seen as neutral (Varela 1999). Hence, the characterization of both the ongoing activity preceding the stimulation and the activity following it is necessary for a complete description of the dynamics of a moment of consciousness. This point is also borne out by the binocular rivalry study, in which the temporal dynamics of fluctuations in perception are constitutive of the experience.

The last point we wish to make in this section concerns the issue of causality. We believe that neurophenomenology can not only produce original dynamic NCC data (in contrast to building block/content NCC data) but can also help illuminate the causal principles involved in conscious activity. For instance, it seems natural to interpret the depth perception study in terms of the dynamic systems principle of circular causality, in which local events self-organize into global coherent patterns that in turn modulate and constrain local events (see Freeman 1999b; Thompson and Varela 2001). Thus, the antecedent and 'rolling' subjective-experiential context modulates the way the perceptual object appears or is experientially lived during the moment of perception, and the content of this momentary conscious state reciprocally affects the flow of experience. At a neurodynamical level, the brain response presumably results from the intertwining of the endogenous brain activity and the peripheral afferent activity evoked by the stimulus. Therefore, the way the stimulus is directly lived by the subject (as indicated by disciplined phenomenological reports using first-person methods) could serve as a heuristic to shed light on the causal principles by which local afferent activity both modulates and is modulated by the contextual influence of the current large-scale neural assembly, such that this local activity either participates in, or is discarded from, the emergence of the following large-scale assembly. According to this perspective, a moment of consciousness is realized in the brain as a dynamic global state variable that constrains local activities (Freeman 1999b; Thompson and Varela 2001). Or as Kelso puts it: 'Mind itself is a spatiotemporal pattern that molds the metastable dynamic patterns of the brain' (Kelso 1995, p. 288).

9 Conclusion

We began this chapter by mentioning a number of issues associated with the explanatory gap. As we have seen, the neurophenomenological strategy for dealing with the epistemological and methodological issues about first-person data is to employ first-person methods in tandem with third-person methods. But what about the explanatory gap in the distinctly philosophical sense of the conceptual gap between subjective experience and neurobiology?

We wish to distinguish between the explanatory gap as a scientific problem and the hard problem of consciousness in the philosophy of mind. The explanatory gap as a scientific problem is the problem of presenting a model that can account for both the phenomenology and neurobiology of consciousness in an integrated and coherent way. The hard problem is an abstract metaphysical problem about the place of consciousness in nature (Chalmers 1996). This problem is standardly formulated as the issue of whether it is conceptually possible to derive subjective experience (or phenomenal consciousness) from objective physical nature. If it is possible, then materialism or physicalistic monism is supposed to gain support; if it is not possible, then property dualism (or panpsychism or idealism) is supposed to gain support.

Although Varela (1996) originally proposed neurophenomenology as a 'methodological remedy for the hard problem', a careful reading of this paper indicates that he did not aim to address the metaphysical hard problem of consciousness on its own terms. The main reason, following analyses and arguments from phenomenological philosophers (e.g., Merleau-Ponty 1962), is that these terms – in particular the Cartesian conceptual framework of the 'mental' versus the 'physical' – are considered to be part of the problem, not part of the solution. From a phenomenological perspective, Cartesian concepts of the mental and the physical cannot be allowed to govern the problem of consciousness. Space prevents further discussion of this point here (see Thompson forthcoming).

With regard to the explanatory gap as a scientific problem, our view is that one of the main obstacles to progress has to do with the way the conscious subject is mobilized in the experimental protocols of neuroscientific research on consciousness. Experimental investigations of the neural correlates of consciousness usually focus on one or another particular feature of experience, and accordingly try both to control as much as possible any variability in the content of subjective experience and to minimize reliance on the subject's verbal reports. Yet this approach seems too limited for investigating the labile, spontaneous, and self-affecting character of conscious processes and brain activity.[27] Thus, it is hard to see how this approach could ever bridge the gap between subjective experience and neurobiology.

We believe that a more fruitful way for the experimentalist to investigate these sorts of complex subjective and neurodynamical processes, and to define and control the variables of interest, is to make extensive use of the first-person perspective, approached in a phenomenologically disciplined way. Hence, neurophenomenology, without denying the validity of trying to control experimentally the subjective context from the outside, favours a complementary 'endogenous' strategy based on using first-person methods. By enriching our understanding of subjective

---

[27] See Varela (1999), Friston (2000a, 2000b), Lutz (2002), Hanna and Thompson (2004), and Cosmelli (2004).

experience through such phenomenological investigations, and using these investigations to cast light on neurodynamics, neurophenomenology aims to narrow the epistemological and methodological distance between subjective experience and brain processes in the concrete context of the working neuroscientist. At a more abstract conceptual level, neurophenomenology aims not to *close* the explanatory gap (in the sense of conceptual or ontological reduction), but rather to *bridge* the gap by establishing dynamic reciprocal constraints between subjective experience and neurobiology. At the present time, neurophenomenology does not claim to have constructed such bridges but only to have proposed a clear scientific research program for making progress on that task.

References

Austin, J. 1998. *Zen and the Brain.* Cambridge, MA: MIT Press.

Baars, B. J. 2002. The conscious access hypothesis: Origins and recent evidence. *Trends in Cognitive Sciences* 6: 47–52.

Bitbol, M. 2002. Science as if situation mattered. *Phenomenology and the Cognitive Sciences* 1: 181–224.

Blake, R. 2001. A primer on binocular rivalry, including current controversies. *Brain and Mind* 2: 5–38.

Blake, R., and Logothetis, N. 2002. Visual competition. *Nature Reviews Neuroscience* 3: 1–11.

Block, N. 1997. On a confusion about a function of consciousness. In N. Block, O. Flanagan, and G. Güzeldere (eds.), *The Nature of Consciousness: Philosophical Debates,* pp. 375–416. Cambridge, MA: MIT Press/A Bradford Book.

Block, N. 2001. Paradox and cross purposes in recent work on consciousness. *Cognition* 79: 197–219.

Botero, J.-J. 1999. The immediately given as ground and background. In J. Petitot, F. J. Varela, B. Pachoud, and J.-M. Roy (eds.), *Naturalizing Phenomenology: Issues in Contemporary Phenomenology and Cognitive Science,* pp. 440–463. Stanford, CA: Stanford University Press.

Burge, T. 1997. Two kinds of consciousness. In N. Block, O. Flanagan, and G. Güzeldere (eds.), *The Nature of Consciousness: Philosophical Debates,* pp. 427–434. Cambridge, MA: MIT Press/A Bradford Book.

Chalmers, D. 1996. *The Conscious Mind.* New York: Oxford University Press.

Chalmers, D. J. 1997. Availability: The cognitive basis of experience? In N. Block, O. Flanagan, and G. Güzeldere (eds.), *The Nature of Consciousness: Philosophical Debates,* pp. 421–424. Cambridge, MA: MIT Press/A Bradford Book.

Chalmers, D. J. 2000. What is a neural correlate of consciousness? In T. Metzinger (ed.), *Neural Correlates of Consciousness,* pp. 18–39. Cambridge, MA: MIT Press.

Church, J. 1997. Fallacies or analyses? In N. Block, O. Flanagan, and G. Güzeldere (eds.), *The Nature of Consciousness: Philosophical Debates,* pp. 425–426. Cambridge, MA: MIT Press/A Bradford Book.

Churchland, P. S. 1986. *Neurophilosophy: Toward a Unified Science of the Mind-Brain.* Cambridge, MA: MIT Press/A Bradford Book.

Churchland, P. S. 2002. *Brainwise: Studies in Neurophilosophy.* Cambridge, MA: MIT Press/A Bradford Book.

Cosmelli, D. 2004. *Des montagnes et des vallées: Perception consciente et structure dynamique de l'intégration cérébrale chez l'être humain dans l'expérience de rivalité binoculaire.* Doctoral thesis. Paris: Ecole Polytechnique.

Cosmelli, D., David, O., Lachaux, J.-P., Martinerie, J., Garnero, L., Renault, B., and Varela, F. J. 2004. Waves of consciousness: Ongoing cortical patterns during binocular rivalry. *Neuroimage* 23: 128–140.

Crick, F., and Koch, C. 1998. Consciousness and neuroscience. *Cerebral Cortex* 8: 97–107.

Crick, F., and Koch, C. 2003. A framework for consciousness. *Nature Neuroscience* 6: 119–126.

Damasio, A. R. 1990. Synchronous activation in multiple cortical regions: A mechanism for recall. *Seminars in the Neurosciences* 2: 287–297.

Damasio, A. R. 1999. *The Feeling of What Happens: Body and Emotion in the Making of Consciousness.* New York: Harcourt Brace.

David, O., Cosmelli, D., and Friston, K. J. 2004. Evaluation of a different measures of functional connectivity using a neural mass model. *Neuroimage* 21: 659–673.

David, O., Cosmelli, D., Hasboun, D., and Garnero, L. 2003. A multitrial analysis for revealing significant corticocortical networks in magentoencaphalography and electroencephalography. *Neuroimage* 20: 186–201.

David, O., Cosmelli, D., Lachaux, J.-P., Baillet, S., Garnero, L., and Martinerie, J. 2003. A theoretical and experimental introduction to the non-invasive study of large-scale neural phase-synchronization in human beings (invited paper). *International Journal of Computational Cognition* (http://www.YangSky.com/yangijcc.htm) 1: 53–77.

David, O., Garnero, L., Cosmelli, D., and Varela, F. J. 2002. Estimation of neural dynamics from MEG/EEG cortical current density maps: Application to the reconstruction of large-scale cortical synchrony. *IEEE Transactions on Biomedical Engineering* 49: 975–987.

Davidson, R. J., Kabat-Zinn, J., Schumacher, J., Rosenkranz, M., Muller, D., Santorelli, S. F., Urbanowski, F., Harrington, A., Bonus, K., and Sheridan, J. F. 2003. Alterations in brain and immune function produced by mindfulness meditation. *Psychosomatic Medicine* 65: 564–570.

Dehaene, S., and Naccache, L. 2001. Towards a cognitive neuroscience of consciousness: Basic evidence and a workspace framework. *Cognition* 79: 1–37.

Dennett, D. C. 1991. *Consciousness Explained.* Boston: Little Brown.

Dennett, D. C. 1997. The path not taken. In N. Block, O. Flanagan, and G. Güzeldere (eds.), *The Nature of Consciousness: Philosophical Debates*, pp. 417–420. Cambridge, MA: MIT Press/A Bradford Book.

Dennett, D. C. 2001. Are we explaining consciousness yet? *Cognition* 79: 221–237.

Dennett, D. C., and Kinsbourne, D. C. 1992. Time and the observer: The where and when of consciousness in the brain. *Behavioral and Brain Sciences* 15: 183–247.

Depraz, N. 1999. The phenomenological reduction as *praxis*. In F. J. Varela and J. Shear (eds.), *The View from Within: First-Person Approaches to the Study of Consciousness*, pp. 95–110. Thorverton, UK: Imprint Academic.

Depraz, N., and Cosmelli, D. 2004. Empathy and openness: Practices of intersubjectivity at the core of the science of consciousness. In E. Thompson (ed.), *The Problem of Consciousness: New Essays in Phenomenological Philosophy of Mind*. Canadian Journal of Philosophy Supplementary Volume. Calgary: University of Alberta Press.

Depraz, N., Varela, F. J., and Vermersch, P. 2000. The gesture of awareness: An account of its structural dynamics. In M. Velmans (ed.), *Investigating Phenomenal Consciousness*, pp. 121–136. Amsterdam and Philadelphia: John Benjamins Press.

Depraz, N., Varela, F. J., and Vermersch, P. 2003. *On Becoming Aware*. Amsterdam and Philadelphia: John Benjamins Press.

Derryberry, D., and Tucker, D. M. 1994. Motivating the focus of attention. In P. M. Niedenthal and S. Kitayama (eds.), *The Heart's Eye: Emotional Influences in Perception and Attention*, pp. 167–196. New York: Academic Press.

Dreyfus, H. 1991. *Being-In-The-World: A Commentary on Heidegger's Being and Time, Division I*. Cambridge, MA: MIT Press.

Dreyfus, H. 2002. Intelligence without representation – Merleau-Ponty's critique of mental representation. *Phenomenology and the Cognitive Sciences* 1: 367–383.

Dreyfus, H., and Dreyfus, S. 1982. *Mind over Machine*. New York: Free Press.

Ellis, R. 2001. Implications of inattentional blindness for 'enactive' theories of consciousness. *Brain and Mind* 2: 297–322.

Engel, A., and Singer, W. 2001. Temporal binding and the neural correlates of sensory awareness. *Trends in Cognitive Sciences* 5: 16–25.

Engel, A. K., Fries, P., and Singer, W. 2001. Dynamic predictions: Oscillations and synchrony in top-down processing. *Nature Reviews Neuroscience* 2: 704–716.

Epstein, M. 1996. *Thoughts Without a Thinker*. New York: Basic Books.

Foreman, R. K. C. (ed.). 1990. *The Problem of Pure Consciousness*. New York: Oxford University Press.

Freeman, W. J. 1999a. *How Brains Make up Their Minds*. London: Weidenfeld and Nicolson.

Freeman, W. 1999b. Consciousness, intentionality, and causality. *Journal of Consciousness Studies* 6: 143–172.

Freeman, W. J. 2000. Emotion is essential to all intentional behaviors. In M. Lewis and I. Granic (eds.), *Emotion, Development, and Self-Organization*, pp. 209–235. Cambridge: Cambridge University Press.

Fries, P., Reynolds, J. H., Rorie, A. E., and Desimone, R. 2001. Modulation of oscillatory neuronal synchronization by selective visual attention. *Science* 291: 1560–1563.

Friston, K. J. 2000a. The labile brain. I. Neuronal transients and nonlinear coupling. *Philosophical Transactions of the Royal Society of London Series B* 355: 215–236.

Friston, K. J. 2000b. The labile brain. II. Transients, complexity, and selection. *Philosophical Transactions of the Royal Society of London Series B* 355: 237–252.

Frith, C. 2002. How can we share experiences? *Trends in Cognitive Sciences* 6: 374.

Gallagher, S. 1986. Lived body and environment. *Research in Phenomenology* 16: 139–170.

Gendlin, E. T. 1981. *Focusing*. New York: Bantam.

Gupta, B. 1998. *The Disinterested Witness*. Evanston, IL: Northwestern University Press.

Gurwitsch, A. 1964. *The Field of Consciousness*. Pittsburgh, PA: Dusquesne University Press.

Hanna, R., and Thompson, E. 2004. Neurophenomenology and the spontaneity of consciousness. In E. Thompson (ed.), *The Problem of Consciousness: New Essays in Phenomenological Philosophy of Mind*. Canadian Journal of Philosophy Supplementary Volume. Calgary: University of Alberta Press.

Heidegger, M. 1996. *Being and Time*. Trans. J. Stambaugh. Albany: State University of New York Press.

Hobson, J. A. 1999. *Consciousness*. New York: W. H. Freeman.

Hurlbert, R. T., and Heavey, C. L. 2001. Telling what we know: Describing inner experience. *Trends in Cognitive Sciences* 5: 400–403.

Hurley, S. L., and Noë, A. 2003. Neural plasticity and consciousness. *Biology and Philosophy* 18: 131–168.

Husserl, E. 2001. *Analyses Concerning Passive and Active Synthesis: Lectures on Transcendental Logic*. Trans. A. J. Steinbock. Dordrecht: Kluwer Academic Publishers.

Jack, A. I., and Roepstorff, A. 2002. Introspection and cognitive brain mapping: From stimulus-response to script-report. *Trends in Cognitive Sciences* 6: 333–339.

Jack, A. I., and Roepstorff, A. (eds.). 2003. *Trusting the Subject? The Use of Introspective Evidence in Cognitive Science*. Volume 1. Thorverton, UK: Imprint Academic.

Jack, A. I., and Shallice, T. 2001. Introspective physicalism as an approach to the science of consciousness. *Cognition* 79: 161–196.

Jonas, H. 1966. *The Phenomenon of Life*. Chicago: University of Chicago Press.

Kabat-Zinn, J. 1990. *Full Catastrophe Living: Using the Wisdom of Your Body and Mind to Face Stress, Pain, and Illness*. New York: Dell.

Kahana, M. J., Seelig, D., and Madsen, J. R. 2001. Theta returns. *Current Opinion in Neurobiology* 11: 739–744.

Kanwisher, N. 2001. Neural events and perceptual awareness. *Cognition* 79: 89–113.

Kelso, J. A. S. 1995. *Dynamic Patterns: The Self-Organization of Brain and Behavior*. Cambridge, MA: MIT Press.

Köhler, W. 1947. *Gestalt Psychology*. New York: Liveright.

Kriegel, U. 2003. Consciousness as intransitive self-consciousness: Two views and an argument. *Canadian Journal of Philosophy* 33: 103–132.

Kuhn, T. 1970. *The Structure of Scientific Revolutions*. Chicago: University of Chicago Press.

Lachaux, J.-P., Chavez, M., and Lutz, A. 2003. A simple measure of correlation across time, frequency and space between continuous brain signals. *Journal of Neuroscience Methods* 123: 175–188.

Lachaux, J.-P., Rodriguez, E., Martinerie, J., and Varela, F. J. 1999. Measuring phase synchrony in brain signals. *Human Brain Mapping* 8: 194–208.

Lambie, J. A., and Marcel, A. J. 2002. Consciousness and the varieties of emotion experience: A theoretical framework. *Psychological Review* 109: 219–259.

Le Van Quyen, M. 2003. Disentangling the dynamic core: A research program for neurodynamics at the large scale. *Biological Research* 36: 67–88.

Le Van Quyen, M., and Petitmengin, C. 2002. Neuronal dynamics and conscious experience: An example of reciprocal causation before epileptic seizures. *Phenomenology and the Cognitive Sciences* 1: 169–180.

Lewis, M. D. 2000. Emotional self-organization at three time scales. In M. D. Lewis and I. Granic (eds.), *Emotion, Development, and Self-Organization: Dynamic Systems Approaches to Emotional Development*, pp. 37–69. Cambridge and New York: Cambridge University Press.

Lewis, M. D. In press. Bridging emotion theory and neurobiology through dynamic systems modeling. *Behavioral and Brain Sciences*.

Llinas, R., Ribary, U., Contreras, D., and Pedroarena, C. 1998. The neuronal basis for consciousness. *Philosophical Transactions of the Royal Society of London Series B* 353: 1801–1818.

Lutz, A. 2002. Toward a neurophenomenology as an account of generative passages: A first empirical case study. *Phenomenology and the Cognitive Sciences* 1: 133–167.

Lutz, A., and Thompson, E. 2003. Neurophenomenology: Integrating subjective experience and brain dynamics in the neuroscience of consciousness. *Journal of Consciousness Studies* 10: 31–52.

Lutz, A., Lachaux, J.-P., Martinerie, J., and Varela, F. J. 2002. Guiding the study of brain dynamics by using first-person data: Synchrony patterns correlate with ongoing conscious states during a simple visual task. *Proceedings of the National Academy of Sciences USA*. 99: 1586–1591.

Maturana, H. R. 1969. The neurophysiology of cognition. In P. Garvin (ed.), *Cognition: A Multiple View*. New York: Spartan Books.

Maturana, H. R. 1970. Biology of cognition. In H. R. Maturana and F. J. Varela, *Autopoiesis and Cognition: The Realization of the Living*, pp. 2–58. Boston Studies in the Philosophy of Science, vol. 43. Dordrecht: D. Reidel.

Maturana, H. R., and Varela, F J. 1980. *Autopoiesis and Cognition: The Realization of the Living*. Boston Studies in the Philosophy of Science, vol. 42. Dordrecht: D. Reidel.

Merleau-Ponty, M. 1962. *Phenomenology of Perception*. Trans. C. Smith. London: Routledge Press.

Metzinger, T. (ed.). 2000. *Neural Correlates of Consciousness*. Cambridge, MA: MIT Press.

Moran, D. 2000. *Introduction to Phenomenology*. London: Routledge Press.

Moustakas, C. 1994. *Phenomenological Research Methods*. Thousand Oaks, London, New Delhi: Sage Publications.

Nagel, T. 1979. What is it like to be a bat? In T. Nagel, *Mortal Questions*, pp. 165–180. New York: Cambridge University Press.

Noë, A., and Thompson, E. 2004a. Are there neural correlates of consciousness? *Journal of Consciousness Studies* 11: 3–28.

Noë, A., and Thompson, E. 2004b. Sorting out the neural basis of consciousness: Authors' reply to commentators. *Journal of Consciousness Studies* 11: 87–98.

Ohl, F. W., Scheich, H., and Freeman, W. J. 2001. Change in pattern of ongoing cortical activity with auditory category learning. *Nature* 412: 733–736.

O'Keefe, J., and Burgess, N. 1999. Theta activity, virtual navigation and the human hippocampus. *Trends in Cognitive Sciences* 11: 403–406.

O'Regan, J. K., and Noë, A. 2001. A sensorimotor account of vision and visual consciousness. *Behavioral and Brain Sciences* 24: 939–1031.

Panksepp, J. 1998a. *Affective Neuroscience: The Foundations of Human and Animal Emotions.* New York: Oxford University Press.

Panksepp, J. 1998b. The periconscious substrates of consciousness: Affective states and the evolutionary origins of self. *Journal of Consciousness Studies* 5: 566–582.

Parvizi, J., and Damasio, A. 2001. Consciousness and the brainstem. *Cognition* 79: 135–159.

Pessoa, L., Thompson, E., and Noë, A. 1998. Finding out about filling-in: A guide to perceptual completion for visual science and the philosophy of perception. *Behavioral and Brain Sciences* 21: 723–802.

Petitmengin, C. 2001. *L'expérience intuitive.* Paris: L'Harmattan.

Petitmengin-Peugeot, C. 1999. The intuitive experience. In F. J. Varela and J. Shear (eds.), *The View from Within,* pp. 43–78. Thorveton, UK: Imprint Academic.

Petitot, J., Varela, F. J., Pachoud, B., and Roy, J.-M. (eds.). 1999. *Naturalizing Phenomenology.* Stanford, CA: Stanford University Press.

Price, D., and Barrell, J. 1980. An experiential approach with quantitative methods: A research paradigm. *Journal of Humanistic Psychology* 20: 75–95.

Price, D., Barrell, J., and Rainville, P. 2002. Integrating experiential-phenomenological methods and neuroscience to study neural mechanisms of pain and consciousness. *Consciousness and Cognition* 11: 593–608.

Rees, G., Krieman, G., and Koch, C. 2002. Neural correlates of consciousness in humans. *Nature Reviews Neuroscience* 3: 261–270.

Rodriguez, E., George, N., Lachaux, J. P., Martinerie, J., Renault, B., and Varela, F. J. 1999. Perception's shadow: Long-distance synchronization of human brain activity. *Nature* 397: 430–433.

Rosenthal, D. M. 1997. A theory of consciousness. In N. Block et al. (eds.), *The Nature of Consciousness,* pp. 729–753. Cambridge, MA: MIT Press.

Roy, J.-M. 2004. Phenomenological claims and the myth of the given. In E. Thompson (ed.), *The Problem of Consciousness: New Essays in Phenomenological Philosophy of Mind.* Canadian Journal of Philosophy Supplementary Volume.

Roy, J.-M., Petitot, J., Pachoud, B., and Varela, F. J. 1999. Beyond the gap: An introduction to naturalizing phenomenology. In J. Petitot, F. J. Varela, B. Pachoud, and J.-M. Roy (eds.), *Naturalizing Phenomenology,* pp. 1–80. Stanford, CA: Stanford University Press.

Rudrauf, D., Lutz, A., Cosmelli, D., Lachaux, J.-P., and Le Van Quyen, M. 2003. From autopoiesis to neurophenomenology. *Biological Research* 36: 27–66.

Sarnthein, J., Petsche, H., Rappelsberger, P., Shaw, G. L., and von Stein, A. 1998. Synchronization between prefrontal and posterior association cortex during human working memory. *Proceedings of the National Academy of Sciences USA* 95: 7092–7096.

Sartre, J.-P. 1956. *Being and Nothingness*. Trans. Hazel Barnes. New York: Philosophical Library.

Schiff, S. J., So, P., Chang, T., Burke, R. E., and Sauer, T. 1996. Detecting dynamical interdependence and generalized synchrony through mutual prediction in a neural ensemble. *Physical Review E* 54: 6706–6724.

Schooler, J. W. 2002. Re-representing consciousness: Dissociations between experience and meta-consciousness. *Trends in Cognitive Sciences* 6: 339–344.

Searle, J. S. 1983. *Intentionality: An Essay in the Philosophy of Mind*. Cambridge and New York: Cambridge University Press.

Searle, J. 2000. Consciousness. *Annual Review of Neuroscience* 23: 557–578.

Searle, J. 2004. Comments on Noë & Thompson 'Are there NCCs?' *Journal of Consciousness Studies* 11: 79–82.

Sellars, W. 1956. Empiricism and the philosophy of mind. In H. Feigl and M. Scriven (eds.), *Minnesota Studies in the Philosophy of Science*. Volume 1. *The Foundations of Science and the Concepts of Psychology and Psychoanalysis*, pp. 253–329. Minneapolis: University of Minnesota Press.

Singer, W. 1999. Neuronal synchrony: A versatile code for the definition of relations? *Neuron* 24: 49–65.

Steinbock, A. J. 1999. Saturated intentionality. In Donn Welton (ed.), *The Body*, pp. 178–199. Oxford: Basil Blackwell.

Thompson, E. 2001. Empathy and consciousness. *Journal of Consciousness Studies* 8: 1–32.

Thompson, E. (ed.). 2004. *The Problem of Consciousness. New Essays in Phenomenological Philosophy of Mind*. Canadian Journal of Philosophy Supplementary Volume. Calgary: University of Alberta Press.

Thompson, E. Forthcoming. *Mind in Life: Biology, Phenomenology, and the Sciences of the Mind*. Cambridge, MA: Harvard University Press.

Thompson, E., and Varela, F. J. 2001. Radical embodiment: Neural dynamics and consciousness. *Trends in Cognitive Sciences* 5: 418–425.

Thompson, E., and Zahavi, D. In press. Contemporary Continental perspectives: Phenomenology. In P. D. Zelazo and M. Moscovitsch (eds.), *The Cambridge Companion to Consciousness*. New York and Cambridge: Cambridge University Press.

Thompson, E., Noë, A., and Pessoa, L. 1999. Perceptual completion: A case study in phenomenology and cognitive science. In J. Petitot, F. J. Varela, B. Pachoud, and J.-M. Roy (eds.), *Naturalizing Phenomenology: Issues in Contemporary Phenomenology and Cognitive Science*, pp. 161–195. Stanford, CA: Stanford University Press.

Thompson, E., Palacios, A., and Varela, F. J. 1992. Ways of coloring: Comparative color vision as a case study for cognitive science. *Behavioral and Brain Sciences* 15: 1–74. Reprinted in Alva Noë and Evan Thompson (eds.), *Vision and Mind: Readings in the Philosophy of Perception*. Cambridge, MA: MIT Press, 2002.

Tononi, G., and Edelman, G. M. 1998. Consciousness and complexity. *Science* 282: 1846–1851.

VanRullen, R., and Koch, C. 2003. Is perception discrete or continuous? *Trends in Cognitive Sciences* 7: 207–213.

Varela, F. J. 1979. *Principles of Biological Autonomy*. New York: Elsevier North Holland.

Varela, F. J. 1991. Organism: A meshwork of selfless selves. In A. Tauber (ed.), *Organism and the Origin of Self*, pp. 79–107. Dordrecht: Kluwer Academic Publishers.

Varela, F. J. 1995. Resonant cell assemblies: A new approach to cognitive functions and neuronal synchrony. *Biological Research* 28: 81–95.

Varela, F. J. 1996. Neurophenomenology: A methodological remedy to the hard problem. *Journal of Consciousness Studies* 3: 330–350.

Varela, F. J. 1997a. Patterns of life: Intertwining identity and cognition. *Brain and Cognition* 34: 72–87.

Varela, F. J. 1997b. The naturalization of phenomenology as the transcendence of nature. *Alter* 5: 355–381.

Varela, F. J. 1999. The specious present: A neurophenomenology of time consciousness. In J. Petitot et al. (eds.), *Naturalizing Phenomenology*, pp. 266–314. Stanford, CA: Stanford University Press.

Varela, F. J., and Bourgine, P. (eds.). 1991. *Toward a Practice of Autonomous Systems: Proceedings of the First European Conference on Artificial Life.* Cambridge, MA: MIT Press.

Varela, F. J., and Shear, J. 1999a. First-person accounts: Why, what, and how. In F. J. Varela and J. Shear (eds.), *The View from Within: First-Person Approaches to the Study of Consciousness*, pp. 1–14. Thorveton, UK: Imprint Academic.

Varela, F. J., and Shear, J. (eds.). 1999b. *The View from Within: First-Person Approaches to the Study of Consciousness.* Thorverton, UK: Imprint Academic.

Varela, F. J., and Thompson, E. 2003. Neural synchrony and the unity of mind: A neurophenomenological perspective. In A. Cleeremans (ed.), *The Unity of Consciousness: Binding, Integration and Dissociation*, pp. 266–287. New York: Oxford University Press.

Varela, F. J., Lachaux, J.-P., Rodriguez, E., and Martinerie, J. 2001. The brainweb: Phase synchronization and large-scale integration. *Nature Reviews Neuroscience* 2: 229–239.

Varela, F. J., Thompson, E., and Rosch, E. 1991. *The Embodied Mind: Cognitive Science and Human Experience.* Cambridge, MA: MIT Press.

von Stein, A., and Sarnthein, J. 2000. Different frequencies for different scales of cortical intergration: From local gamma to long range alpha/theta synchronization. *International Journal of Psychophysiology* 38: 301–313.

von Stein, A., Chiang, C., and König, P. 2000. Top-down processing mediated by interareal synchronization. *Proceedings of the National Academy of Sciences USA* 97: 14748–14753.

Wallace, A. 1998. *The Bridge of Quiescence: Experiencing Tibetan Buddhist Meditation.* La Salle, IL: Open Court.

Wallace, A. 1999. The Buddhist tradition of *shamatha*: Methods for refining and examining consciousness. In F. J. Varela and J. Shear (eds.), *The View from Within: First-Person Approaches to the Study of Consciousness*, pp. 175–188. Thorverton, UK: Imprint Academic.

Watt, D. F. 1999. Emotion and consciousness: Implications of affective neuroscience for extended reticular thalamic activating system theories of consciousness. Electronic publication of the Association for the Scientific Study of Consciousness. Available at: http://server.philvt.edu/assc/watt/default.htm.

Wider, K. V. 1997. *The Bodily Basis of Consciousness: Sartre and Contemporary Philosophy of Mind.* Ithaca, NY: Cornell University Press.

Williams, P. 1998. *The Reflexive Nature of Awareness.* London: Curzon Press.

Zahavi, D. 1999. *Self-Awareness and Alterity.* Evanston, IL: Northwestern University Press.

Zahavi, D. 2002. First-person thoughts and embodied self-awareness: Some reflections on the relation between recent analytic philosophy and phenomenology. *Phenomenology and the Cognitive Sciences* 1: 7–26.

Zahavi, D. 2003. *Husserl's Phenomenology.* Stanford, CA: Stanford University Press.

Zahavi, D. 2004. Intentionality and phenomenality: A phenomenological take on the hard problem. In E. Thompson (ed.), *The Problem of Consciousness: New Essays in Phenomenological Philosophy of Mind.* Canadian Journal of Philosophy Supplementary Volume. Calgary: University of Alberta Press.

Zahavi, D., and Parnas, J. 1998. Phenomenal consciousness and self-awareness: A phenomenological critique of representational theory. *Journal of Consciousness Studies* 5: 687–705.

3

Out of the Mouths of Autistics: Subjective Report and Its Role in Cognitive Theorizing

Victoria McGeer

The theoretical work that emerges from a study on the work of memory, learning, and other higher functions, such as consciousness, is this: if the psychological (functional) taxonomy is ill-defined, then the search for neural substrates for those functions will be correspondingly ill-defined. There are, certainly, remarkable data, found at all levels in psychology and neuroscience, but precisely how to interpret the data in terms of a theory of neurobiological capacities, representations and processes is yet to be discovered.

– Patricia Churchland, *Neurophilosophy*

My primary concern in this chapter is with subjective report, and in particular, first-person reports of abnormal sensory and/or perceptual experiences. This topic raises interesting questions at two distinct levels: First, there are philosophical questions about the nature of subjective experience, subjective awareness of experience, and subjects' capacity to articulate what they are experiencing. Secondly, there are questions about how philosophical theories of such matters interact with empirical theories of – and research into – abnormal neurocognitive conditions. To focus my discussion of these questions, I will be considering, in particular, the phenomenon of subjective report in high-functioning individuals with autism.

1 Understanding Autism: Some Methodological and Substantive Concerns

Autism presents a highly complex challenge for researchers trying to negotiate between neurological and cognitive levels of theorizing. As with other neurodevelopmental disorders, such research involves working along two dimensions at once. As T. W. Robbins explains:

The challenge to research into childhood autism lies in relating what appears to be a set of apparently somewhat independent symptoms . . . to corresponding deficits in brain systems. This research approach can be viewed as one of 'vertical integration' in which theories in one domain are strengthened by structural congruences in another. . . .

However, autism generally presents as a set of symptoms, the precise significance of which, relative to one another, remains to be established. . . . Analogously, the main neurobiological findings in autism seem to implicate many possible foci which are probably affected to varying extents in different individuals. These are problems of 'horizontal integration' and it seems evident that a complete understanding of the disorder has to achieve an orderly account in both the 'vertical' and the 'horizontal dimensions'. (Robbins, 1997)

There is one further dimension of complexity to add to this challenge: the problem of 'temporal integration'. Autism is a neuro*developmental* disorder, unfolding in time and hence affecting individuals (and their brains) in cascading complicated ways throughout their development. As Helen Tager-Flusberg notes,

. . . neurodevelopmental disorders are more often associated with *diffuse* cortical damage, which suggests that the impact of such disorders is more widespread, affecting complex neural systems rather than simple localized areas. Furthermore, across a range of developmental syndromes, we find that not only are particular cortical systems affected but often associated atypical subcortical structures are involved as well. For example, in autism both the cerebellum and the limbic system show significant abnormalities. These finding suggest deviations in brain development that begin early in embryology and cannot be easily classified and interpreted as later acquired focal lesions. (Tager-Flusberg, 1999)

One suggestion here is that the range of symptoms found at the cognitive/behavioural level are not related to one another at all but derive simply from abnormalities in diffuse underlying neural systems. Another possibility, not excluded by this first, is that abnormalities in specific cortical areas make for a primary cognitive deficit – for instance, a hypothesized 'theory of mind' deficit (discussed later in this section) – and this primarily deficit is responsible for (i.e., functionally related to) at least a subset of the cognitive/behavioural symptoms. A third 'neuroconstructivist' possibility proposes a developmentally more complicated set of relations among the various symptoms: Early neurological abnormalities, with typically low-level (noncognitive) behavioural manifestations, affect the developing child's interactions with its environment, leading to atypical inputs for the developing brain. These atypical inputs themselves encourage atypical development in a number of high-level cortical structures, leading to an overall profile of abnormal brain functioning. Symptoms at the cognitive/behavioural level will both speak to and reflect this pattern of atypical development, making the relations among them impossible to sort out in purely atemporal terms (for further discussion of these alternatives, see Gerrans and McGeer, 2003). As Annette-Karmiloff Smith explains:

> The neuroconstructivist modification in perspective crucially influences the way in which atypical development is considered. In this approach, the deletion, reduplication or mispositioning of genes will be expected to subtly change the course of development pathways, with stronger effects on some outcomes and weaker effects on others. A totally specific disorder will, *ex hypothesi*, be extremely unlikely, thereby changing the focus of research in pathology. Rather than solely aiming to identify a damaged module at the cognitive level, researchers are encouraged to seek more subtle effects beyond the seemingly unique one, as well as to question whether successful behaviour (the presumed 'intact' part of the brain) is reached by the same processes as in normal development. This change in perspective means that atypical development should not be considered in terms of a catalogue of impaired and intact functions, in which non-affected modules are considered to develop normally, independently of the others. Such claims are based on the static, adult neurophysiological model which is inappropriate for understanding the dynamics of developmental disorders. (Karmiloff-Smith, 1998)

With these methodological considerations in mind, consider now the range of symptoms with which researchers in autism must contend. To begin with, autism is a spectrum disorder, with individuals (usually designated 'low-functioning') that are severely retarded at one end of the spectrum and those (usually called 'high-functioning') that have normal to above average IQ at the other. (A substantial proportion of high-functioning autistics are sometimes given the differentiating diagnosis of 'Asperger's syndrome', but for the purposes of this chapter, I will consider these two high-functioning groups together, using the terms 'Asperger's syndrome' and 'high-functioning' autistic interchangeably.)[1]

---

[1] In the latest editions of the American Psychiatric Association Diagnostic and Statistical Manual (DSM-IV) and the World Health Organization. International Classification of Diseases (ICD-10), 'Asperger's Disorder' is listed as a distinct nosological entity under the category of Pervasive Development Disorders (along with 'Austistic Disorder', 'Rett's Disorder' and "Childhood Disintegrative Disorder'). There is still substantial disagreement amongst clinicians and researchers about whether Asperger's syndrome, in particular, really constitutes a distinct disorder from autism or merely lies on the milder

The core cognitive deficits, according to which diagnosis is made, involve characteristically impaired social, communicative and imaginative skills – the last, for instance, typically identified through an absence of pretend play in early childhood. Communicative deficits are both prelinguistic and linguistic, involving, for instance, an absence of proto-declarative pointing (to call attention to things in a shared environment), an absence of normal gestural speech, and abnormal prosody. There is limited or absent awareness of conversational cues, or the give and take of normal conversation. Pronoun reversals – for example, 'I' for 'you' – are also a common feature. As for social skills, even high-functioning autistics – those with relatively good or even superior cognitive skills – show distinctive impairments in these respects. If there is curiosity about the world, it is manifestly not directed towards other people, who seem to occupy a predominantly instrumental role in the autistic's environment. (Those who interact with autistics often report feeling like 'objects'.) Autistics show little or no understanding of the give and take of social life, and no capacity to engage with others in anything like a normal way. They generally occupy the world as if on their own, often developing and adhering to strict routines that make little sense from our nonautistic point of view. Similarly, their activities and interests seem to us repetitive, highly restricted and often deeply odd. Memory skills amongst autistics can sometimes be extraordinary, and a small proportion also show 'savant talents' in music, calculating, drawing and so on.

While these cognitive features are certainly distinctive, they do not encompass the totality of autistic abnormalities. Motor stereotypies (hand flapping, spinning, rocking) are a common feature throughout the autistic range, and motor clumsiness is also sometimes apparent, particularly amongst high-functioning autistics (perhaps distinctively associated with Asperger's syndrome (Green et al., 2002)). Extreme emotional

---

end of a continuum (Klin et al., 2000; Ozonoff et al., 2000; Mayes et al., 2001; Mayes and Calhoun, 2001). The distinguishing features of Asperger's syndrome are no evidence of general language delay or of delay in reasoning skills outside the social domain. However, other symptoms characteristic of autism, including abnormalities in language *use*, are generally present. Moreover, individuals who do have language and/or cognitive delay – those who fall within the current categorization of 'Autistic Disorder' (DSM-IV) or 'Childhood Autism' (ICD-10) – may develop skills in both these respects that lead to their being designated 'high-functioning' (For further discussion, see 'A Note on Nosology' in Baron-Cohen et al., 2000). Given the substantial overlap of symptoms between these disorders, it seems unlikely that the points I want to make in this chapter are affected by the diagnostic distinction. However, this is a matter for further research.

reactions, moodiness, tantrums and symptoms of anxiety disorder are also frequently observed.

Of particular interest is a further diagnostic item listed in connection with both autism and Asperger's syndrome in DSM-IV – namely, a 'persistent preoccupation with parts of objects'. Perceptual abnormalities that involve a marked preference for featural over contextual processing have been interestingly explored, with some research even indicating that autistics are less susceptible than normal or Downs syndrome control subjects to low-level perceptual illusions induced by an embedding context (e.g., the Ebbinghaus [or Titchener circles] illusion, the Ponzo illusion, the Poggendorf illusion and so on) (Happé, 1996). These abnormalities may indicate a more general cognitive tendency towards what Uta Frith calls 'weak central coherence' – a relative inability to process information, whatever its form (e.g., visual, syntactic, semantic and so on), in context (Frith, 1989). How such a tendency may be related to the particular social-cognitive difficulties described here is still unclear (Frith and Happé, 1994; Happé, 2000). However, what is clear is that such perceptual abnormalities constitute a significant element of the autistic phenotype, importantly shaping how individuals understand the world.

In addition to these perceptual abnormalities, there is another prominent aspect of the disorder – at least as revealed in subjective report – that has received surprisingly scant attention from cognitive researchers, either in their theoretical accounts or in their empirical research – namely, the prevalence of sharply abnormal sensory experiences. As Uta Frith herself remarks:

One mysterious feature that is not currently given much importance may hold further clues. Some Asperger individuals give first-hand accounts of sharply uncomfortable sensory and strong emotional experiences, often including sudden panic. From autobiographical accounts, we learn that again and again the Asperger individual's interpretation of perceptions by ear, eye, or touch tends to be extremely faint or overwhelmingly strong. There can be hyper- as well as hyposensitivity. Feeling scratchy clothes, for example, is not merely uncomfortable, but agonizing. On the other hand, pain may be tolerated to a surprising degree. (Frith, 1991)

While these reports are certainly interesting, revealing aspects of the disorder that seem otherwise overlooked, it is worth asking how much attention researchers should pay to these first-hand accounts. After all, as Frith goes on to observe, these kinds of subjective reports are restricted to a fairly small subset of the wider autistic population (about 20–25%) – a subset that is significantly reduced if those with Asperger's syndrome are excluded. That is, they are gleaned only from individuals who are linguistically able, who tend to have normal or above-average IQs, and who very often have a late-developed capacity to pass standard 'theory of mind' tests, such as the now famous false-belief task (though it must be stressed that such a capacity does not, in these cases, indicate anything like a capacity for normal social reasoning). All in all, such individuals are unusual relative to the general autistic population; and while they may have a milder form of the same underlying disorder affecting more seriously disabled autistics, how far we can generalize from these reports remains an open question.

Nevertheless, despite exercising caution in taking verbally adept individuals to speak for the experiences of autistics in general, their reports do seem to give us a rare glimpse of what it is like to be autistic, or what the world is like from an autistic point of view. As such, they provide a unique and important set of data for research in autism, data that underscores the existence of phenomena that may not be so salient from the third-person point of view: for example, the extent and impact of autistic children's sensory abnormalities in coming to learn about and negotiate the world around them. Indeed, these abnormalities and their consequences may be peculiarly invisible from the third-person point of view – at least if we rely on it exclusively – unduly constraining proposed hypotheses for explaining some aspect of autistic behaviour.

Take, for instance, the well-known phenomenon of autistics' failing to make normal eye contact when interacting with other people. (In fact, this is one of the targeted features of social interaction in which autistics are often given explicit training: 'When you first meet people, look them in the eyes.') How might we explain this characteristic behavioural abnormality?

Here's one possibility (explored in Baron-Cohen, 1996): Although autistics have no trouble understanding what the eyes are for (i.e., looking at things), and although they are able to understand that someone is looking at them or at something else, depending on the direction of that person's gaze (as judged from their interpretation of photographs or cartoons), they take no interest in what others are looking at. They have no drive towards sharing attention – towards looking at what the other is looking at, or to securing joint attention; and they don't seem to particularly care whether or not someone is looking at them. Simply put, the eyes are not a particularly interesting or information-rich stimulus for autistic individuals; they don't make eye-contact because they're not motivated to do so.

Contrast this hypothesis with a second possibility: Far from the eyes constituting a relatively impoverished stimulus for autistics, it may be that normally sustained eye contact is too arousing for them to bear. After all, if the eyes are normally such a salient feature to human beings, and if autistics are hypersensitive in certain sensory domains, then it is at least possible that the eyes fall into this domain. Consequently, autistics avoid eye contact, not because they find other people's gaze informationally uninteresting but because they can't make use of the information available, given how sensorily overwhelming it is to look directly into another's eyes.

Why would such a hypothesis occur to us? Perhaps only by paying more heed to subjective report. For example, on being asked to make eye contact with people when speaking to them, one resistant young autistic complained, 'Their eyes are on fire!' Or consider another report from Therese Jolliffe:

Looking at people's faces, particularly into their eyes, is one of the hardest things for me to do. When I do look at people, I have nearly always had to make a conscious effort to do so and then I can usually only do it for a second.... People do not appreciate how unbearably difficult it is for me to look at a person. It disturbs my quietness and is terribly frightening – though the fear decreases with increasing distance away from the person. (Jolliffe et al., 1992)

These reports certainly suggest that sensory disturbances have something to do with this standardly cited example of an autistic social-cognitive deficit, making this second hypothesis a live, but overlooked, research option. Hence, I want to suggest there is a blind spot in autism research. On the one hand, there is a great deal of effort and attention devoted to understanding autistic abnormalities and the connection among them, as these are or have become evident from the third-person point of view. On the other hand, there is a growing body of (anecdotal) evidence from autistic self report – mostly in the form of autobiographical writings – that sensory abnormalities are a central feature of autistic experience, and perhaps central to understanding the nature of the disorder. Yet this possibility seems to be generally ignored or at least substantially underexplored by cognitive theorists. The question is: why? My speculative answer is that subjective reports of abnormal experiences are liable to fall prey to two methodological cum philosophical objections that ordinary subjective reports do not; and for this reason, they tend to be ignored or discounted – explained away, rather than explained in terms of the phenomena they straightforwardly express.

Of course, dealing with subjective report in cognitive research is always a tricky business, and there are two questions that really ought to be posed in connection with it: (1) In any particular case or kind of case, are these reports reliable? – that is, do they give us accurate information about what individual subjects are (really) experiencing? and (2) how can we explain what the subjects are saying? Notice that if the answer to the first question is 'no – subjects are not reliable in making these reports', then the type of answer given to the second question is significantly altered: Instead of trying to explain subjects' experience as they report it to be, the researcher must now account for why they *say* what they do, given that what they say doesn't accurately represent how they are experiencing things. For example, in cases of hysterical blindness, subjects with no apparent physiological damage may *say* that they are blind. But researchers would be justified in doubting the credibility of these reports if, for instance, these subjects do significantly *worse* than chance in forced-choice visual tests, thereby showing their capacity to use visual information to support the contention of blindness (discussed in Dennett, 1991, pp. 326–327).

Clearly, then, the first 'reliability' question is a critical one to pose. Still, I will be arguing that there are certain background views about subjective report that may bias the answer towards false negatives when it comes to assessing their credibility in cases where what's reported is especially strange or unusual. These are the two methodological cum philosophical concerns I referred to: The first is not as serious as the second, though it sets the stage for examining a pervasive way researchers in both philosophy and psychology tend to regard subjective report – namely, as a sort of testimony to the nature of something that's observed from only one point of view. I call this first source of bias a Humean resistance to miraculous testimony. The second, more significant, source of bias, which I discuss in section 3, involves an adherence to a 'neo-perceptual' model of introspective awareness and self-report. The important and distinctive thing about this model is that it makes self-awareness dependent on a properly functioning cognitive mechanism for tracking our mental life; and, of course, the breakdown of such a mechanism is just what might be expected in pathological cases.

These are both rather general considerations in the sense that if they count at all, they count against crediting all anomalous subjective reports – subjective reports made in a variety of pathological or abnormal cases. In my final section 4, I will consider a third, seemingly additional, reason for discounting subjective report in the particular case of autism.

It stems from what many researchers consider to be a central – perhaps, *the* central – deficit in this disorder: namely, a domain-specific inability to conceptualize, attribute or reason about specifically *mental* states – whether of *self* or other (for discussion, see Baron-Cohen et al., 1993; Baron-Cohen et al., 2000). This deficit is usually characterized as an inability to develop or deploy a 'theory of mind' – or 'ToM'. However, as we shall see, not all researchers who subscribe to this view think theory of mind is theorylike in structure, or learned in a way that theories ordinarily are.

I will not here review the empirical research that supports the idea of such a domain-specific disability in autism. I take it to be rather compelling. My concern instead is with how this disability has been conceptualized by certain influential researchers in the field – and, specifically, with how this conceptualization affects researchers' attitudes towards autistic self-knowledge and self-report. In particular, my concern is this: If having such a disability can be equated with *lacking normal introspective awareness* (Frith and Happé, 1999), then it seems reasonable to discount the reliability of autistic subjective report in general. In this case, anything unusual in these reports would be explained away in terms of the global introspecting disability, rather than explained in terms of the phenomena directly reported. Or to put this another way, the 'valuable data' of autistic subjective report would be valuable just in the sense of providing more evidence for a specific ToM disability. So my goal, in this final section, is to examine whether the equation between a ToM disability and lack of introspective awareness makes good theoretical or empirical sense. I will close by considering some implications of this discussion for future research in autism.

2 Subjective Report and the Humean Resistance to Miraculous Testimony

If we begin with the naive question – what is subjective report? – a commonsense answer immediately suggests itself: It is a kind of testimony to the states or events that constitute individuals' subjective condition. That is, it is a testimony to how individuals find they are experiencing the world (including their own bodies); or it is a testimony to what they believe, desire, hope or fear – a testimony to the nature and existence of their own intentional states.

Now with regard to testimony generally, we may ask a further question – namely, when is it rational to credit another person's reports of states and events that we do not witness ourselves? Again, a common-sense answer suggests itself: when these states and events fall within the range of normal, expectable, or at least *explicable* occurrences. But what about reports that fall quite outside this normal range? Is it rational to credit such testimony then? In the *Enquiries Concerning Human Understanding*, David Hume considers this question with respect to testimony about the occurrence of 'miracles': that is, events that violate laws of nature and thereby stand outside of any normal experience. His answer there is unequivocally 'no'. He writes:

> No testimony is sufficient to establish a miracle unless the testimony be of such a kind, that its falsehood would be more miraculous, than the fact which it endeavours to establish; and even in that case, there is a mutual destruction of arguments, and the superior only gives the assurance suitable to that degree of force, which remains, after deducting the inferior. (Hume, 1975, Section X, I, 91)

To put Hume's conclusion in modern parlance: *the less probable the reported event, the less credible the report.* Here is Hume again:

> Suppose that the fact which the testimony endeavours to establish, partakes of the extraordinary and the marvellous; in that case, the evidence, resulting from the testimony, admits of a diminution, greater or less, in proportion as the fact is more or less unusual. (Section X. I. 89)

Consider now the following autistic subjective reports:

> I had – and always had had, as long as I could remember – a great fear of jew-ellery. That terror also included hairclips and metal buttons. I thought they were frightening, detestable, revolting. If I was made to touch jewellery, I felt a sharp whistling metallic noise in my ears, and my stomach turned over. Like a note falsely electrified, that sound would creep from the base of my spine upwards until it rang in my ears, tumbled down into my throat and settled like nausea into my stomach. (Gerland, 1997, pp. 54, 157)

> As much as I loved to chew scratchy and gritty textures, I often found it impossible even to touch some objects. I hated stiff things, satiny things, scratchy things, things that fit me too tightly. Thinking about them, imagining them, visualizing them . . . any time my thoughts found them, goose bumps and chills and a general sense of unease would follow. I routinely stripped off everything I had on even if we were in a public place. I constantly threw my shoes away, often as we were driving in the car. I guess I thought I would get rid of the nasty things forever! . . .
> I also found many noises and bright lights nearly impossible to bear. High frequencies and brassy, tin sounds clawed my nerves. . . . Bright lights, mid-day sun, reflected lights, strobe lights, flickering lights, fluorescent lights; each seemed to sear my eyes. Together, the sharp sounds and bright lights were more than enough to overload my senses. My head would feel tight, my stomach would

churn, and my pulse would run my heart ragged until I found a safety zone. (Willey, 1999)

I perceived sound and visual information directly and consciously only at the cost of its cohesion. I could interpret the part but lost the whole. I saw the nose but lost the face, saw the hand but continued to see the body but would not know what it was except piece by piece. I'd get the intonation but lose the meaning of the words or get the meaning of the words but only at the cost of tuning out the intonation, as though independent of the words. (Williams, 1999)

How far do we credit such reports? After all, they certainly sound 'extraordinary and marvellous' in a Humean sense. Or consider another example: subjective reports of so-called synaesthetic experience, where subjects purport to have experiences in one modality as a consequence of stimulation in a quite different modality. For instance, they report seeing particular graphemes as tinged always with particular colors (5s as red, for instance), or hearing particular musical tones as colored in particular ways (C#s as purplish-blue). Falling outside our own experience – that is to say, *normal* human experience (in a statistical sense), it may be tempting to apply Hume's principle: Such reports are simply incredible. In fact, as V. S. Ramachandran and E. M. Hubbard comment in a recent article on synaesthesia, 'despite a century of research, the phenomenon is still sometimes dismissed as bogus' (Ramachandran and Hubbard, 2001b, p. 4). Could this resistance be due in part to an application of Hume's principle? If so, I think it is misguided for two distinct reasons.

In the first place, Hume's reasoning depends on assessing the improbability of the reported event. But how do we go about making these assessments when it comes to reports of subjective experience? Using 'normal' experience as a guide will be reasonable only insofar as we are quite convinced that no organic abnormality could account for the experiences *as described*. But this must come after a thoroughgoing investigation of the phenomena concerned, rather than providing a priori grounds for undercutting such investigations. After all, given the complexity of the neural systems realizing our experience and given how little we still understand about these neural systems, the possibility of such untoward phenomenon is not something we can reasonably assess simply in light of normal experience.

A second, more important, reason is that applying Hume's principle in the case of subjective report means taking the situation of testimony that Hume envisaged far too literally. In effect, it is to imagine an old-fashioned picture of the mind in which subjects themselves are the only witness to the putative events reported – only they have direct perceptual access to their own subjective condition; and cognitive scientists are 'on the outside', depending exclusively or primarily on what their subjects say.

In fact, this is not the position of cognitive – or consciousness – scientists. (Whether or not it's the position of subjects themselves, I'll return to in section 3.) As Ramachandran and Hubbard have elegantly demonstrated in a recent set of experiments on synaesthesia, scientists do have ways of directly verifying what their test subjects say about themselves. More, their methods allow them to explore the mechanisms that explain why subjects' experiences take the shape that they do. For instance, after noting that certain of their grapheme-color synaesthetes described their visual experience in ways that suggested a fairly low-level visuo-perceptual effect, Ramachandran and Hubbard reasoned that these subjects should experience certain phenomena under experimental conditions that are generally accepted as indicative of low-level perceptual processes – phenomena such as perceptual grouping and 'pop-out'.[2]

Thus, in one such experiment designed to test for synaesthetic pop-out, they asked their subjects to identify a geometric shape notionally inscribed by the 2s embedded in a matrix of 5s. The matrix was presented in black and white, as shown in Plate 1 (following p. 310). Their responses were then compared with nonsynaesthetic control subjects. As might be expected, control subjects took some time to identify a triangle. The synaesthetic subjects, by contrast, 'were significantly better as detecting the embedded shape', demonstrating a genuine pop-out effect (such as displayed in Plate 2, following p. 310) (Ramachandran and Hubbard, 2001b, p. 7; see also Ramachandran and Hubbard, 2001a). These and other experiments demonstrate quite conclusively that these subjects' synaesthetic experiences were very much in keeping with how they reported them to be.

The upshot is plain. Though ordinary folk may not be in a position to verify what others say about their experience, scientists certainly are

---

[2] It's worth noting that in order to get this far, Ramachandran and Hubbard acknowledge a need for 'probing the introspective phenomenological reports of these subjects, even though this strategy is unpopular in conventional psychophysics. For instance, to the question: "Do you literally *see* the number 5 as red or does it merely remind you of red – the way a black-and-white half-tone photo of a banana reminds you of yellow?" our first two subjects replied with remarks like, "Well, that's hard to answer. It's not a memory thing. I do *see* the colour red. But I *know* it's just black. But with the banana I can also imagine it to be a different colour. It's hard to do that with the 5s", all of which further suggests that synaesthesia is a sensory phenomenon' (Ramachandran and Hubbard, 2001b, note 3, p. 7).

if they are clever enough to figure out what other discriminating behavioural manifestations there will be of the nature and quality of a person's experience beyond verbal report. Careful scientific investigation at
the behavioural level can give us – in fact, already has given us – good
reason to think that abnormal experiences, unlike miracles, do occur
after all. Moreover, we may have such reason even in the absence of direct physical evidence, or understanding, of the underlying organic cause
of various particular abnormalities. These observations point us towards
three related conclusions: First, in the case of strange or unusual sensory/
perceptual reports, we have no prima facie grounds for doubting the reliability of the subjects in question, especially when their reports are systematic and robust; their reliability or unreliability must be established
by further investigation. Secondly, if subjects are somehow miscuing us
about the nature of their own experience, there is no reason to think that
such miscuings cannot be detected from the third-person point of view –
or indeed that what they say cannot be objectively verified. Thirdly, this
very fact casts some doubt on the commonsense conception of subjective
report with which we began. I turn now to a fuller examination of this
final point.

3 Subjective Report as Grounded in Competing
Philosophical Models of Introspection

3.1 The Neoperceptual Model

As indicated earlier, Hume's reasoning about miraculous testimony fits
particularly well with a testimonial conception of self-report according
to which persons are testifying to something that they witness but that
their interlocutor does not – namely, their own experience. This way
of putting things has a rather perceptual flavour not to be taken literally anymore – that is, not in the sense of persons casting an inner eye
over their experience. Nevertheless, many philosophers and cognitive
scientists are drawn towards thinking of reflective self-awareness in quasi-
perceptual terms. In fact, they endorse what I will call a 'neoperceptual'
(NP) model of introspection, which replaces the 'inner eye' with a respectable subpersonal cognitive mechanism.[3] The key features of the

---

[3] The contemporary locus classicus for this view is Armstrong (1968, 1993). For critical discussion, see Shoemaker's account of the 'broad' perceptual model in Shoemaker (1996,
essays 10 and 11).

neoperceptual model are these:

i. The mind/brain is composed of what I will call *first-order* states and processes: for example, sensations, perceptions, emotions, beliefs, desires and so forth – states and processes that, insofar as they have representational content at all, represent conditions in the world (including conditions of our own bodies).

ii. We are aware of some of these states/processes courtesy of a (subpersonal) cognitive mechanism that causally produces in us independently existing *second-order* or *'meta-representational' beliefs:* beliefs whose representational content is about these first-order (mental) states and processes. This causal process constitutes the special faculty of 'introspection' or 'introspective judgement'.

iii. This causal process is generally reliable, at least in normal individuals under normal conditions.

iv. Subjective reports are normally the direct linguistic expression of the *second-order beliefs* thus generated.

Notice that in this model, there are two distinct sources of error in self-report: First, persons may not be giving adequate linguistic expression to their second-order beliefs, because of linguistic incompetence, confusion, distraction, memory problems and so forth. And, second, persons may not be expressing higher-order beliefs that are properly formed by this mechanism at all, because the mechanism is broken down, malformed or just not deployed. Some philosophers – for example, Alvin Goldman – may regard this second source of error as an advantage of the model, since it suggests that abnormal reports in pathological conditions could well be explained in terms of this sort of breakdown (Goldman, 1997). However, since this kind of breakdown is always possible on the NP model, it also suggests that researchers should not simply help themselves to the assumption, stipulated in iii, that our introspective mechanisms are generally reliable. Thus, as Goldman himself worries, if researchers want to use subjective reports as evidence for what their subjects are experiencing, it seems that they must first establish the reliability of their subjects' introspective mechanisms, even in cases where the first source of error is ruled out – that is, there is no evidence of linguistic incompetence, confusion, distraction and so on (Goldman, 1997; Goldman, 2000).

As I indicated in section 2, this task of verifying and even exploring what subjects claim about their experiences is not beyond the reach of cognitive scientists. Still, the NP model introduces, by its very structure, a source of error or unreliability in these reports that may cast inappropriate doubt on the prima facie credibility of subjective report in both normal and pathological conditions. To see why this doubt is inappropriate, I introduce an alternative model of self-report and reflective self-knowledge that simply rules out this putative source of error.

3.2 *The Reflective-Expressivist Model – Self-Knowledge of Intentional States*

The reflective-expressivist (RE) model builds on a feature already present in the NP model of self-knowledge and self-report. Recall that on the NP model, we report our first-order states by directly expressing our second-order beliefs about them: That is, we directly express what is believed (e.g., that we have a certain belief or experience). On pain of regress, we don't have to form any third-order beliefs in order to express the contents of our second-order beliefs – namely, that we have a certain belief or experience. We simply express these contents directly. In fact, advocates of the NP model should accept the fact that we also very often just directly express our first-order states: If I say 'that's a latte', then under the usual provisos of sincerity and linguistic competence, I am taken to directly express the content of my belief that there's a latte before me. It only seems necessary to depart from this basic expressivist structure when we're asked to engage in a certain mode of self-reflective activity – that is, when we're asked to pay attention to ourselves, as against (so it seems) the world – and say how it is with our own subjective condition.

But now let us examine this self-reflective task more closely. Suppose I spy a drink in front of me and judge (or take) it to be a latte. Phenomenologically, as all sides agree, it seems that when I'm asked whether I believe – that is, really *believe!* – that there's a latte before me, all I can do to make my answer is attend more closely to the apparent qualities of the drink itself: Does it look like a latte? Smell like a latte? Do I have background info that would call my conclusions into doubt? Does the drink still command my unswerving latte judgement? Moreover, if I perceptually 'retake' the drink in this way – that is, as a sophisticated folk-psychologist with background knowledge, however implicit, of the norms governing belief ascription – then I know that my positive reassessment of the evidence that leads me to say there's a latte before me is sufficient for my saying that I truly believe there's a latte before me. In other words, I don't seem to answer the belief question about myself by scrutinizing or tracking my own internal states at all. I simply *speak out of* my (first-order) latte-believing condition. (For this way of putting the point and a congenial defense of this type of expressivist view, see Bar-On, 2000; Bar-On and Long, 2001)

Of course, this phenomenological evidence is not conclusive; it can be accounted for in a number of ways. Even so, it makes vivid the conceptual point that knowing what I believe really only depends on two things: (1) having background knowledge of what believing in general requires – namely, a robust inclination to judge something is the case; and (2) the capacity to make and express judgements that report particular features of the world. Hence, there is no barrier in principle to embracing the comparatively minimal RE model of self-reflective activity. The key feature of this model is that such activity does not depend on a capacity for tracking one's own first-order states in addition to tracking what those first-order states are about. One's first-order states are not the direct target of cognitive or even subcognitive activity. Instead, such activity depends simply on a capacity to look at the world again, prompted by the recognition that it may not be as it seems – that is, as one initially judges it to be – and trying to rule out possible sources of error.

There are two important things to note about this minimalist approach to subjective knowledge and subjective report. First, what makes self-reflective activity distinctively *reflective* is that a person's intentionally expressed judgements – that is, judgements expressed with intention and in intentional language – depend on (a) knowing about folk-psychological concepts and the norms that govern them – that is, what counts as believing, hoping, desiring and so forth; and (b) knowing how the norms apply in particular cases (for instance, is my hesitation in reaffirming that this drink is a latte due to something detectably strange about the drink itself? – in which case I may register my doubt by hedging what I claim about myself. Or am I hesitating because I'm surprised by your question? I look again, I see it's a latte; of course, I believe it's a latte!). The second thing to note is that if a person's knowledge of these norms is limited or abnormal, the capacity to express (intentional) self-reflective judgements will be correspondingly compromised. This will be relevant to the case of autism, which I return to in section 4 below.

3.3 Extending the Reflective-Expressivist Model – Self-Knowledge of Experiential States

So far I've only talked about subjective reports, or better subjective expressions, of first-order intentional states. But what about subjective expressions of our own sensory/perceptual experiences, expressions of how things look, taste, smell and so on? Are there any special problems for the reflective expressivist here?

Suppose I catch a whiff of that irresistible latte smell, or better, have a close encounter with that irresistible latte taste. If I say, 'that smells (or tastes) like a latte', then on the NP model, I must be scanning or tracking (perhaps subcognitvely) some internal mental state of mine – a latte smell or taste experience. On the RE model, by contrast, I just directly express how I am experiencing things; I directly express how I take things to be, according to one sense modality or another.

But now suppose I'm asked to be more reflective about my experience, to say more precisely what my experience is *really* like. How do I go about answering this question? Once again, it seems that the only way to respond is by focussing my attention on the world, just as I do when I'm answering the belief question. Only now I focus on how in this moment things in the world, including states of my own body, really smell, taste, feel and so forth. According to the reflective expressivist, then, self-reflection in the sensory case is the same as in the belief case: It does not involve any special, introspective sort of mentalistic scanning; it does involve a special – that is, attentive, focussed – redeployment or reengagement of my sensory systems towards some aspect of the world. It involves my perceptually retaking the world in a certain modality, and it is this retaking that I express in confirming self-report: 'It smells like a rich, dark, smoothly roasted latte (to me)' (cf. Harman, 1990).

Briefly, then, and schematically, there are three things to note about making experiential reports on the RE model: First, I am able to *know* that I'm having a certain olfactory experience, say, just by virtue of being able to focus on *how* things smell in the world. Secondly, I am able to communicate that I have such an experience to others simply by virtue of knowing how to say how things smell in words (e.g., it smells bitter, sweet, like roses and so on). Thirdly, and most significantly, knowing how to say in more elaborate detail how things smell, taste, feel and so forth does not require any skilled folk-psychological understanding of the norms governing the ascription of belief and other intentional states. Again, this matters for assessing autistic subjective reports, and I'll come back to it in section 4.

3.4 Comparing the Reflective-Expressivist and Neoperceptual Models
There is much more to say in defence and elaboration of the RE model of self-awareness and self-report.[4] For present purposes, my aim is simply

---

[4] For further discussion, see Bar-On (2000), Bar-On and Long (2001); McGeer (in preparation).

to articulate a coherent alternative to the NP model that is worth exploring in its own right. In my view, what makes it worth exploring are the
following salient points of comparison:

i. The RE model is more economical than the NP model. Like the NP
model, it requires an account of how subjects are linguistically en-
abled to express the contents of their mental states without forming
(higher-order) beliefs about them. The RE model only differs with
the NP model over whether it is first- or second-order states that are
directly expressed in self-report. However, unlike the NP model,
the RE model obviates the positing of a tracking mechanism or
process responsible for generating second-order representations
of particular first-order states. This economy leads to a straightforward empirical prediction: The search for such a mechanism
and/or process in the brain will be forlorn. Correspondingly, at
the theoretical level, the very idea that such a mechanism or process could be selectively damaged, either through trauma or in the
course of development, will have no place in cognitive theory.

ii. The RE model rules out a certain kind of error in self-report –
namely, error by way of mistaking or misrepresenting one's own
conscious occurrent first-order states – for instance, one's own occurrent sensory experiences, as they putatively are in themselves
as against how they merely *seem* to a person to be. Experiences –
or better experienc*ings* – are firmly located back on the subjective
side of the subjective-objective divide. This has the further advantage of repudiating the strange metaphysics of 'real seeming', what
Daniel Dennett has called 'the bizarre category of the objectively-
subjective – the way things actively objectively seem to you even if
they don't seem to seem that way to you!' (Dennett, 1991, p. 132).

But is this really such an advantage? After all, one important
motivation for retaining the category of the objectively-subjective
is precisely to account for certain types of error in self-report –
namely, the sorts of error for which we seem to have very good reason to believe that individuals are misreporting (misexpressing)
how they are *really* experiencing things. Cases of neglect seem to
provide a good example. For instance, in Anton's syndrome, involving visual anosognosia (or blindness neglect), cortically blind
subjects fail to acknowledge that they cannot see. Is this not a
case of subjects simply mistaking what their own experiences are
really like?

In partial response to this challenge, it's important to recognize that the RE model does not rule out *all* sources of error or unreliability in subjective report. Uncontroversial sources of error or unreliability include, in either pathological or normal forms: limitations of language, limitations of memory, self- and other-deception and, of course, inattentiveness or neglect to features of the world, including, of course, features of one's own body. For example, for interesting pathological reasons yet to be fully understood in neurocognitive terms, the RE model suggests that what visual anosognosics are failing to notice is not something about the quality of their own experiences due specifically to a lack of inner-directed attention; rather, what they are failing to notice is the absence of a visible world due to a peculiar loss of outer-directed attention that accompanies their loss of visual perception. In other words, it's not just that visual anosognosics fail to see the world – this is simply blindness; it's that they *overlook* the absence of a visual world. Further, by overlooking the absence of the visual world, visual anosognosics fail to notice a problem with themselves. On this model, blindness neglect, or any pathology of neglect, is not to be viewed in terms of mis*perceiving* something inner – how one's sensory experiences really are in themselves (see, for example, Goldman, 1997); rather, neglect (as the terms suggests) is more like *forgetting* something outer: a usually temporary and specific *amnesia* concerning how they normally interact with the world (for a similar line of argument, see Dennett, 1991).

iii. Advocates of the RE model can also agree with neoperceptualists that insofar as ordinary subjects have opinions about this, they may be quite dramatically wrong about what accounts for their experiences being a certain way. For these are questions not about the way their experiences are (i.e., how things are, according to how they experience them); rather, these are questions about why they experience things as they do (i.e., how things work to make it the case that they experience things a certain way). Notoriously, those who reason from the first-person point of view may have pretty wild and woolly views about such things, at least from the point of view of science, if not from the point of view of folksy commonsense. For instance, the phenomenon of *change blindness* seems to indicate that we don't recreate and store somewhere in the brain a fully detailed imagistic representation of the scene before us. And yet, given the rich detail we can perceive in the world (as soon as we saccade to it), it may seem to us as perceiving subjects that the visual scene is recreated in our experience in all the high-resolution detail that appears before us. This seemingly plausible story is not unmasked by deeper first-person reflection on our experience, but rather by scientists experimenting with our abilities to act and react perceptually under carefully controlled conditions (Dennett, 1992; Churchland et al., 1994; Rensink, 2000).

iv. The crucial difference between the RE and NP models of self-attribution and self report is finally this: The RE model asserts, whereas the NP model denies, that within the restricted domain of individuals saying how things *seem* to them to be – that is, of their directly expressing how they are *experiencing* things – they cannot, subject to various qualifications, be reasonably doubted. The qualifications are that they are sincere and that uncontroversial sources of error are ruled out: linguistic incompetence, confusion, distraction, neglect and so on.

4 Are There Special Reasons to Doubt the Reliability of Autistic Self-Report?

So far I've argued that there are two background considerations that might lead researchers to underestimate the probable reliability of abnormal experiential or other subjective reports, thereby losing or misreading valuable data in their exploration of these phenomena.

The first is a misplaced adherence to the Humean principle: *the less probable the reported event, the less credible the report.* While this principle is not unreasonable in itself, I think it is misapplied in the case of subjective report (as against testimony to the occurrence of miracles) on two distinct counts: (a) Most importantly, we don't just have a person's verbal report to go on (there are ways of objectively exploring what the person reports); and (b) we really don't know enough at this stage about how the brain works to judge the probability or improbability of strangely abnormal experiences.

The second background consideration involves an adherence to the neoperceptual model of introspection. Researchers who endorse the idea that self-awareness – and, hence, self-report – is mediated by a mechanism for generating metarepresentations of (first-order) mental states and processes will always be confronted by questions of the reliability of this mechanism. Moreover, in cases of abnormal or unusual reports, the possible breakdown of this mechanism must seem a reasonable hypothesis (particularly when combined with the Humean reasoning reviewed earlier). Nevertheless, while this model of introspection is still very popular in philosophical and psychological circles, it is at least optional. Other models, such as the one I've sketched, show promise of accounting for the phenomena we wish to explain in a more economical and finally less mysterious fashion when it comes to conceptualizing the subject of experience, let alone experiences themselves. For embracing an RE model of self-reflective activity means forgoing the analogy between observing the world and observing (or even tracking) inner conscious states, as if these states can and do exist independently of reflecting, judgemental, verbally adept individuals as they are (somehow) in themselves. The RE model is therefore congenial with a more straightforward way of characterizing how subjects are in themselves: They *are* simply their rich and changing conscious experiences – their consciously attended ways of taking the world. Moreover, when they give expression to these ways of taking the world, they show how it is with them subjectively. Their verbal articulations – 'reports', if you like – thus constitute a rich source of *data* for consciousness research, to be explained in terms of what individuals are consciously experiencing, rather than constituting *evidence for* what is actually experienced, to be assessed in terms of how well a given individual is thought to introspect what is putatively an object for them as much as it is for researchers 'on the outside'.

In this last section, I turn to what may seem to be an additional reason for discounting subjective reports in the particular case of autism. My aim here is partly to demonstrate that particular philosophical views are not innocent in the way they enter into and affect empirical research programs.

This additional reason concerns the empirically demonstrable difficulties that autistics have in understanding *mental* states and properties – perceiving them in others, let alone reasoning about them – and so explaining and/or predicting others' behaviour by attributing such states. If autistics cannot attribute/perceive/understand the mental states of others, then surely it's reasonable to expect they'll have equal difficulty with their own mental lives. As cognitive psychologists Uta Frith and Francesca Happé have recently suggested:

The logical extension of the ToM deficit account of autism is that individuals with autism may know as little about their own minds as about the minds of other people. This is not to say that these individuals lack mental states, but that in an important sense they are unable to reflect on their mental states. Simply put, they lack the cognitive machinery to represent their thoughts and feelings *as* thoughts and feelings. (Frith and Happé, 1999)

What *kind* of cognitive machinery is this? According to Frith and Happé, it is the kind of cognitive machinery that would allow individuals to reflect on their inner experience. In particular, it is the kind of machinery that generates representations of mental states – hence, second-order representations – that are about first-order states and processes, themselves about physical things in the world. As the authors elaborate:

It seems plausible that the mechanism that keeps (second-order) representations of mental states separate from (first-order) representations of physical states is the same for self and other attribution. Even if the appreciation of others' mental states results in representations that are more error prone than the representations of own mental states, this difference becomes trivial if one is unable to represent mental states at all. (Frith and Happé, 1999, pp. 4–5)

In reviewing the details of Frith and Happé's proposal, it seems clear that they embrace a version of the neoperceptual model of introspective self-awareness where the unspecified cognitive mechanism or process that produces second-order representations of first-order states (whether of self or other) is replaced by the so-called theory of mind (or ToM) module. As they say themselves, '[I]f the mechanism which underlies the computation of mental states is dysfunctional, then self-knowledge is likely to be impaired just as is the knowledge of other minds' (ibid., p. 7). Moreover, since they propose that this mechanism is what allows for the representation of any kind of mental state – sensory, as well as intentional (e.g., beliefs and desires) – we are left with an error theory to account for abnormal – or even absent – sensory reports: It is the result of autistics' failure to form adequate second-order beliefs about their own sensory experience. As Frith and Happé explain:

If low-functioning autistics are unable to reflect on their inner experiences, then they would be unable to develop over time the richly connected semantic and experiential associations which normally pervade our reflective consciousness. Observation by parents suggests that the *awareness* of sensations and experiences may be peculiar in children with autism. Anecdotal reports of abnormal sensory and pain experiences are on occasion quite extreme.... Abnormal response to heat and cold, as well as hypo- and hyper-sensitivity to sound, light or touch are frequently reported.... Such responses might be expected if there was an inability to reflect on inner experiential states. Of course, normal pain perception is greatly affected by attribution and expectation. *These individuals might feel immediate pain in the same way as everyone else* [my emphasis], but would not be able to attribute to themselves the emotional significance that normally accompanies pain. This might explain why they do not complain about it. (Frith and Happé, 1999)

Compare this, now, with an autistic subjective report:

My insensitivity to pain was now as good as total . . . nothing hurt at all. And yet I felt – my actual feelings were not shut off – because when I was aware that I had injured myself somewhere, I could *sense something* [my emphasis], a non-pain, which branched out into my body from the place where the injury was. But the fact was, it didn't hurt. (Gerland, 1997, pp. 54, 157)

Clearly, in this autistic individual, there is *some* kind of awareness of the experience she calls 'pain' – a term, understood neutrally here, to designate that experiential state (whatever it may be) produced by bodily injury. What kind of awareness is this? It does not seem to be behavioural (i.e., based on how the person finds herself reacting to injury), since Gerland claims that she 'senses' something radiating out from the point of injury. So it must be some kind of inner experiential awareness. In light of this fact, does it make sense to say that Gerland 'feel[s] immediate pain in the same way as everyone else', but lacks the capacity to introspect, and so to become properly aware of her perfectly ordinary pain experience? The explanation seems a bit forced. At the very least, we need to rule out the more straightforward possibility that her actual 'pain' experience is abnormal – that is, that what's abnormal for her is her direct and immediate experience of bodily injury (this possibility is also suggested by Raffman, 1999).

However, this option is not so easily available to Frith and Happé since it involves significant modification to their account of introspective awareness. If autistic individuals are capable of being *fully, reflectively self-aware* of their sensory states, their relative incompetence with self- and other attribution of *intentional* states (like beliefs and desires) cannot be accounted for in terms of the breakdown of some general mechanism for generating second-order representational states – states that are, in their words, about mental, as opposed to physical, states of affairs. What, then, might Frith and Happé say?

I see two possibilities. The first is to hold onto a neoperceptual model of introspective awareness and multiply cognitive mechanisms for generating second-order representations of first-order states. One of these mechanisms, ToM_{sen}, is responsible for producing second-order representations of sensory-perceptual states and is spared in autism; but the other, ToM_{int}, responsible for producing second-order representations of intentional states like belief and desire, is absent or malformed. (Note that this means both mechanisms must be operative in normal cognition to explain normal competence in the full range of self-reports.)

There are, however, some disadvantages to taking this option: In the first place, complicating a functional theory by multiplying mechanisms is an expensive step, requiring one of two justifications: either (a) there is no other way to explain the relevant phenomena, or (b) there is some independent evidence providing collateral support for the proposed innovation. I think neither of these conditions is met in this case. A second disadvantage concerns the overall explanatory value of introducing the ToM$_{sen}$ mechanism to account for abnormal sensory/perceptual reports in autism. While such reports are now explained directly in terms of abnormal sensory/perceptual experiences, there is no explanation of how such experiences might relate to the social-cognitive deficits in autism. Indeed, this account would suggest that they are quite unconnected. Of course, it's always possible to explain their comorbidity in terms of diffuse brain abnormalities, where the pattern of spared and affected capacities has a purely neuromechanical explanation. However, it seems unlikely that such unusual and preoccupying sensory experiences would have no impact on autistic cognitive functioning. In any case, and this is a third disadvantage for Frith and Happé, in taking this option they defeat one of their main objectives in making the proposal that autistics have impaired introspective awareness: namely, to strengthen the claim that a so-called theory of mind deficit is in fact the central deficit in autism, capable of accounting for a range of seemingly unrelated – and yet comorbid – autistic abnormalities.

There is a second possibility that Frith and Happé have open to them. They can take the more radical theoretical step of jettisoning the neo-perceptual model of introspective awareness altogether and embracing something like the reflective-expressivist model instead. Why is this recommended?

Notice, first, that establishing the reliability of autistic self-report in the case of their sensory/perceptual experiences is not so difficult for proponents of the RE model: On this approach, if there are no signs of failure of understanding or attention, gross linguistic difficulties, confusion and so forth, then there is no reason to doubt that self-reporting autistics are directly and reliably expressing how they are experiencing things. Looking again at the now quite extensive sample of autistic autobiographical writings, it's hard to find any signs of these sources of error. While there are certainly oddities of style and content, the capacity these individuals have to write about themselves – and, in particular their sensory experience – in such vivid and articulate detail belies the idea of confusion, linguistic incompetence, lack of understanding or attention and so forth. Hence, their reports should not be written off as unreliable.

If the evidence suggests that autistic self-reports are reliable in the case of sensory-perceptual matters, then adopting the RE model means that there is no need to postulate a *second* mechanism, also putatively operative in normal cognition, for generating second-order beliefs that are exclusively about first-order sensory-perceptual states. Since no such mechanism is required to explain a person's self-reporting competence in general, there is no need to multiply mechanisms to account for why autistics are capable in one domain while failing in another (the domain of self- and other attribution of intentional states). Hence, this option avoids the disadvantage of introducing any further complications into the functional theory.

Nevertheless, there are difficulties for this approach. For instance, we still need to explain the relative *incompetence* that autistics show with intentional mentalistic attribution and reasoning – in particular, as this involves (false) beliefs and (complex) emotional states. How is this hurdle to be overcome?

Recall that on the RE model, competence in intentional self- (and other) attribution does not depend on a capacity to form second-order beliefs about oneself (or others) that are then expressed in intentional reports – reports such as 'I believe that p' or 'she believes that p'. It does not depend on any sort of mechanism for generating such beliefs. Rather, it depends on knowing how and when the norms governing folk-psychological concepts (belief, desire, hope, fear) apply to our dispositions to interact with the world: For instance, if someone is disposed to un-hesitatingly judge that p, then, according to the norms of folk-psychology, this amounts to someone's believing that p. Hence, if I'm aware of this norm, and find myself or anyone else unhesitatingly disposed to judge that p (whether in word or in deed), then I know we each believe that p, and can competently attribute this state to either one of us.

Of course, learning and applying such norms takes a lot of practice; they are much more nuanced than this crude sketch implies, and their conditions of correct application depend on many subtle features of the way humans interact with the world, constituting myriad patterns of normalcy and exception. For instance, if someone confidently insists that p while breaking into a light sweat and nervously tapping a foot, we tend to think, 'Aha, this person probably does not believe that p, but wants me to believe that p' – or even more subtly, 'This person wants me to think that *he* or *she* believes that p', or even more subtly still, 'This person wants me to think that he or she is *lying* about p', and so on. How many day-to-day interactions with others, both humdrum and exceptional, does it take to acquire this kind of dedicated expertise, in knowing both the norms of folk-psychology and in knowing the ways these norms may be kept and broken in various linguistic and nonlinguistic ways?

Acquiring this knowledge may, of course, be crude and partial, and one's capacity to understand and use folk-psychological attribution and explanation correspondingly circumscribed. Much will depend on one's motivation and opportunity for learning, especially as this capacity begins to develop through childhood. So, for instance, congenitally blind and congenitally deaf children (particularly deaf children of hearing parents) are delayed on standard theory of mind tests, presumably not because they lack some special neuro-cognitive mechanism for generating second-order beliefs about intentional states but because their sensory deficits deprive them of the opportunity of engaging in rich, reason-giving interactions with parents and peers, possibly affecting their motivation for engaging in such interactions as well. Hence, they are slow to develop normal folk-psychological expertise (Hobson, 1993; Brown et al., 1997; Peterson and Siegal, 1998; Peterson and Siegal, 1999; Peterson et al., 2000). Could something like this be true for autistic individuals? Could their very profound sensory-perceptual abnormalities sufficiently inhibit their interactions with others, both through development and in an ongoing way, so as to seriously compromise their ability to acquire folk-psychological expertise? I think it could and argue extensively for this possibility elsewhere (McGeer, 2001).

If this explanatory hurdle can be overcome in the manner just suggested, then embracing the RE model of reflective self-awareness and self-report, far from leaving us with no account of autistic social-cognitive problems, opens up new explanatory possibilities and creates new research opportunities. In particular, in keeping with Frith and Happé's original motivation, it suggests that there *is* an important connection between reported sensory/perceptual abnormalities and social-cognitive problems in autism, thereby promising a unified account of these abnormalities. However, against Frith and Happé's dysfunctional ToM hypothesis, it suggests a different kind of connection and an alternative developmental possibility to what they propose. In particular, it suggests that abnormal sensory *reports* are not the contemporary consequence of a selectively damaged innate capacity for cognizing mind. Rather, it suggests that deep and extensive sensory/perceptual abnormalities are both developmentally prior to and importantly implicated in the failure to acquire normal social-cognitive expertise. In other words, it is to suggest that research in autism is best approached from a neuro-constructivist perspective. As Karmiloff-Smith writes:

It is clear that disorders like autism and SLI [Specific Language Impairment] have a genetic origin and that evolutionary pressures have contributed to whatever is innately specified. This is a truism. The question is whether, on the one hand, the deficit results from damage to a domain-specific starting point at the cognitive level, as a result of evolution pre-specifying dedicated processing systems for grammar, theory of mind, and so forth, or whether, on the other hand, evolution has specified more general constraints for higher-level cognition and there is a more indirect way for genetic defects to result in domain-specific outcomes as a function of development. (Karmiloff-Smith, 1998)

In support of this broader developmental approach, consider one final passage from Therese Joliffe's "personal account" of autism:

Reality to an autistic person is a confusing interacting mass of events, people, places, sounds and sights. There seem to be no clear boundaries, order or meaning to anything. A large part of my life is spent just trying to work out the pattern behind everything. . . .

Objects are frightening. Moving objects are harder to cope with because of the added complexity of movement. Moving objects which also make a noise are even harder to cope with because you have to try to take in the sight, movement and further added complexity of the noise. Human beings are the hardest of all to understand because not only do you have to cope with the problem of just seeing them, they move about when you are not expecting them to, they make varying noises and along with this, they place all different kinds of demands on you which are just impossible to understand. . . .

It is the confusion that results from not being able to understand the world around me which I think causes all the fear. This fear then brings a need to withdraw. Anything which helps reduce the confusion . . . has the effect of reducing the fear and ultimately reduces the isolation and despair, thus making life a bit more bearable to live in. If only people could experience what autism is like just for a few minutes, they might then know how to help! (Jolliffe et al., 1992)

Experiencing what autism is like is out of the question for those of us not similarly afflicted. But I do think that paying close attention to autistic self-report can yield substantial insights. In particular, I think it lends credence to the developmental hypothesis I have suggested. But, of course, this is just a hypothesis, requiring further investigation and empirical support. My hope is that once any philosophical biases towards thinking of self-attribution and self-report in broadly perceptual terms have been removed, this line of research will become tempting for cognitive scientists to explore in detail.

Acknowledgments

This work was supported by the McDonnell Foundation through a grant received from the McDonnell Project in Philosophy and the Neurosciences. I would also like to thank Daniel Dennett, Philip Gerrans, Philip Pettit and especially Kathleen Akins for helpful comments on earlier drafts of this chapter.

References

Armstrong, D. (1968, 1993). *A Materialist Theory of the Mind*. London, Routledge.

Bar-On, D. (2000). Speaking my mind. *Philosophical Topics* 28(2): 1–34.

Bar-On, D., and D. C. Long (2001). Avowals and first-person privilege. *Philosophy and Phenomenological Research* 62(2): 311–335.

Baron-Cohen, S. (1996). *Mindblindness: An Essay on Autism and Theory of Mind*. Cambridge, MA, MIT Press.

Baron-Cohen, S., H. Tager-Flusberg, and D. J. Cohen, Eds. (1993). *Understanding Other Minds: Perspectives from Autism*. Oxford, Oxford University Press.

Baron-Cohen, S., H. Tager-Flusberg, and D. J. Cohen, Eds. (2000). *Understanding Other Minds: Perspectives from Developmental Cognitive Neuroscience*. Oxford, Oxford University Press.

Brown, R., R. P. Hobson, and A. Lee (1997). Are there 'autistic-like' features in congenitally blind children? *Journal of Child Psychology and Psychiatry* 38(6): 693–703.

Churchland, P. S., V. S. Ramachandran, and T. J. Sejnowski (1994). A critique of pure vision. In C. Koch and J. Davis (Eds.), *Large Scale Neuronal Theories of the Brain*. Cambridge, MA, MIT Press: 23–60.

Dennett, D. C. (1991). *Consciousness Explained*. Boston, MA, Little, Brown and Company.

Dennett, D. C. (1992). 'Filling in' versus finding out: A ubiquitous confusion in cognitive science. In H. L. Pick, P. van den Broek, and D. C. Knill (Eds.), *Cognition: Conceptual and Methodological Issues*. Washington, DC, American Psychological Association: 33–49.

Frith, U. (1989). *Autism: Explaining the Enigma*. Oxford, Blackwell.

Frith, U., Ed. (1991). *Asperger and his Syndrome*. Cambridge, Cambridge University Press.

Frith, U., and F. Happé (1994). Autism: Beyond 'theory of mind'. *Cognition* 50: 115–132.

Frith, U., and F. Happé (1999). Theory of mind and self-consciousness: What is it like to be autistic? *Mind and Language* 14(1): 1–22.

Gerland, G. (1997). *A Real Person: Life on the Outside*. London, Souvenir Press.

Gerrans, P., and V. McGeer (2003). *Theory of Mind in Autism and Schizophrenia*. Brighton and New York, Psychology Press.

Goldman, A. I. (1997). Science, publicity and consciousness. *Philosophy of Science* 64: 525–545.

Goldman, A. I. (2000). Can science know when you're conscious? Epistemological foundations of cognitive research. *Journal of Consciousness Studies* 7(5): 3–22.

Green, D., G. Baird, A. L. Barnett, L. Henderson, J. Huber, and S. E. Henderson (2002). The severity and nature of motor impairment in Asperger's syndrome: A comparison with Specific Developmental Disorder of Motor Function. *Journal of Child Psychology and Psychiatry and Allied Disciplines* 43(5): 655–668.

Happé, F. (1996). Studying weak central coherence at low levels: Children with autism do not succumb to visual illusions: A research note. *Journal of Child Psychology and Psychiatry* 37(7): 873–877.

Happé, F. (2000). Parts and wholes, meanings and minds: Central coherence and its relation to theory of mind. In S. Baron-Cohen, H. Tager-Flusberg, and D. J. Cohen (Eds.), *Understanding Other Minds: Perspectives from Developmental Cognitive Neuroscience*. Oxford, Oxford University Press: 203–221.

Harman, G. (1990). The Intrinsic Quality of Experience. In J. Tomberlin (Ed.), *Philosophical Perspectives*. Atascadero, CA, Ridgeview 4: 31–52.

Hobson, R. P. (1993). Understanding persons: The role of affect. In S. Baron-Cohen, H. Tager-Flusberg, and D. J. Cohen (Eds.), *Understanding Other Minds: Perspectives from Autism*. Oxford, Oxford University Press: 204–227.

Hume, D. (1975). *Enquiries Concerning Human Understanding and Concerning the Principles of Morals*. Oxford, Clarendon Press.

Jolliffe, T., R. Lansdown, and C. Robinson (1992). Autism: A personal account. *Communication* 26(3): 12–19.

Karmiloff-Smith, A. (1998). Development itself is the key to understanding developmental disorders. *Trends in Cognitive Science* 2(10): 389–398.

Klin, A., F. R. Volkmar, and S. S. Sparrow, Eds. (2000). *Asperger Syndrome*. New Haven, CT, Yale University Medical Center.

Mayes, S. D., and S. L. Calhoun (2001). Non-significance of early speech delay in children with autism and normal intelligence and implications for DSM-IV Asperger's disorder. *Autism* 5(1): 81–94.

Mayes, S. D., S. L. Calhoun, and D. L. Crites (2001). Does DSM-IV Asperger's disorder exist? *Journal of Abnormal Child Psychology* 29(3): 263–271.

McGeer, V. (2001). Psycho-practice, psycho-theory and the contrastive case of autism. *Journal of Consciousness Studies* 8(5–7): 109–132.

McGeer, V. (in preparation). *Mind in Time: The Developmental Character of Human Intentionality*, under contract to MIT Press.

Ozonoff, S., M. South, and J. N. Miller (2000). DSM-IV-defined Asperger syndrome: Cognitive, behavioral and early history differentiation from high-functioning autism. *Autism* 4(1): 29–46.

Peterson, C. C., J. C. Peterson, and J. Webb (2000). Factors influencing the development of a theory of mind in blind children. *British Journal of Developmental Psychology* 18(3): 431–447.

Peterson, C. C., and M. Siegal (1998). Changing focus on the representational mind: Concepts of false photos, false drawings and false beliefs in deaf, autistic and normal children. *British Journal of Developmental Psychology* 16: 301–320.

Raffman, D. (1999). What autism may tell us about self-awareness: A commentary on Frith and Happé. *Mind and Language* 14(1): 23–31.

Ramachandran, V. S., and E. M. Hubbard (2001a). Psychophysical investigations into the neural basis of synaesthesia. *Proceedings of the Royal Society of London, B* 268: 979–983.

Ramachandran, V. S., and E. M. Hubbard (2001b). Synaesthesia: A window into perception, thought and language. *Journal of Consciousness Studies* 8(12): 3–34.

Rensink, R. A. (2000). The dynamic representation of scenes. *Visual Cognition* 7(1–3): 17–42.

Robbins, T. W. (1997). Integrating the neurological and neuropsychological dimensions of autism. In J. Russell (Ed.), *Autism as an Executive Disorder*. Oxford, Oxford University Press: 21–53.

Shoemaker, S. (1996). *The First Person Perspective and Other Essays*. Cambridge, Cambridge University Press.

Tager-Flusberg, H., Ed. (1999). *Neurodevelopmental Disorders*. Boston, MA, MIT Press/Bradford.

Willey, L. H. (1999). *Pretending to Be Normal: Living with Asperger Syndrome*. London, Jessica Kingsley Publishers.

Williams, D. (1999). *Autism and Sensing: The Unlost Instinct*. London, Jessica Kingsley Publishers.

NEURAL REPRESENTATION

4

Moving Beyond Metaphors: Understanding the Mind for What It Is

Chris Eliasmith

1 Introduction

In the last 50 years, there have been three major approaches to understanding cognitive systems and theorizing about the nature of the mind: symbolicism, connectionism, and dynamicism. Each of these approaches has relied heavily on a preferred metaphor for understanding the mind/brain. Most famously, symbolicism, or classical cognitive science, relies on the "mind as computer" metaphor. Under this view, the mind is the software of the brain. Jerry Fodor,[1] for one, has argued that the impressive theoretical power provided by this metaphor is good reason to suppose that cognitive systems have a symbolic "language of thought," which, like a computer programming language, expresses the rules that the system follows. Fodor claims that this metaphor is essential for providing a useful account of how the mind works.

Similarly, connectionists have relied on a metaphor for providing their account of how the mind works. This metaphor, however, is much more subtle than the symbolicist one; connectionists presume that the functioning of the mind is like the functioning of the brain. The subtlety of the "mind as brain" metaphor lies in the fact that connectionists, like symbolicists, are materialists. That is, they also hold that the mind is the brain. However, when providing psychological descriptions, it is the metaphor that matters, not the identity. In deference to the metaphor, the founders of this approach call it "brain-style" processing and claim to

---

[1] *The Language of Thought* (New York: Crowell, 1975).

be discussing "abstract networks."[2] This is not surprising since the computational and representational properties of the nodes in connectionist networks bear little resemblance to neurons in real biological neural networks.[3]

Proponents of dynamicism also rely heavily on a metaphor to understand cognitive systems. Most explicitly, Timothy van Gelder[4] employs the Watt Governor as a metaphor for mind. It is through his analysis of the best way to characterize this dynamic system that he argues that cognitive systems, too, should be understood as nonrepresentational, low-dimensional, dynamical systems. Like the Watt Governor, van Gelder argues, cognitive systems are essentially dynamic and can only be properly understood by characterizing their state changes through time. The "mind as Watt Governor" metaphor suggests that trying to impose any kind of discreteness, either temporal or representational, will lead to a mischaracterization of minds.

Notably, each of symbolicism, connectionism, and dynamicism rely on metaphor not only for explanatory purposes but also for developing their conceptual foundations in understanding the target of the metaphor, that is, the mind. For symbolicists, the properties of Turing machines become shared with minds. For connectionists, the character of representation changes dramatically. Mental representations are taken to consist of "subsymbols" associated with each node, while "whole" representations are real-valued vectors in a high-dimensional property space.[5] Finally, for the dynamicists, because the Watt Governor is best described by dynamic systems theory, which makes no reference to computation or representation, our theories of mind need not appeal to computation or representation either.

In this chapter, I want to suggest that it is time to move beyond these metaphors. We are in the position, I think, to understand the mind for what it is: the result of the dynamics of a complex, physical, information processing system, namely, the brain. Clearly, in some ways this

[2] J. McClelland and D. Rumelhart, "Future directions," in McClelland, J. and D. Rumelhart, eds., *Parallel Distributed Processing: Explorations in the Microstructure of Cognition,* Vol. 2 (Cambridge, MA: MIT Press, 1986), pp. 547–552.

[3] As discussed in chapter 10 of W. Bechtel and A. Abrahamsen, *Connectionism and the Mind: Parallel Processing, Dynamics, and Evolution in Networks* (2d ed., Oxford: Blackwell, 2001).

[4] "What Might Cognition Be, If Not Computation?" *Journal of Philosophy,* XCI, 7 (1995): 345–381.

[5] See, for example, Paul Smolensky's "On the Proper Treatment of Connectionism," *Behavioral and Brain Sciences,* XI, 1 (1988): 1–23.

is a rather boring thesis to defend. It is just a statement of plain old "monistic materialism" or "token identity theory," call it what you will. It is, in essence, just the uncontroversial view that, so far as we know, you do not have a mind without a brain. But I further want to argue that the best way to understand this physical system is by using a different set of conceptual tools than those employed by symbolicists, connectionists, and dynamicists individually. That is, the right toolbox will consist in an extended subset of the tools suggested by these various metaphors.

The reason we need to move beyond metaphors is because, in science, analogical thinking can sometimes constrain available hypotheses. This is not to deny that analogies are incredibly useful tools at many points during the development of a scientific theory. It is only to say that, sometimes, analogies only go so far. Take, for instance, the development of the current theory of the nature of light. In the 19th century, light was understood in terms of two metaphors: light as a wave and light as a particle. Thomas Young was the best-known proponent of the first view, and Isaac Newton was the best-known proponent of the second. Each used his favored analogy to suggest new experiments and develop new predictions.[6] Thus, these analogies played a role similar to that played by the analogies discussed in contemporary cognitive science. However, as we know in the case of light, both analogies are false. Hence, the famed "wave-particle duality" of light: sometimes it behaves like a particle, and sometimes it behaves like a wave. Neither analogy by itself captures all the phenomena displayed by light, but both are extremely useful in characterizing some of those phenomena. So, understanding what light *is* required moving beyond the metaphors.

I want to suggest that the same is true in the case of cognition. Each of the aforementioned metaphors has some insight to offer regarding certain phenomena displayed by cognitive systems. However, none of these metaphors is likely lead us to all of the right answers. Thus, my project in trying to move beyond these metaphors is a synthetic one. I want to provide a way of understanding cognitive systems that draws on the strengths of symbolicism, connectionism, and dynamicism. The best way of doing this is to understand minds for what they are. To phrase this as a conditional, if minds are the behavior of complex, dynamic,

---

[6] For a detailed description of the analogies, predictions, and experiments, see C. Eliasmith and P. Thagard, "Particles, Waves and Explanatory Coherence," *British Journal of the Philosophy of Science*, XLVIII (1997): 1–19.

information-processing systems, then we should use the conceptual tools that we have for understanding such systems when trying to understand minds. In this chapter, I outline a general theory that describes representation and dynamics in neural systems (R&D theory) that realizes the consequent of this conditional. I argue that R&D theory can help unify neural and psychological explanations of cognitive systems and that the theory suggests a need to reevaluate standard functionalist claims.

First, however, it is instructive to see how R&D theory does not demand the invention of new conceptual tools; the relevant tools are already well tested. So, in some ways, the theory is neither risky nor surprising. What is surprising, perhaps, is that our most powerful tools for understanding the kinds of systems that minds are have yet to be applied to minds. I suggest that this surprising oversight is due to an overreliance on the mind-as-computer metaphor.

2 A Brief History of Cognitive Science

While the main purpose of this chapter is clearly not historical, a brief perusal of the relevant historical landscape helps situate both the theory and the subsequent discussion.

2.1 Prehistory

While much is sometimes made of the difference between philosophical and psychological behaviorism, there was general agreement on this much: internal representations, states, and structures are irrelevant for understanding the behavior of cognitive systems. For psychologists, like John Watson and B. F. Skinner, this was true because only input/output relations are scientifically accessible. For philosophers such as Gilbert Ryle, this was true because mental predicates, if they were to be consistent with natural science, must be analyzable in terms of behavioral predicates. In either case, looking inside the "black box" that was the object of study was prohibited for behaviorists.

It is interesting to note that engineers of the day respected a similar constraint. In order to understand dynamic physical systems, the central tool they employed was (classical) control theory. Classical control theory, notoriously, only characterizes physical systems in terms of their input/output relations in order to determine the relevant controller. Classical control theory was limited to designing nonoptimal, single-variable, static controllers, and it depended on graphical methods and rules of thumb and did not allow for the inclusion of noise.[7] While the limitations of classical controllers and methods are now well known, they nevertheless allowed engineers to build systems of kinds they had not systematically built before: goal-directed systems.

While classical control theory was useful, especially in the 1940s, for building warhead guidance systems, some researchers thought it was clearly more than that. They suggested that classical control theory could provide a theoretical foundation for describing living systems as well. Most famously, the interdisciplinary movement founded in the early 1940s known as "cybernetics" was based on precisely this contention.[8] Cyberneticists claimed that living systems were also essentially goal-directed systems. Thus, closed-loop control, it was argued, should be a good way to understand the behavior of living systems. Given the nature of classical control theory, cyberneticists focused on characterizing the input/output behavior of living systems, not their internal processes. With the so-called cognitive revolution of the mid-1950s, interest in cybernetics waned due in part to its close association with, and similar theoretical commitments to, behaviorism.

2.2 *The Cognitive Revolution*
In the mid-1950s, with the publication of a series of seminal papers,[9] the "cognitive revolution" took place. One simplistic way to characterize this shift from behaviorism to cognitivism is that it became no longer taboo to look inside the black box. Quite the contrary: Internal states, internal processes, and internal representations became standard fare when thinking about the mind. Making sense of the insides of that black box was heavily influenced by concurrent successes in building and programming computers to perform complex tasks. Thus, many early cognitive scientists saw, when they opened the lid of the box, a computer. As explored in

---

7 For a succinct description of the history of control theory, see Frank Lewis's *Applied Optimal Control and Estimation* (New York: Prentice-Hall, 1992).

8 For a statement of the motivations of cybernetics, see Arturo Rosenblueth, Norbert Wiener, and Julian Bigelow, "Behavior, Purpose, and Teleology," *Philosophy of Science*, X (1943): 18–24.

9 These papers include, but are not limited to: A. Newell, C. Shaw, and H. Simon, "Elements of a Theory of Human Problem Solving," *Psychological Review*, LXV (1958): 151–166; G. Miller "The Magical Number Seven, Plus or Minus Two: Some Limits on Our Capacity for Processing Information," *Psychological Review*, LXIII (1956): 81–97; J. Bruner, J. Goodnow, and G. Austin, *A Study Of Thinking* (New York: Wiley, 1956).

detail by Jerry Fodor,[10] "[c]omputers show us how to connect semantical with causal properties for *symbols*"; thus, computers have what it takes to be intentional minds. Once cognitive scientists began to think of minds as computers, a number of new theoretical tools became available for characterizing cognition. For instance, the computer's theoretical counterpart, the Turing machine, suggested novel philosophical theses, including functionalism and multiple realizability, about the mind. More practically, the typical architecture of computers, the von Neumann architecture, was thought by many to be relevant for understanding our cognitive architecture.

However, adoption of the von Neumann architecture for understanding minds was seen by many as poorly motivated. As a result, the early 1980s saw a revival of the so-called connectionist research program. Rather than adopting the architecture of a digital computer, these researchers thought that an architecture more like that seen in the brain would provide a better model for cognition.[11] As a result of this theoretical shift, connectionists were very successful at building models sensitive to statistical structure and could begin to explain many phenomena not easily captured by symbolicists (e.g., object recognition, generalization, learning, etc.).

For some, however, connectionists had clearly not escaped the influence of the mind-as-computer metaphor. Connectionists still spoke of representations and thought of the mind as a kind of computer. These critics argued that minds are not essentially computational; they are essentially physical, dynamic systems.[12] They suggested that if we want to know which functions a system can actually perform in the real world, we must know how to characterize the system's dynamics. Furthermore, since cognitive systems evolved in dynamic environments, we should expect evolved control systems, like brains, to be more like the Watt Governor – dynamic, continuous, coupled directly to what they control – than like a discrete-state Turing machine that computes over "disconnected" representations. As a result, these "dynamicists" suggested that dynamic systems theory, not computational theory, was the right quantitative tool for

---

[10] *Psychosemantics* (Cambridge, MA: MIT Press, 1987), p. 18.

[11] As discussed in both Smolensky, op. cit., and the introduction to D. Rumelhart and J. McClelland, eds., *Parallel Distributed Processing: Explorations in the Microstructure of Cognition*, Vol. 1 (Cambridge, MA: MIT Press, 1986).

[12] See the various contributions to R. Port and T. van Gelder, eds., *Mind as Motion: Explorations in the Dynamics of Cognition* (Cambridge, MA: MIT Press, 1995), especially the editors' introduction.

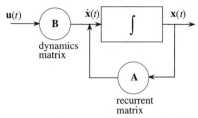

FIGURE 4.1. A modern control theoretic system description. The vector $\mathbf{u}(t)$ is the input to the system. $\mathbf{A}$ and $\mathbf{B}$ are matrices that define the behavior of the system, $\mathbf{x}(t)$ is the system state variable (generally a vector), and $\dot{\mathbf{x}}(t)$ is the derivative of the state vector. The standard transfer function in control theory, as shown in the rectangle, is integration. (I have simplified this diagram for a generic linear system from the typical, truly general one found in most control theory texts by excluding the feedthrough and output matrices. Nothing turns on this simplification in this context.)

understanding minds. They claimed that notions like "chaos," "hysteresis," "attractors," and "state-space" underwrite the conceptual tools best suited for describing cognitive systems.

2.3 A Puzzling Oversight

In some ways, dynamicists revived the commitments of the predecessors of the cognitive revolution. Notably, the Watt Governor is a standard example of a classical control system. If minds are to be like Watt Governors, they are to be like classical control systems, just what the cyberneticists had argued. One worry with this retrospective approach is that the original problems come along with the original solutions. The limitations of classical control theory are severe, so severe that they will probably not allow us to understand a system as complex as the brain.

However, an important series of theoretical advances in control theory went completely unnoticed during the cognitive revolution. During the heyday of the computer, in the 1960s, many of the limitations of classical control theory were rectified with the introduction of what is now known as "modern" control theory.[13] Modern control theory introduced the notion of an "internal system description" to control theory. An internal system description is one that includes *system state variables* (i.e., variables describing the state of the system itself) as part of the description (see Figure 4.1). It is interesting that with the cognitive revolution,

---

[13] This introduction is largely credited to R. Kalman in his "A New Approach to Linear Filtering and Prediction Problems," *ASME Journal of Basic Engineering*, LXXXII (1960): 35–45.

researchers interested in the behavior of living systems realized that they needed to "look inside" the systems they were studying, and at about the same time, researchers interested in controlling engineered systems began to look inside as well. Both, nearly simultaneously, opened their black box. However, as already discussed, those interested in cognitive behavior adopted the computer as a metaphor for the workings of the mind. Unfortunately, the ubiquity of this metaphor has served to distance the cognitive sciences from modern control theory. Nevertheless, I argue in section 3.3 that modern control theory offers tools better suited than computational theory for understanding biological systems as fundamentally physical, dynamic systems operating in changing, uncertain environments.

This is not to suggest that each of the dominant metaphors should be taken as irrelevant to our understanding of minds. Both connectionism and dynamicism highlight important limitations of the original mind-as-computer metaphor. Connectionism challenged the symbolicist conception of representation, noting how important statistical considerations are for capturing certain kinds of cognitive phenomena. Dynamicist critiques of symbolicism focused on its lack of a principled account of the temporal properties of cognitive systems.[14] Nevertheless, it was the symbolicists, armed with their metaphor, who rightly justified opening the black box. And, furthermore, both connectionism and dynamicism introduced their own misleading metaphors.

3 Representation and Dynamics in Neural Systems: A Theory

Moving beyond metaphors, that is, taking seriously the view that minds are complex, physical, dynamic, and information-processing systems, means using our best tools for describing systems with these properties. In the remainder of this section, I propose and defend a theory of representation and dynamics in neural systems (R&D theory) that takes precisely this approach. R&D theory relies on modern control theory, information theory, and recent results from neuroscience to provide an account of what minds are.[15]

I have broken this account into three parts. The first defines representation, the second describes computation, and the third section, on

---

[14] T. van Gelder, op. cit.

[15] For an in-depth technical description of this approach, see Chris Eliasmith and Charles Anderson, *Neural Engineering: Computation, Representation and Dynamics in Neurobiological Systems* (Cambridge, MA: MIT Press, 2003).

dynamics, shows how the preceding characterizations of representation and computation can be merged with control theory to provide an account of neural and cognitive function. The result, I argue, is a theory that avoids the weaknesses and capitalizes on the strengths of past approaches.

3.1 Representation

A central tenet of R&D theory is that we can adapt the information theoretic account of *codes* to understanding representation in neural systems. Codes, in engineering, are defined in terms of a complementary encoding and decoding procedure between two alphabets in such systems. Morse code, for example, is defined by the one-to-one relation between letters of the Roman alphabet and the alphabet composed of a standard set of dashes and dots. The encoding procedure is the mapping from the Roman alphabet to the Morse code alphabet, and the decoding procedure is its inverse.

In order to characterize representation in a cognitive/neural system, we can identify each of these procedures and their relevant alphabets. The encoding procedure is quite easy to identify: it is the transduction of stimuli by the system resulting in a series of neural "action potentials," or "spikes." The precise nature of this encoding has been explored in depth via quantitative models.[16] So, encoding is what neuroscientists typically talk about. When I show a cognitive system a stimulus, some neurons or other "fire." Unfortunately, neuroscientists often stop here in their characterization of representation, but this is insufficient. We also need to identify a decoding procedure; otherwise there is no way to determine the relevance of the encoding for the system. If no information about the stimulus can be extracted from the spiking neuron, then it makes no sense to say that it represents the stimulus. Representations, at a minimum, must potentially be able to "stand in" for their referents.

Quite surprisingly, despite typically nonlinear encoding, a good linear decoding can be found.[17] And there are several established methods for determining linear decoders, given the statistics of the neural populations that respond to certain stimuli.[18] Notably, these decoders are sensitive both to the temporal statistics of the stimuli and to what other

---

[16] See J. Bower and D. Beeman, *The Book of GENESIS: Exploring Realistic Neural Models with the GEneral NEural SImulation System* (Berlin: Springer Verlag, 1998) for a review of such models.

[17] As demonstrated by F. Rieke, D. Warland, R. de Ruyter van Steveninick, and W. Bialek, *Spikes: Exploring the Neural Code* (Cambridge, MA: MIT Press, 1997), pp. 76–87.

[18] As discussed in Eliasmith and Anderson, op. cit.

elements in the population encode. Thus, if you have multiple neurons involved in the (distributed) representation of a time-varying object, they can "cooperate" to provide a better representation.

Having specified the encoding and decoding procedures, we still need to specify the relevant alphabets. While the specific cases will diverge greatly, we can describe the alphabets generally: neural responses (encoded alphabet) code physical properties (decoded alphabet). In fact, it is possible to be a bit more specific. Neuroscientists generally agree that the basic element of the neural alphabets is the neural spike. However, there are many possibilities for how such spikes are used: average production rate of neural spikes (i.e., a rate code); specific timings of neural spikes (i.e., a timing code); population-wide groupings of neural spikes (i.e., a population code); or the synchrony of neural spikes across neurons (i.e., a synchrony code). Of these possibilities, arguably the best evidence exists for a combination of timing codes and population codes.[19] For this reason, let us take the combination of these basic coding schemes to define the alphabet of neural responses. Thus, the encoded alphabet is the set of temporally patterned neural spikes over populations of neurons.

It is much more difficult to be specific about the nature of the alphabet of physical properties. Of course, we can begin by looking to the physical sciences for categories of physical properties that might be encoded by nervous systems. Indeed, we find that many of the properties that physicists traditionally use do seem to be represented in nervous systems, for example, displacement, velocity, acceleration, wavelength, temperature, pressure, mass, and so on. But there are many physical properties not discussed by physicists that also seem to be encoded in nervous systems, for example, red, hot, square, dangerous, edible, object, conspecific, and so on. Presumably, all of these "higher-level" properties are inferred on the basis of representations of properties more like those that physicists talk about. In other words, encodings of "edible" depend, in some complex way, on encodings of "low-level" physical properties like wavelength, velocity, and so on. While R&D theory itself does not determine precisely what is involved in such complex relations, there is reason to suppose that R&D theory provides the necessary tools for describing such relations. To see why this is so, let us consider a simple example.

---

[19] For an overview of this evidence, see F. Rieke et al., op. cit.; Eliasmith and Anderson, op. cit.; and L. Abbott, "Decoding Neuronal Firing and Modelling Neural Networks," *Quarterly Review of Biophysics*, XXVII, 3 (1994): 291–331.

It is clearly important for an animal to be able to know where various objects in its environment are. As a result, in mammals, there are a number of internal representations of signals that convey and update this kind of information. One such representation is found in parietal areas, particularly in lateral intraparietal cortex (LIP). For simplicity, let us consider the representation of only the horizontal position of an object in the environment. As a population, some neurons in this area *encode* an object's position over time. This representation can be understood as a scalar variable, whose units are degrees from midline (decoded alphabet), that is, encoded into a series of neural spikes (encoded alphabet). Using the quantitative tools mentioned earlier, we can determine the relevant decoder. Once we have such a decoder, we can then estimate what the actual position of the object is, given the neural spiking in this population. Thus, we can determine precisely how well (or what aspects of) the original property (in this case, the actual position) is represented by the neural population. We can then use this characterization to understand the role that the representation plays in the cognitive system as a whole.

As mentioned, this is a simple example. But notice that it not only describes how to characterize representation but also shows how we can move from talking about neurons to talking about "higher-level" variables, like object position. That is, we can move from discussing the "basic" representations (i.e., neural spikes) to higher-level representations (i.e., mathematical objects with units). This suggests that we can build up a kind of "representational hierarchy" that permits us to move further and further away from the neural-level description, while remaining responsible to it. For instance, we could talk about the larger population of neurons that encodes position in three-dimensional space. We could dissect this higher-level description into its lower-level components (i.e., horizontal, vertical, and depth positions), or we could dissect it into the activity of individual neurons. Which description we employ will depend on the kind of explanation we need. Notably, this hierarchy can be rigorously and generally defined to include scalars, vectors, functions, vector fields, and so on.[20] The fact that all of the levels of such a hierarchy can be written in a standard form suggests that this characterization provides a unified way of understanding representation in neurobiological systems.

Note that the focus of this chapter is on how to characterize representational states, computations over these states, and the dynamics of

---

[20] This generalization is made explicit in Eliasmith and Anderson, op. cit., pp. 79–80.

these states. As a result, I do not generally address concerns related to content determination. This is largely because such a discussion would lead me far afield. Nevertheless, it is interesting to note that R&D theory is suggestive of a particular approach to content determination. Notice, first, that both the encoding and decoding are essential for determining the identity of a representation. This means that both what causes a neural state and how that state is used by the system are likely to play a role in content determination. This suggests that some kind of two-factor theory is consistent with R&D theory. As well, a single neural state may play a role in multiple contents concurrently, because distinct, yet related, representations (and hence, contents) can be identified at different levels of organization at the same time. This may initially seem problematic, but because the relation between levels of representation is quantitatively defined (hence, we know precisely what role a single neural state is playing in each of the representations defined over it), we should expect the parallel content relations also to be well defined. Of course, such comments only provide a hint of a theory of content; they do not constitute one.[21]

In any case, there is no reason to consider such a theory of content if its underlying theoretical assumptions are not appropriate to cognitive systems. So, we should notice that the strength of the previous characterization of representation lies in its generality. That is, regardless of what the higher-level representations look like (i.e., what kind of mathematical objects with units they are), R&D theory will apply. So R&D theory, while having definite consequences for what constitutes a good representational story, is silent as to which particular one is correct for a given neural system. This is desirable for a theory of mind because higher-level representations are clearly theoretical postulates (at least at this point in the development of neuroscience). Although we can directly measure the voltage changes of individual neurons, making claims about how they are grouped to represent the world is not easily confirmable. Presumably, the right representational story will be the most coherent and predictively successful one.

---

[21] Note that in order to understand the relation between representations at a given organizational level, we need to consider computational relations, as discussed in the next section. For an in-depth but preliminary discussion of a theory of content that is consistent with R&D theory, see my *How Neurons Mean: A Neurocomputational Theory of Representational Content* (St. Louis, MO: Washington University PhD dissertation in Philosophy, 2000).

3.2 *Computation*

Of course, no representational characterization will be justified if it does not help us understand how the system functions. Luckily, a good characterization of neural representation paves the way for a good understanding of neural computation. This is because, like representations, computations can be characterized using decoding. But rather than using the "representational decoder" discussed earlier, we can use a "transformational decoder." We can think of the transformational decoder as defining a kind of biased decoding. That is, in determining a transformation, we extract information *other than* what the population is taken to represent. The bias, then, is away from a "pure," or representational, decoding of the encoded information. For example, if we think that the quantity x is encoded in some neural population, when defining the representation we determine the representational decoders that estimate x. However, when defining a computation, we identify transformational decoders that estimate some function, $f(x)$, of the represented quantity. In other words, we find decoders that, rather than extracting the signal represented by a population, extract some transformed version of that signal. The same techniques used to find representational decoders are applicable in this case and result in decoders that can support both linear and nonlinear transformations.[22]

Given this understanding of neural computation, there is an important ambiguity that arises in the preceding characterization of representation. It stems from the fact that information encoded into a population may now be decoded in a variety of ways. Suppose we are again considering the population that encodes object position. Not surprisingly, we can decode that population to provide an estimate of object position. However, we can also decode that same information to provide an estimate of some function of object position (e.g., the square). Since representation is defined in terms of encoding and decoding, it seems that we need a way to pick which of these possible decodings is the relevant one for defining the representation in the original population. To resolve this issue, we can specify that what a population represents is determined by the decoding that results in the quantity that all other decodings are functions of. Thus, in this example, the population would be said to represent object position (since both object position and its square are decoded). Of course, object position is also a function of the square of object position (i.e., $x = \sqrt{x^2}$).

[22] As demonstrated in Eliasmith and Anderson, op. cit., pp. 143–160.

This further difficulty can be resolved by noticing that the right physical quantities (i.e., the decoded alphabet) for representation are those that are part of a coherent, consistent, and useful theory. In other words, we characterize cognitive systems as representing positions because we characterize the world in terms of positions, and cognitive systems represent the world.

Importantly, this understanding of neural computation applies at all levels of the representational hierarchy and accounts for complex transformations. So, for example, it can be used to define inference relations, traditionally thought necessary for characterizing the relations between high-level representations. Again consider the specific example of determining object position. Suppose that the available data from sensory receptors make it equally likely that an object is in one of two positions (represented as a bimodal probability distribution over possible positions). However, also suppose that prior information, in the form of a statistical model, favors one of those positions (perhaps one is consistent with past-known locations given current velocity, and the other is not). Using the notion of computation defined earlier, it is straightforward to build a model that incorporates transformations between and representations of 1) the top-down model, 2) the bottom-up data, and 3) the actual inferred position of the object (inferred on the basis of Bayes' rule, for example). As expected, in this situation, the most likely position given the a priori information would be the one consistent with the top-down model. However, if the bottom-up data is significantly stronger in favor of an alternate position, this will influence the preferred estimate, and so on.[23] So, although simple, performing linear decoding can support the kinds of complex transformations needed to articulate descriptions of cognitive behavior. Statistical inference is just one example.

Before moving on to a consideration of dynamics, it is important to realize that this way of characterizing representation and computation does not demand that there are "little decoders" inside the head. That is, this view does not entail that the system itself needs to decode the representations it employs. In fact, according to this account, there are no directly observable counterparts to the representational or transformational decoders. Rather, they are embedded in the synaptic weights

---

[23] For the technical details and results of the model described here, see ibid., pp. 275–283. For a brief discussion of more logic-like inference on symbolic representations, see section 4.1.

between neighboring neurons. That is, coupling weights of neighboring neurons indirectly reflect a particular population decoder, but they are not identical to the population decoder. This is because connection weights are best characterized as determined by both the decoding of the incoming signal and the encoding of the outgoing signal. Practically speaking, this means that changing a connection weight changes both the transformation being performed and the tuning curve of the receiving neuron. As is well known from both connectionism and computational neuroscience, this is exactly what happens in such networks. In essence, the encoding/decoding distinction is not one that neurobiological systems need to respect in order to perform their functions, but it is extremely useful in trying to understand such systems and how they do, in fact, manage to perform those functions.

3.3 Dynamics
While it may be understandable that dynamics were initially ignored by those studying cognitive systems as computational (theoretically, time is irrelevant for successful scomputation), it would be strange, indeed, to leave dynamics out of the study of minds as physical, neurobiological systems. Even the simplest nervous systems performing the simplest functions demand temporal characterizations (e.g., locomotion, digestion, sensing). It is not surprising, then, that single neural cells have almost always been modeled by neuroscientists as essentially dynamic systems. In contemporary neuroscience, electrophysiologists often analyze cellular responses in terms of "onsets," "latencies," "stimulus intervals," "steady states," "decays," and so on – these are all terms describing temporal properties of a neuron's response. The fact is, the systems under study in neurobiology are dynamic systems and as such they make it very difficult to ignore time.

Notably, modern control theory was developed precisely because understanding complex dynamics is essential for building something that works in the real world. Modern control theory permits both the analysis and synthesis of elaborate dynamic systems. Because of its general formulation, modern control theory applies to chemical, mechanical, electrical, digital, or analog systems. As well, it can be used to characterize nonlinear, time-varying, probabilistic, or noisy systems. As a result of this generality, modern control theory is applied to a huge variety of control problems, including autopilot design, spacecraft control, design of manufacturing facilities, robotics, chemical process control, electrical systems design, design of environmental regulators, and so on. It should not be surprising, then, that it proves useful for the analysis of the dynamics of cognitive neurobiological systems as well.

Having identified quantitative tools for characterizing dynamics, and for characterizing representation and computation, how do we bring them together? An essential step in employing the techniques of control theory is identifying the system state variable ($\mathbf{x}(t)$ in Figure 4.1). Given the preceding analysis of representation, it is natural to suggest that the state variable *just is* the neural representation.

However, things are not quite so simple. Because neurons have intrinsic dynamics dictated by their particular physical characteristics, we must adapt standard control theory to neurobiological systems. Fortunately, this can be done without loss of generality.[24] As well, all of the computations needed to implement such systems can be implemented using transformations as defined earlier. As a result, we can directly apply the myriad techniques for analyzing complex dynamic systems that have been developed using modern control theory to this quantitative characterization of neurobiological systems.

To get a sense of how representation and dynamics can be integrated, let us revisit the simple example introduced previously, object position representation in area LIP. Note that animals often need not just to know where some object currently is but also to remember where it was. Decades of experiments in LIP have shown that neurons in this area have sustained responses during the interval between a brief stimulus presentation and a delayed "go" signal.[25] In other words, these neurons seem to underlie the (short-term) memory of where an interesting object is in the world. Recall that I earlier characterized this area as representing $x(t)$, the position of an object. Now we know the dynamics of this area, namely, stability without subsequent input. According to R&D theory, we can let the representation be the state variable for the system whose dynamics are characterized in this manner.

Mathematically, these dynamics are easy to express with a differential equation: $x(t) = \int u(t)\,dt$. In words, this system acts as a kind of integrator. In fact, neural systems with this kind of dynamics are often called "neural integrators" and are found in a number of brain areas, including

---

[24] For the relevant derivations see ibid., pp. 221–225.

[25] For a detailed description and review of these experiments and their results, see C. Colby and M. Goldberg, "Space and Attention in Parietal Cortex," *Annual Review of Neuroscience*, XX (1999): 319–349; and R. Andersen, L. Snyder, D. Bradley, and J. Xing, "Multimodal Representation of Space in the Posterior Parietal Cortex and Its Use in Planning Movements," *Annual Review of Neuroscience*, XX (1997): 303–330.

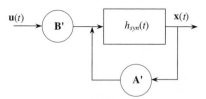

FIGURE 4.2. A control theoretic description of neurobiological systems. All variables are the same as in Figure 4.1. However, the matrices $\mathbf{A}'$ and $\mathbf{B}'$ take into account that there is a different transfer function, $h_{syn}(t)$, than in Figure 4.1. As well, $\mathbf{x}(t)$ is taken to be represented by a neural population.

brainstem, frontal lobes, hippocampus, and parietal areas. Neural integrators act like memories because when there is no input (i.e., $u(t) = 0$), the change in the output over time is 0 (i.e., $\dot{x} = dx/dt = 0$). Thus, such systems are stable with no subsequent inputs. Looking for the moment at Figure 4.1, we can see that the desired values of the $\mathbf{A}$ and $\mathbf{B}$ matrices will be 0 and 1, respectively, in order to implement a system with these dynamics. Since we have a means of "translating" this canonical control system into one that respects neural dynamics, we can determine the values of $\mathbf{A}'$ and $\mathbf{B}'$ in Figure 4.2; they turn out to be 1 and τ (the time constant of the intrinsic neural dynamics), respectively. We can now set about building a model of this system at the level of single spiking neurons, which gives rise to these dynamics – originally described at a higher level. In fact, the representation in LIP is far more complex than this, but the representational characterization of R&D theory is a general one, and so such complexities are easily incorporated. As well, more complex dynamics are often necessary for describing neural systems, but again, the generality of R&D theory allows these to be incorporated using similar techniques.[26] So, while the neural integrator model is extremely simple, it shows how R&D theory provides a principled means of explaining a cognitive behavior (i.e., memory) in a neurally plausible network.

3.4 Three Principles
R&D theory is succinctly summarized by three principles:

1. Neural representations are defined by the combination of nonlinear encoding (exemplified by neuron tuning curves) and weighted linear decoding.

---

[26] For various examples, see Eliasmith and Anderson, op. cit., especially chapters 6 and 8.

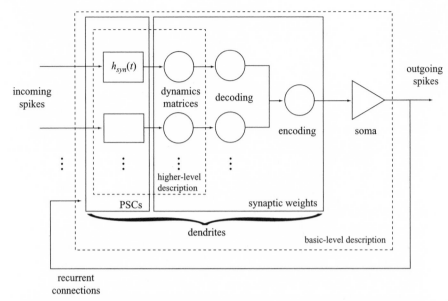

FIGURE 4.3. Generic neural subsystem. A synthesis of the preceding characterizations of representation (encoding/decoding), computation (biased decoding), and dynamics (captured by $h_{syn}(t)$ and the dynamics matrices). Dotted lines distinguish basic-level (i.e., neural) and higher-level (i.e., mathematical objects with units) descriptions.

2. Transformations of neural representations are functions of variables that are represented by neural populations. Transformations are determined using an alternately weighted linear decoding.

3. Neural dynamics are characterized by considering neural representations as control theoretic state variables. Thus, the dynamics of neurobiological systems can be analyzed using control theory.

To summarize these principles, Figure 4.3 shows a "generic neural subsystem." This figure synthesizes the previous characterizations of representation, computation, and dynamics across multiple levels of description.

While recent, this approach has been successfully used to characterize a number of different systems, including the vestibular system, the lamprey locomotive system, the eye stabilization system, working memory,[27]

---

[27] These examples can all be found in ibid.

and the limb control system.[28] As well, a rigorous formulation of each of these principles is available.[29]

4 Comparison to Current Approaches

Notice that R&D theory is not a description of cognitive systems as "being like" anything; cognitive systems are neurobiological systems that are best described by certain well-established quantitative tools for describing physical systems. That is, I am not proposing an analogy of "the mind as neural integrator" or even "the mind as a control system." While we may notice that certain control structures mimic some behaviors of neurobiological systems, we have to build the detailed neurobiological model and then determine if the mind really does work that way. In other words, R&D theory in no way suggests that we should stop at the analogy. Rather, it gives us principled means of comparing a full-blown, neuron-level implementation with the actual neurobiological system. Connectionists, symbolicists, and dynamicists typically build models that begin and end with metaphors; in particular, they do not specify how their models relate to the mind as a physical system. But the devil is in the details.

To see what is gained by moving away from the various metaphors that have dominated theorizing about the mind, let me briefly compare R&D theory to past approaches. In doing so, I describe several of what I take to be the central strengths and weaknesses of each approach. These lists are not intended to be exhaustive, but representative. So long as I am right about at least one of each, my claim that R&D theory should be preferred follows.

4.1 Symbolicism
The central problem for symbolicists, which I have already mentioned, is that time is largely ignored, or considered only after the fact.[30] Other typical concerns about symbolic models include their brittleness (i.e.,

---

[28] As described in Z. Nenadic, C. Anderson, and B. Ghosh, "Control of Arm Movement Using a Population of Neurons," in J. Bower, ed., *Computational Neuroscience: Trends In Research 2000* (Amsterdam: Elsevier Press, 2000).

[29] This formulation can be found in Eliasmith and Anderson, op. cit., pp. 230–231.

[30] Perhaps the most concerted effort to include time in such models is in Allen Newell's *Unified Theories Of Cognition* (Cambridge, MA: Harvard University Press, 1990). However, this attempt is both after the fact and largely inconsistent, as discussed in my "The Third Contender: A Critical Examination of the Dynamicist Theory of Cognition," *Philosophical Psychology*, IX, 4 (1996): 441–463.

lack of ability to survive partial destruction of the system or its representations), high-level discreteness,[31] and unconvincing descriptions of low-level perceptual processes.[32] R&D theory suffers from none of these limitations due to its essential inclusion of time and its responsibility to the underlying neural architecture.

The central strengths of symbolicism are demonstrated by its many past successes (e.g., ACT, SOAR, etc.). These are largely due to its espousal of cognitivism, that is, its willingness to peer inside the black box. Doing so has made representation an essential ingredient for providing good explanations of cognitive systems. As well, symbolicism is supported by a powerful and general theory of computation. Together, the commitment to representation and the quantitative theory of computation make for a unified understanding of cognitive systems. R&D theory shares similar strengths. While computational theory is bolstered by including control and information theory, and the notion of representation is sharpened to relate directly to neural systems, the ability to provide unified representational explanations remains.

That being said, a typical concern of symbolicists regarding approaches that are concerned with neural implementations is that the demonstrated symbol-manipulating abilities of cognitive systems are lost in the concern for neural detail. In one sense, this issue is easily addressed in the context of R&D theory. This is because the numeric representations in R&D theory are just another kind of syntax. While it is not a typical syntax for the logic used to describe cognitive function by symbolicists, the syntax itself does not determine the kinds of functions computable with the system.[33] Given the discussion in section 3.2, we know that this theory supports quite general computation, that is, linear and nonlinear functions. As a result, most, if not all, of the functions computed with standard symbolicist syntax can be computed with the numerical syntax adopted by R&D theory. More to the point, perhaps, past work using numerical distributed representations has shown that structure-sensitive processing of

---

[31] As discussed in E. Smith, "Concepts and Induction," in M. Posner, ed., *Foundations of Cognitive Science* (Cambridge, MA: MIT Press, 1989), pp. 501–526.

[32] This is made clear in the case of visual processes by P. Churchland, V. Ramachandran, and T. Sejnowski, "A Critique of Pure Vision," in C. Koch and J. Davis, eds., *Large-Scale Neuronal Theories of the Brain* (Cambridge, MA: MIT Press, 1994).

[33] This general point has been argued in J. Girard, "Proof-Nets: The Parallel Syntax for Proof-Theory," in Aldo Ursini and Paolo Agliano, eds., *Logic and Algebra* (New York: Marcel Dekker, 1996); and J. Girard, "Linear Logic," *Theoretical Computer Science*, L, 1 (1987): 1–102.

the kind demanded by Jerry Fodor and Zenon Pylyshyn[34] can be achieved in such a system.[35] Furthermore, this kind of representational system has been used to model high-level cognitive functions, like analogical mapping.[36] As a result, structured symbol manipulation is not lost by adopting R&D theory. Rather, a precise neural description of such manipulation is gained.

4.2 Connectionism

Of past approaches, connectionism is probably the most similar to R&D theory. This raises the question: Is R&D theory merely glorified connectionism? A first response is to note that glorifying connectionism (i.e., making it more neurally plausible) is no mean feat. The neural plausibility of many connectionist models leaves much to be desired. Localist models are generally not neurally plausible at all. But even distributed models seldom "look" much like real neurobiological systems. They include neurons with continuous, real-valued inputs and outputs, and often have purely linear or generic sigmoid response functions. Real neurobiological networks have highly heterogeneous, nonlinear, spiking neurons. Connectionist themselves are seldom certain precisely what the relation is between their models and what goes on the brain.[37]

Of course, such connectionist models are far more neurally plausible than symbolicist ones. As a result, they are not brittle like symbolic systems, but rather degrade gracefully with damage. As well, they are supremely statistically sensitive and are thus ideal for describing many perceptual and cognitive processes that have eluded symbolicists. And finally, connectionist models do, on occasion, consider time to be central to neural processing.[38] Again, R&D theory shares each of these strengths and, in fact, improves on a number of them (e.g., neural plausibility and the integration of time).

But more importantly, R&D theory also improves on connectionism. Connectionism has been predominantly a bottom-up approach to

---

[34] "Connectionism and Cognitive Architecture: A Critical Analysis," *Cognition*, XXVIII (1988): 3–71.

[35] As demonstrated in T. Plate, *Distributed Representations and Nested Compositional Structure* (Toronto: University of Toronto PhD dissertation in Computer Science, 1994).

[36] As in Eliasmith and P. Thagard, "Integrating Structure and Meaning: A Distributed Model of Analogical Mapping," *Cognitive Science*, XXV, 2 (2001): 245–286.

[37] See the various discussions by the editors in J. McClelland and D. Rumelhart, eds., *Parallel Distributed Processing: Explorations in the Microstructure of Cognition*, Vol. 2.

[38] See the selection of the examples in P. Churchland and T. Sejnowski, *The Computational Brain* (Cambridge, MA: MIT Press, 1992).

cognitive modeling. The basic method is straightforward: connect simple nodes together and train them to compute complex functions. While this approach can provide some useful insights (e.g., determining what kinds of statistical structure can be detected in the training set), it is unlikely to lead to a useful model of a brain that consists of billions of neurons. Connecting 10 billion nodes together and training them will probably not result in much. So, one of the main difficulties that connectionism suffers from is the lack of a principled method.

Progress in decomposing complex physical systems often necessitates an integration of bottom-up and top-down information.[39] So, in the case of neurobiology, it is essential to be able to test top-down hypotheses regarding brain function that are consistent with known lower-level facts. That is, we must be able to relate high-level characterizations of psychological processes (e.g., "working memory") to more specific implementational claims (e.g., that networks of certain kinds of neurons can realize such processes). Connectionists, unfortunately, have no principled method for incorporating top-down constraints on the design and analysis of their models. R&D theory, in contrast, explicitly combines both higher- and lower-level constraints on models.

The third principle of R&D theory captures this synthesis. It is with this principle that the analyses of representation, computation, and dynamics come together (see Figure 4.3). As an example, consider the recent proposal by Rajesh Rao and Dana Ballard[40] that the visual systems acts like a dynamic, optimal linear estimator (i.e., a linear control structure known as a Kalman filter).[41] Using R&D theory, we can build a large-scale, complex network to test this hypothesis. This is because the hypothesis is a precise high-level description, there is a significant amount of neural data available regarding the visual system, and principle 3 tells us how to combine these. Using the tools of connectionism, we simply cannot test this kind of high-level claim. It is not at all evident how we can train a network to realize an optimal estimator, or what an appropriate network

---

[39] As argued by W. Bechtel and R. Richardson in their *Discovering Complexity: Decomposition and Localization as Strategies in Scientific Research* (Princeton, NJ: Princeton University Press, 1993).

[40] "Predictive Coding in the Visual Cortex: A Functional Interpretation of Some Extra-Classical Receptive-Field Effects," *Nature Neuroscience*, II, 1 (1999): 79–87.

[41] Another high-level hypothesis regarding the use of the Kalman filter has been made in the context of the construction and use of cognitive maps in hippocampus in K. Balakrishnan, O. Bousquet, and V. Honavar, "Spatial Learning and Localization in Animals: A Computational Model and Its Implications for Mobile Robots," *Adaptive Behavior*, VII, 2 (1999): 173–216.

architecture would be. So, R&D theory is able to test high-level hypotheses in ways not available to connectionists. This makes R&D theory much better able to bridge the gap between psychological and neural descriptions of behavior than connectionism.

Another way of making this point is to contrast the kind of characterization of dynamics offered by principle 3 with that typical of connectionism. While connectionists often consider the importance of time and, in particular, have introduced and explored the relation between recurrent networks and dynamic computation, they do not have a systematic means of analyzing or constructing networks with these properties. Principle 3, by adopting control theory, makes such analyses possible within R&D theory. That is, control theory has a large set of well-established quantitative tools for both analyzing and constructing control structures. And because R&D theory provides a means of intertranslating standard and "neural" control structures, such tools can be used in a neurobiological context. This is extremely important for understanding the dynamic properties, and for otherwise predicting the overall behavior of a network constructed using the R&D approach. In other words, R&D relates rather imprecise connectionist notions like "recurrence" to a specific understanding of dynamics of physical systems that is subject to well-known analytical tools. This makes it possible to rigorously design networks with highly complex dynamics (like the Kalman filter mentioned earlier), a task left mostly to chance with connectionism.

The previous discussion shows how principle 3 supports building networks that have the complex dynamics demanded by a higher-level hypothesis. In addition, principle 3 supports building networks that have the complex representations demanded by a higher-level hypothesis. For example, Eliasmith and Anderson[42] describe a model of working memory that accounts for representational phenomena not previously accounted for. Specifically, this model employed complex representations to demonstrate how working memory could be sensitive not only to spatial properties (i.e., position) but also to other properties concurrently (e.g., shape). In addition, the model gives rise to both neural predictions (e.g., connectivity and firing patterns), and psychological predictions (e.g., kinds of error and conditions for error). This was possible only because R&D theory provides a means of determining the detailed connection weights given high-level descriptions of the system. Again, it is unclear how such

---

[42] "Beyond Bumps: Spiking Networks that Store Smooth N-Dimensional Functions," *Neurocomputing*, XXXVIII (2001): 581–586.

a network could have been learned (that this is not an easy task is a good explanation for why it had not been previously done).

In both of these examples, R&D theory is distinguished from connectionism because it does not share the same heavy reliance on learning for model construction. Unfortunately, getting a model to learn what you want it to can be extremely challenging, even if you build in large amounts of innate information (and choosing what to build in tends to be something of an art). But connectionists have little recourse to alternative methods of network construction, and so the severe limitations and intrinsic problems with trying to learn complex networks are an inherent feature of connectionism. R&D theory, in contrast, allows for high-level characterizations of behavior to be imposed on the network that is constructed. As a result, connection weights can be analytically determined, not learned.

Nevertheless, R&D theory is also able to incorporate standard learning rules.[43] And more than this, R&D theory can provide new insights regarding learning. This is because being able to analytically construct weights also provides some insight into methods for deconstructing weights. So, given a set of learned weights, the techniques of R&D theory can be used to suggest what function is being instantiated by the network.[44] Often, when some input/output mapping has been learned by a connectionist network, it is very difficult to know exactly which function has been learned because the testable mappings will always be finite. Using R&D to determine which linear decoders combine to give a set of provided connection weights can be used to give an exhaustive characterization of what higher-level function is actually being computed by the network.

Such connection weight analyses are possible because R&D theory, unlike connectionism, explicitly distinguishes the encoding and decoding processes when defining representation and computation. While the relation between this distinction and the observable properties of neurobiological systems is subtle, as noted in section 3.2, the theoretical payoff is considerable.

To summarize, R&D theory should be preferred to connectionism for two main reasons. First, R&D provides for a better understanding of neural connection weights, no matter how they are generated. There is much

---

[43] As discussed in chapter 9 of Eliasmith and Anderson, *Neural Engineering: Computation, Representation and Dynamics in Neurobiological Systems.*

[44] For a simple example, see Eliasmith and Anderson, op. cit., pp. 294–298.

less mystery to a network's function if we have a good means of analyzing whatever it is that determines that function. While learning is powerful, and biologically important, it cannot be a replacement for understanding what, precisely, a network is doing. Second, R&D theory provides a principled means of relating neural and psychological data. This makes representationally and dynamically complex cognitive phenomena accessible to neural-level modeling. Given a high-level description of the right kind, R&D theory can help us determine how that can be realized in a neural system. Connectionists, in contrast, do not have a principled means of relating these two domains. As a result, high-level hypotheses can be difficult to test in a connectionist framework.

So, unlike connectionism, R&D theory carefully relates neural and psychological characterizations of behavior to provide new insights into both. And while it is possible that certain hybrid models (either symbolicist/connectionist hybrids or localist/distributed hybrids) may make up for some of the limitations of each of the components of the hybrid alone, there is an important price being paid for that kind of improvement. Namely, it becomes unclear precisely what the cognitive theory on offer is supposed to be. R&D theory, in contrast, is highly unified and succinctly summarized by three simple, yet quantifiable, principles. To put it simply, Occam's razor cuts in favor of R&D theory. But it should be reiterated that this unification buys significantly more than just a simpler theory. It provides a unique set of conceptual tools for relating, integrating, and analyzing neural and psychological accounts of cognitive behavior.

4.3 Dynamicism

Of the three approaches, dynamicism, by design, is the most radical departure from the mind-as-computer metaphor. In some ways, this explains both its strengths and its weaknesses. Having derided talk of representation and computation, dynamicists have put in their place talk of "lumped parameters" and "trajectories through state-space." Unfortunately, it is difficult to know how lumped parameters (e.g., "motivation" and "preference")[45] relate to the system that they are supposed to help

---

[45] These are two of the lumped parameters included in the model of feeding described in J. Busemeyer and J. Townsend, "Decision Field Theory: A Dynamic-Cognitive Approach to Decision Making in an Uncertain Environment," *Psychological Review*, C, 3 (1993): 432–459.

describe. While we can measure the arm angle of the Watt Governor, it is not at all clear how we can measure the "motivation" of a complex neurobiological system. But this kind of measurement is demanded by dynamicist models. As well, some dynamicists insist that cognitive models must be low-dimensional, in order to distinguish their models from those of connectionists.[46] But insistence on low-dimensionality greatly reduces the flexibility of the models and does not seem to be a principled constraint.[47] Finally, because the Watt Governor, a standard example of a classical control system, has been chosen as a central exemplar of the dynamicists approach, the well-known limitations of classical control theory are likely to plague dynamicism. Clearly, these limitations are not ones shared by R&D theory.

What the replacement of the mind-as-computer metaphor by the "mind as Watt Governor" metaphor gained for dynamicists was an appreciation of the importance of time for describing the behavior of cognitive systems. No other approach so relentlessly and convincingly presented arguments to the effect that cognitive behaviors were essentially temporal.[48] If, for example, a system cannot make a decision before all of the options have (or the system has) expired, there is little sense to be made of the claim that such a system is cognitive. Furthermore, there is evidence that perception and action, two clearly temporal behaviors, provide the foundation for much of our "more cognitive" behavior.[49] While dynamicists have done a good job of making this kind of argument, the consequences of such arguments need not include the rejection of representation and computation that dynamicists espouse. R&D theory, which essentially includes precisely these kinds of dynamics, shows how representation, computation, and dynamics can be integrated in order to tell a unified story about how the mind works.

[46] T. van Gelder, op. cit.

[47] These points are discussed in detail in my "Commentary: Dynamical Models and van Gelder's Dynamicism: Two Different Things," *Behavioral and Brain Sciences*, XXI, 5 (1998): 616–665; and my "The Third Contender: A Critical Examination of the Dynamicist Theory of Cognition."

[48] Such arguments are prominent in T. van Gelder, op. cit.; T. van Gelder, "The Dynamical Hypothesis In Cognitive Science," *Behavioral and Brain Sciences*, XXI, 5 (1998): 615–665; and the various contributions to R. Port and T. van Gelder, op. cit.

[49] This view, associated variously with the terms "embodied," "embedded," or "dynamicist," has been expressed in, for example, F. Varela, E. Thompson, and E. Rosch, *The Embodied Mind: Cognitive Science and Human Experience* (Cambridge, MA: MIT Press, 1991); and more recently in D. Ballard, M. Hayhoe, P. Pook, and R. Rao, "Deictic Codes for the Embodiment of Cognition," *Behavioral and Brain Sciences*, XX, 4 (1997): 723–767.

4.4 Discussion

So, in short, R&D theory adopts and improves upon the dynamics of dynamicism, the neural plausibility of connectionism, and the representational commitments of symbolism. As such, it is a promising synthesis and extension of past approaches to understanding cognitive systems because it includes the essential ingredients. Of course, it is not clear whether R&D theory combines those ingredients in the right way. Because it is a recent proposal for explaining cognitive systems, its current successes are few. While it has been used to effectively model perceptual (e.g., the vestibular system), motor (e.g., eye control), and cognitive (e.g., working memory) processes, these particular examples of perceptual, motor, and cognitive behavior are relatively simple. So, while the resources for constructing neurally plausible models of phenomena that demand complex dynamics over complex representations are available, it remains to be clearly demonstrated that such complexity can be incorporated into R&D theoretic models.

As well, R&D theory does not, in itself, satisfactorily answer questions regarding the semantics of representational states. As Fred Dretske[50] has noted, coding theory does not solve the problem of representational semantics. Thus, R&D theory needs to be supplemented with a theory of meaning, as mentioned in section 3.1. In fact, I think R&D theory suggests a novel theory of meaning that avoids the problems of past theories.[51] Nevertheless, this remains to be clearly demonstrated.

However, even in this nascent form, R&D theory has some important theoretical implications for work in philosophy of mind and cognitive science. For example, functionalism regarding the identity of mental states may need to be reconceived. If, as R&D theory entails, the function of a mental state must be defined in terms of its time course, and not just its inputs and outputs, it is unlikely that functional isomorphism of the kind that H. Putnam[52] envisioned will be sufficient for settling the identity of mental states. If the dynamics of some aspects of mental life are central to their nature, then an atemporal functionalism is not warranted. Standard functionalism in philosophy of mind is clearly atemporal. And, I take it, some (if not many) aspects of mental life have their character in virtue of their dynamics (e.g., shooting pains, relaxed conversations,

---

[50] *Knowledge and the Flow of Information* (Cambridge, MA: MIT Press, 1981).

[51] For an attempt at articulating such a theory, see my *How Neurons Mean: A Neurocomputational Theory of Representational Content.*

[52] "Philosophy and Our Mental Life," in H. Putnam, ed., *Mind, Language and Reality: Philosophical Papers* (Cambridge: Cambridge University Press, 1975), pp. 291–303.

and recognizing friends). So, a "temporal" functionalism is necessary for properly characterizing minds. In other words, input, outputs, and their time course must all be specified to identify a mental state. While some mental functions may not be especially tied to dynamics (e.g., addition), others will be (e.g., catching a ball). Specifying ranges of dynamics that result in the successful realization of that function will allow us to determine if some mind or other can really be in a given mental state.

These considerations have further consequences for the role of the Turing machine in cognitive science.[53] While cognitive functions will still be Turing computable, they will not be realizable by every universal machine. This is because computing over time (i.e., with time as a variable in the function being computed) is different from computing in time (i.e., arriving at the result in a certain time frame). When this difference is acknowledged, it becomes clear that Turing machines as originally conceived (i.e., under the assumption of infinite time) are relevant theoretically, but much less so practically (i.e., for understanding and identifying real minds). I take it that more argument is needed to establish such conclusions but that, at the very least, adopting R&D theory shows how such a position is plausible.

5 Conclusion

Perhaps, then, R&D theory or something like it can help rid us of the constraints of metaphorical thinking. Such an approach holds promise for preserving many of the strengths, and avoiding many of the weaknesses, of past approaches to understanding the mind. But more than this, it is also suggestive of new perspectives that we might adopt on some of the central issues in philosophy of mind and cognitive science.

Because cognitive science is interdisciplinary, it should not be surprising that a cognitive theory has consequences for a variety of disciplines. I have suggested some of the consequences of R&D theory for neuroscience (e.g., careful consideration of decoding), psychology (e.g., quantitative dynamic descriptions of cognitive phenomena), and philosophy (e.g., reconsidering functionalism). These are consequences that should be embraced in order to improve our understanding of cognitive systems. In other words, the time is ripe for moving beyond the metaphors.

---

[53] These consequences are more fully explored in my "The Myth of the Turing Machine: The Failings of Functionalism and Related Theses," *Journal of Experimental and Theoretical Artificial Intelligence*, XIV (2002): 1–8.

Acknowledgements

Reprinted with permission from *Journal of Philosophy* 10(2003): 493–520.

Special thanks to Charles H. Anderson. Thanks as well to William Bechtel, Ned Block, David Byrd, Rob Cummins, Brian Keeley, Brian McLaughlin, William Ramsey, Paul Thagard and Charles Wallis for comments on earlier versions. Funding has been provided in part by the Mathers Foundation, the McDonnell Center for Higher Brain Function, and the McDonnell Project for Philosophy and the Neurosciences.

5

Brain Time and Phenomenological Time

Rick Grush

As 'tis from the disposition of visible and tangible objects we receive the idea of space, so from the succession of ideas and impressions we form the idea of time.

– David Hume, *A Treatise of Human Nature*

... space and time ... are ... pure intuitions that lie *a priori* at the basis of the empirical. ... [T]hey are mere forms of our sensibility, which must precede all empirical intuition, or perception of actual objects.

– Immanuel Kant, *Prolegomena to Any Future Metaphysics*

... there are cases in which on the basis of a temporally extended content of consciousness a unitary apprehension takes place which is spread out over a temporal interval (the so-called specious present). ... That several successive tones yield a melody is possible only in this way, that the succession of psychical processes are united "forthwith" in a common structure.

– Edmund Husserl, *The Phenomenology of Inner Time-Consciousness*[1]

1 Introduction

The topic of this chapter is temporal representation. More specifically, I intend to provide an outline of a theory of what it is that our *brains do* (at the subpersonal level) such that *we experience* (at the personal level) certain aspects of time in the way that we do. A few words about both sides of this relation are in order.

[1] This particular passage is one in which Husserl (1905, p. 41) is articulating, with approval, a view he attributes to L. W. Stern (1897).

First, the brain. I will actually be making little substantive contact with neurophysiology. The main thrust of my strategy on the brain side is to articulate an information-processing structure that accounts for various behavioral and phenomenological facts. The neurophysiological hypothesis is that the brain implements this information-processing structure. The amount of neurophysiology won't be nil, but at this stage of the game, our understanding of the brain's capacities for temporal representation is incredibly slim. The experimental side of neurophysiology is in need of some theoretical speculations to help it get going in earnest.

Now to personal-level phenomenology. There are only a few central aspects of this that I will be addressing. It will help to mention some of the aspects I will *not* be addressing. I won't be addressing memory, including how it is that our memories come to us with the conviction that they concern events that happened in the past. Nor will I be concerned with what might be called objective temporal representation. My belief that Kant's *Critique of Pure Reason* was published less than 100 years after Locke's *Essay Concerning Human Understanding* does depend on my ability to represent objective temporal relations and objective units of time, such as years. But the capacities involved in such temporal representation are not my concern.

Rather, I am directly interested in what I will call *behavioral time*. This is the time that is manifest in our immediate perceptual and behavioral goings on, and in terms of which these goings on unfold. I will expand on this shortly, but first an analogy with spatial representation may prove helpful. In philosophical, psychological, and cognitive neuroscientific circles, it is common to distinguish allocentric/objective from egocentric spatial representation. The contrast is between an ability to represent the spatial relations between things in a way that does not depend on my own whereabouts, and an ability to represent the spatial relations and properties of things in relation to myself. My ability to represent the Arc de Triomphe as being between the Obelisk and La Grande Arche de la Defense in no way depends on my own location in space or even my own beliefs about my location in space, whereas my (egocentric) belief that there is a pitcher's mound 60 feet 6 inches west of me relies on my own location as a sort of reference point.

But there are *two* senses in which a spatial representation can be nonobjective. In the sort of case that I have just called egocentric, the location of objects is represented in relation to oneself, rather than as being at some objectively specifiable spot in an objective framework. But the units of magnitude and axes of such a specification might yet be objective. My belief that the pitcher's mound is 60 feet 6 inches west of me is not fully objective because it makes reference to my own location as a reference point. But objective elements are present in this belief state, such as the axes (east–west), and the units of magnitude (feet, inches). These axes and units would be the same as those involved in my entirely objective belief that the pitcher's mound at Yankee Stadium is 60 feet 6 inches west of home plate.

I will use the expression *behavioral space* for a kind of spatial representation that is fully subjective – in which not only the reference point but also axes and magnitudes that define the space are nonobjective.[2] My representation of the coffee cup as being *right there* when I see it and reach out for it is an example of a behavioral spatial representation. First, this representation clearly invokes, even if only implicitly, my own body as the origin of the space within which the cup is located. Second, the axes of this space are to a good first approximation defined by the axial asymmetries of up/down, left/right and front/back, and these axes are defined by my own behavioral capacities.[3] Finally, the magnitudes involved – the difference between the cup that is *right there* and the sugar bowl that is *over there* – are also imbued with content via their connections to my own behavior. I may have no clear idea how far the coffee cup is from me in inches, feet, or meters, but I have a very precise representation of its distance specified in behavioral terms, as is evident from the fact that I can quickly and accurately grasp it.

Back to time. Analogues of allocentric/objective, egocentric and behavioral space are readily specifiable in the temporal domain. Allocentric/objective temporal representation is exploited by my belief that Kant's masterwork was published 90 years after Locke's; and also in my belief that the numeral "4" always appears in the seconds position of my watch one second after the numeral "3" appears there. Egocentric temporal representation is involved in my belief that Kant's first *Critique* was published 222 years ago (i.e., back from *now*); and it is also manifested when I see the numeral 3 appear in the seconds spot of my stopwatch

---

[2] Robin Le Poidevin (1999) also makes a case for the parallelism of space and time on this issue, but, it seems, fails to clearly distinguish between what I am calling "egocentric" and "behavioral" space and time.

[3] I say this is a first approximation mostly because there is good reason to think that another axis defined by *gravitational* up/down is integral to the behavioral space. This is still not objective, for what counts as gravitational up/down is defined not by the actual gravitational field but by the felt and anticipated force. The axis is lost in the behavioral space during free fall, *even if the free fall is happening in earth's gravitational field.*

and I come to believe that the numeral 4 will appear one second from *now*. Egocentric temporal representation uses my current time, *now*, as a temporal reference point, much like egocentric spatial representation uses my current location, *here*, as a spatial reference point. But both the objective and egocentric time employ objective units, such as years and seconds.

But the *behavioral* time specifies the temporal dimension and magnitudes not in terms of such objective units but in terms of behavioral capacities. When I move to intercept and hit a racquetball that is moving quickly through the court, I may have no accurate idea, in terms of seconds or milliseconds, of how far in the future the point of impact between my racquet and the ball will be. But I am nevertheless quite aware in behavioral terms. My movements and planning reveal an exquisite sensitivity to the temporal features of the event as it unfolds. A more common example might be moving one's hand down to catch a pencil that has just rolled off the edge of a table. One's attention is palpably focused on a spatiotemporal point – *just there* and *just then* (a half meter or so from the torso and a few hundred millliseconds in the future, though the units are not in terms of feet or milliseconds but are behaviorally defined) – at which the hand will intercept the pencil.

It should be clear that the distinction between these three kinds of representation is not a matter of magnitude. I can represent quite large and quite small spatial and temporal intervals both objectively and egocentrically, especially since the units in both cases are objective units. My belief that Kant's masterwork was published 222 years ago, and my belief that the numeral 4 will appear in the seconds position of my watch one second after the numeral 3 appears there are examples. In the case of *behavioral* space and time, large and small magnitudes can be represented, although because the units derive their content from perception and behavior, discriminatory abilities are best within a certain perceptually and behaviorally relevant range and degrade as we exceed the limits of what typically becomes behaviorally manifested, both with very large spatial and temporal magnitudes, as well as very small spatial and temporal magnitudes.

One apparent disanalogy between space and time concerns the axes. In the case of space, it is clear that there is a difference between objective axes, such as north and south, and behavioral axes, such as left and right. It might not appear that there is an analogous distinction with respect to time. The objective past and future seem to be the same as the behavioral past and future. But the cases are, in fact, more analogous than it might seem. Note that what makes left/right different from north/south is not that they are not always aligned. Being misalignable is often a *symptom* of axes being distinct, but it is not criterial. I might, in fact, be situated such that my left and right are exactly aligned with north and south. They would be distinct nevertheless because the difference between the axes does not lie in any lack of alignment, but rather in their content ("north" means in the direction of the intersection of the earth's Arctic surface and its axis of rotation; "left" means "what I could reach/look toward, etc., by moving thus"). And this would remain true even if, for whatever reason, I *always* faced due east so that the alignment was never lost. Similarly, the behavioral temporal axis of future/past and the objective temporal axis of future/past are distinct because of their content. The fact, supposing it is one, that they are always aligned is interesting, but it does not render the two axes identical, at least not for the purposes of this inquiry.[4]

The point is that the coextension of the axes, even if this coextension is ubiquitous, does not render them identical in content. We can perfectly well understand what it would be for them to misalign, even if they in fact never do. Notice that representational media allow us to tinker with how the relation between the behavioral future/past axis and the objective future/past axis is experienced. It is plausible to assume that at least part of what is going on with representational media is that they allow us to have a sort of surrogate phenomenal experience that is similar to what we would experience in some counterfactual situation: Looking at a photograph gives us a visual experience similar to what we would see if we were looking out from the point of view of the camera, and so on. The surrogate experiential import of seeing a photograph is one in which (represented) objective time is stopped with respect to the behavioral time – behavioral time keeps moving, but the represented objective time does not. The effect is more phenomenologically pronounced when the event represented is one that isn't (objectively) temporally static – a photograph of a bullet halfway through a glass of water is more effective in bringing this out than a photograph of relatives standing and smiling. Time-lapse movies keep the alignment but alter the usual relative pace

---

4 Christoph Hoerl (1998) takes this apparent disanalogy between space and time to be quite important, but it seems to me that he makes far too much of the fact that behavioral and objective time are in fixed alignment in a way unlike behavioral and objective space. Indeed, their ability to misalign seems to be *criterial* for Hoerl (1998, especially pp. 164–165) of the difference between what he calls egocentric and objective representation. But if I am right, this is a mistake.

of the units. Most relevant for the present point are movies watched in reverse, in which the axes of behavioral future/past and the represented objective future/past are misaligned – the behavioral future marches forward into the (represented) objective past: One can anticipate that in just a (behavioral) moment, one will see what happened an (objective) moment earlier. Perhaps to a person who had always faced due east, a photograph of familiar objects taken from a camera facing *west* would have a phenomenological novelty and queerness similar to our viewing of movies in reverse.

Not only will I be limiting myself to behavioral time in this chapter but I will also be focusing attention on a very limited but special region of behavioral time, what I will call the *behavioral now*. There is reason to believe that a small temporal interval, spanning a few hundred milliseconds into the past and future, has a special kind of status both psychologically, philosophically and neurophysiologically. To make brief mention of some of the more famous examples: The locus classicus in psychology is William James' "specious present," though the behavioral now that I am interested in is of much shorter duration.[5] In the philosophical tradition, Edmund Husserl was the first to examine time consciousness in great detail. He posited a structure of primary protentions and retentions as brief temporal regions in which the mind is actively synthesizing and interpreting temporally conditioned representations: anticipations of what will be unfolding in the immediate future as well as retentions of what has just happened. For Husserl, protentions and retentions are substantive features of what is being experienced *now*. Daniel Dennett has pointed out a number of pitfalls in our understanding of the representation of time, and he has argued that within some small interval,[6] the brain is a

---

[5] James attributes the expression to E. R. Clay, but I have been unable to locate a definitive reference. Nevertheless, James' account is the locus of substantive development. My own limitation of the "specious present" to a fraction of a second is more in line with C. D. Broad (1923) and J. D. Mabbot (1951), and quite unlike what James has in mind. And like Broad and Mabbot, and unlike James, perception and action are more central to the specious present I am interested in than short-term memory. The considerations discussed in section 7 of this chapter should be taken to answer the main challenge of Gilbert Plumer (1985), though it is also arguably the case that I am here discussing what he takes to be "perception" rather than "sensation." Making substantive contact with philosophical discussions of temporal representation is beyond the scope of this chapter.

[6] Dennett is not as clear as one might like, but the implication seems to be that the temporal window of interest is an interval such that at temporal scales smaller than this interval, the distinction between "Orwellian" and "Stalinesque" temporal interpretation breaks down. See Dennett 1991; Dennett and Kinsbourne 1992. See section 7 for more discussion.

very active interpreter of temporal experience, and not a passive "Cartesian theatre." In cognitive neuroscience, there has been a recent flurry of activity on the temporality of perception and perception of temporality that suggests that the brain is an active temporal interpreter. For example, David Eagleman and Terrence Sejnowski have shown that what one perceives as having happened at t depends in part on what happens with the stimulus within some small interval after t. (Dennett and Eagleman et al. will be discussed in more detail in section 7.)

The next five sections construct the theory. Sections 2 and 3 concern not time but space. There are two reasons for the spatial detour. The first is that the account of temporal representation is parallel in many respects to an account of spatial representation that I have articulated elsewhere, and various of its features are more perspicuous in the case of space. Section 2 outlines some theories of how neural states can come to carry spatial information, and section 3 provides an account of how states that carry spatial information can come to have spatial import for the subject.

In sections 4 and 5, I develop a theory of how neural states can carry temporally conditioned information. Section 4 recaps the emulation theory of representation (Grush 2002, 2004), which is an information-processing framework according to which the core of the brain's representational machinery involves forward models (also called *emulators*), especially as used in Kalman-filter control schemes. Section 5 extends that framework from mere filtering to a combination filter-smoother-predictor that maintains information about a represented system over some temporal interval, rather than a temporal point as the bare filter does. Section 6 then offers some speculations concerning what is involved in such states not only being able to carry temporal information but also having temporal import for the subject.

In the final section 7, I will return to a general discussion. First, I will compare the theory presented here with some recent work in temporal perception by Eagleman and Sejnowski (Eagleman and Sejnowski 2000) and its relation to some of Dennett's views on temporal representation (Dennett 1991; Dennett and Kinsbourne 1992); and later on I will remark on a proposal by T. van Gelder (1996). Second, I make a few speculations concerning why such an information-processing scheme as I describe might have evolved, and why it might have the specific temporal dimensions that it has. And finally, I will make some remarks meant to show how the theoretical stance of this chapter can be seen as part of a larger program in cognitive science that embraces a Kantian approach to core elements of the content of experience, rather than the more typical Humean approach.

2 Spatial Information

There are many kinds of spatial representation, and even when the scope is narrowed to what neuroscientists and psychologists call egocentric space (by which they probably mean something more like what I have described as behavioral space than egocentric space), the neural areas implicated are many and complex. Nevertheless, the posterior parietal cortex (PPC) has emerged as perhaps the most important cortical area involved in the representation of behavioral space. Damage to this area, especially the right PPC in humans, leads to a characteristic pattern of deficits that, to a first approximation, can be described as an impairment of the capacity to represent contralateral behavioral space. But although the behavioral deficit data suggest that this is what the PPC is doing, single-cell recordings from the area do not reveal anything like a map of the behavioral space. Given that there are topographic maps of many other sorts throughout the brain, one might have expected there to be a topographic map of the behavioral space. But instead, neurons in this area appear to correspond to things like sensory stimulation at a point on the retina, or the stretch of eye muscles, or head orientation with respect to the torso. What the connection is between things like this and a representation of the behavioral space might not be obvious. But as in all things, the font of clarity is good theory.

The major breakthrough in understanding the PPC came in work by David Zipser and A. Anderson (Zipser and Andersen 1988). These researchers trained a connectionist network to solve the following toy sensorimotor problem: An artificial two-dimensional organism with one rotatable eye was to determine the location of a stimulus that projected onto its retina. The difficulty is that because the eye can rotate, simply knowing the location of stimulation on the retina does not determine a unique stimulation location. The same stimulation location on the retina might have been caused by any of a number of stimulation locations, depending on the orientation of the eye. However, given the location of stimulation of the retina *and* the orientation of the eye with respect to the head, the location of the stimulus relative to the head can be determined. With this in mind, the neural network was trained by being given two inputs: location of retinal stimulation and eye orientation (both kinds of information are readily available in real nervous systems). And from these it was to correctly ascertain the location of the external stimulus relative to the head.

After it learned to solve the problem, the network could be analyzed in detail to see how it ticked – one advantage of artificial networks over their biological counterparts. The analysis revealed that the "hidden" units (the units between the input and output units that, in many connectionist architectures, are where the interesting work happens) appeared to be acting as linear gain fields. That is, a given hidden unit would have its activity determined by two factors: i) a Gaussian function of distance between the location of the stimulus's projection on the retina and some preferred spot on the retina, such that if the stimulus falls on this preferred spot, the activity is highest and the activity falls off with the stimulus's distance from that location; and ii) a linear function of preferred eye orientation– highest if the eye is oriented toward the preferred direction (left or right) and lower as it is oriented in the opposite direction. The net result is that a given hidden unit's activity is most strong for a certain combination of retinal location of stimulation and eye orientation, and such a combination corresponds to a location in head-centered space. (The details are a bit more complicated, but these are the essentials.)

Neurophysiological studies verified that actual PPC neurons have response profiles that fairly closely match those of the artificial units, enough to support the supposition that perhaps the PPC neurons were processing information in more or less the same way. And not only does the gain field hypothesis explain how sensory signals can be combined with signals concerning eye orientation to determine a location relative to the head, but the mechanism also generalizes to providing information about spatial location with respect to the head or torso, given the necessary postural signals concerning the orientation of the eyes with respect to the head, the head with respect to the torso. So the answer to the question: How can neural systems carry information about the location of an object in egocentric space? One answer, suggested by the Zipser and Andersen model, is by combining information about the location of the stimulation on the sensory periphery together with information concerning the orientation and location of the sensory platforms (such as the retina) relative to the head, and the head relative to the torso, and so on.

Of course, one might ask: Why should the brain care about developing units (neurons or groups of neurons) whose activity corresponds to the location of a stimulus relative to the head or torso? It might be thought that positing spatial representations of this sort is unnecessarily complicated. The brain can implement coordinate transformations from sensation directly to action without having to construct this kind of intermediary. Such coordinate transformations would map points in a high-dimensional sensory space directly to points in a high-dimensional motor space. The point in sensory space would carry information about, for example, where on the retina a stimulus is occurring, or what the orientation of the eyes relative to the head and head relative to the torso are. The points in motor space would specify patterns of activity (dynamic or kinematic, perhaps in the form of flexor-extensor tension equilibrium points, whatever) that would result in some sort of interaction with the stimulus. And the proposal is that all of this might happen without the intermediary of units whose activity is correlated with points in head- or torso-centered space.

This is true, and if one restricts attention to one sensation-action episode, then it would seem that constructing a space-representing intermediary is more work than needs to be done. But when one looks at the bigger picture and realizes that there are many sensory channels, many effectors, and coordination between the effectors is sometimes required, it becomes clear that having a common representational format is the most economical way to do things. The overall computational load is tremendously reduced by determining location relative to the head and/or torso. A torso-centered spatial representation allows for a stimulus to be placed at a location in the behavioral space regardless of whether it is seen or felt, providing a common framework for action whether the action will involve the left hand, right hand, head and mouth, or anything else. Representation of spatial location with respect to the torso provides for a common coordinate system, so to speak, within which all perception and action can be coordinated, and lets one space-representing state do the work of perhaps many thousands of special-purpose coordinate transformations.

Next, it might be objected that while having a common representational format is useful, the idea that that format must exist in terms of any sort of *space* does not follow. In particular, it might be suggested that something along the lines of Alexandre Pouget's basis function model will do the job (Pouget and Sejnowski 1997). I can't go into this here, but there is no headlong conflict between the basis function model and the idea that the common representational format is spatial. The three axial asymmetries of up/down, left/right, and front/back, which span behavioral space, are not only adequate but indeed natural bases for any kind of basis-function representation of this sort to employ. And indeed, Pouget's model is one quite elegant example of how the construction of an intermediary (in this case, sets of basis functions) has advantages over an indefinitely large set of specific coordinate transformations.

3 Spatial Import

The last section gave a very brief explanation of how it is possible for states of neural systems to carry information about locations in space relative to the torso. But carrying information about such things cannot be the whole story of our spatial perceptual experience, for surely we can have perceptual or experiential states that carry spatial information but lack spatial import, that is, are such that we don't experience them *as* spatial.

To help me explain what I mean here, I will refer to the *sonic guide,* a sensory substitution device designed to provide blind subjects with an artificial distance sense that is in some ways analogous to vision (Heil 1987, p. 236). The sonic guide features a transmitter worn on the head with a speaker that produces a constant ultrasonic probe tone. A stereophonic microphone, also mounted on the headgear, picks up the echoes produced by the reflection of this probe tone off objects in the subject's vicinity. A processor takes the information gained through these echoes and translates it into audible sound profiles that the subject hears through stereo earphones. There are four relevant aspects to the translation:

1. Echoes from a distance are translated into higher pitches than echoes from nearby. E.g. as a surface moves toward the subject, the sound it reflects will be translated into a tone which gets lower in pitch.
2. Weak echoes are translated into lower volumes. Thus as an object approaches the subject, the subject will hear a tone which increases in volume (as the object gets closer, it will ceteris paribus reflect more sound energy, resulting in a stronger echo), and gets lower in pitch (because it is getting closer, as in (1) above). An object which grows, but stays at a constant distance, will get louder, but stay at the same pitch.
3. Echoes from soft surfaces – e.g. grass and fur – are translated into fuzzier tones, while reflections from smooth surfaces – e.g. glass and concrete – are translated into purer tones. This allows subjects to distinguish the lawn from the sidewalk, for example.
4. The left–right position of the reflecting surface is translated into different arrival times of the translated sound at each ear. (Note that it is not required that this coding exploit the same inter-aural differences which code direction in normal subjects. In fact, if the differences are

exaggerated substantially by the guide, then one would expect a better ability to judge angle than we typically have.) (Heil 1987, p. 237).

As John Heil (1987, p. 238) describes it, the

sonic guide taps a wealth of auditory information ordinarily unavailable to human beings, information that overlaps in interesting ways with that afforded by vision. Spatial relationships, motions, shapes, and sizes of objects at a distance from the observer are detectable, in the usual case, only visually. The sonic guide provides a systematic and reliable means of hearing such things.

The first lesson I want to draw from the sonic guide has to do with the distinction between spatial information and spatial import. If you or I were to don the guide, we would be in receipt of a great deal of information about the spatial characteristics of and relations between objects and surfaces in our vicinity. For example, a bowling ball on the ground a few meters in front of us might cause us to hear middle C at 35 dB, say. Our experience of this sound carries spatial information, in that the features of this sound covary in the required ways with the location of the ball. Not only would *our experience* (describable as, say, hearing middle C at 35 dB) carry such information, but neural states (describable in terms of firing frequencies and patterns in some pool of neurons) would also, of course, be carrying this information. The pattern of neurons on the auditory pathways and cortical areas that fire when and only when middle C at 35 dB is heard will be a neural pattern that carries information to the effect that a roughly bowling-ball-sized object is 2 meters ahead on the ground. So at both the personal experiential level and subpersonal neural levels, there are states that carry spatial information about the location of the ball in egocentric space. And yet for all that, you or I would not experience anything with any significant spatial *import*. It would be merely a particular pattern of sound heard through earphones and wouldn't have the phenomenology of experiencing an object in the behavioral space. This should make the distinction between carrying spatial information and having spatial import for the subject clear enough.

Nevertheless, and this leads to the second lesson to be drawn from the sonic guide, it is possible for subjects who are sufficiently practiced with the guide to be such that through it, they do enjoy perceptual states with spatial import. I will suppose what seems to be plausible, that subjects who have been using the device for a while (perhaps since birth) and are competent with it are actually *perceiving* the objects in their environment directly, rather than *reasoning out* what the environment must be like on the basis of pitches and volumes. This seems to be accepted by Heil who, in discussing the sonic guide and another sensory substitution device, P. Bach-Y-Rita's TVSS (Bach-Y-Rita 1972), notes:

> Devices like the sonic guide and the TVSS prove useful only after the sensations they produce become transparent. In employing a TVSS, for instance, one ceases to be aware of vibrations on one's skin and becomes aware, rather, of objects and events scanned by the camera. Similarly, successful use of the sonic guide requires one to hear things and goings-on rather than the echoes produced by the device.... [Children] less than about 13 months...do this quite naturally, while older children do so only with difficulty. (Heil 1987, p. 237)

The question is, what is the difference between you and me, on the one hand, and the blind subject practiced with the guide and for whom its deliverances have spatial import, on the other? The quick and surely correct answer is that the subject is used to the guide, and we are not. But what does this mean?

I have elsewhere articulated and defended a theory of spatial content that I call the *skill theory*, which was first expressed in more or less the form I defend it by Gareth Evans.[7] The skill theory maintains that a sensory episode becomes imbued with spatial content for an organism if that sensory episode disposes or enables the organism to behave in certain spatially significant ways. To hear a sound as coming from the left *just is* for the sound to dispose me to turn to the left in order to orient toward it, or to run to the right in order to distance myself from it (and similarly for an indefinite number of other possible behaviors). As Evans put it:

> The subject hears the sound as coming from such and such a position, but how is this position to be specified? We envisage specifications like this: he hears the sound *up*, or *down*, *to the right* or *to the left*, *in front* or *behind*, or *over there*. It is clear that these terms are *egocentric* terms: they involve the specification of the position of the sound in relation to the observer's own body. But these egocentric terms derive their meaning from their (complicated) connections with the actions of the subject. (Evans 1985, p. 384)

> Auditory input, or rather the complex property of auditory input which codes the direction of the sound, acquires a spatial *content* for an organism by being linked with behavioral output. (Evans 1985, p. 385)

> ...we must say that having the perceptual information at least partly consists in being disposed to do various things. (Evans 1985, p. 383)

---

[7] For Evans' view, see "Understanding demonstratives" (Evans 1981), chapter 6 of *The Varieties of Reference*, and "Molyneux's question." For my own development of this view, see Grush 1998, 2000.

What does this mean in terms of neural infrastructure? The Zipser and Anderson model implicitly involves an action element in that the network is trained to make correct assessments of stimulus location, given only retinal location and eye orientation. In biological practice, such correct assessments would be manifest only via some behavior or other: trying to grasp the stimulus or foveate on it, for instance. In more sophisticated models, such as Pouget's basis function model, the basis functions that extract the spatial information from sensory signals are essentially linked to behavior via sets of coefficients that, together with the perceptually determined basis functions, poise the organism to execute any of a range of spatial behaviors – or as Evans put it, dispose the organism to do various things. (I discuss all of this in much more detail in Grush 1998, 2000).

At the subpersonal level, appropriate experience, perhaps in conjunction with such favorable conditions as starting at an early enough age, effects a coordination between the relevant discriminable sensory states and spatial behavior via the right sort of connections, presumably in the PPC, between those neural pools whose job it is to put sensory information into a format capable of guiding behavior by stabilizing it with respect to a coordinate frame centered on the torso (in the central case). This is what the gain fields and basis functions do. The manifestation of this subpersonal machinery at the personal level is the experiencing of the stimulus as being located somewhere in the behavioral space.

Though we all sensorily discriminate, via hearing, middle C at 35 dB, the sonic guide user's auditory cortex mainlines this channel of input to the PPC in a way that extracts the spatial information from the input into a format that is of immediate use in cuing and guiding motor action. The subject is in a position, without cognitive preliminary, to walk toward the ball, or orient toward it, or to throw a dart at it. At this point, it is crucial to recall clearly what we are trying to provide an account of. We are trying to provide an account of how we have experience of objects as being located in behavioral space. And the behavioral space, recall, just is the space that has the subject's body as the implicit center, and whose axes and magnitudes are given through the subject's behavioral potentialities. So having a perceptual episode in which the PPC extracts the information from the episode in such a way as to be able to cue and guide behavior *just is* imbuing that personal-level experience with behavioral spatial import.[8]

I have indulged in this excursion into a theory of the spatial content of perceptual experience in order to lay the groundwork for a theory of the

---

[8] Again, this has been a very brief discussion of a very complicated topic. For much more detail on all of the issues broached in this section, see Grush 1998, 2000.

temporal content of experience. This theory will parallel the account of spatial representation in that first, I will describe a kind of information-processing framework that maintains states that carry temporal information. This information-processing framework is a generalization of the *emulation framework*, which I have articulated at length elsewhere. The next section, section 4, provides a very brief introduction to this framework (readers familiar with the Kalman-filter-based emulation framework may safely skip this section). Section 5 briefly outlines the generalization of this framework that allows for the production of temporally thick-state estimates. Section 6 puts forth a proposal concerning the conditions under which states that carry temporal information, as described in section 5, have behavioral-temporal *import*.

4 Representation: The Emulation Theory

In this section, I will provide a *very* quick introduction to a theory of neural representation that I have developed in much more detail elsewhere.[9] The *emulation framework* is an information-processing architecture based on constructs from control theory and signal processing. The basic idea is that the brain constructs entities, most likely implemented in systems of neural circuits, that act as models of the body and/or environment. During normal sensorimotor activity, the brain sends motor commands to the body, and the body and environment conspire to produce sensory signals sent back to the brain. Circuits that observe the outgoing efferent commands and subsequent incoming afferent signals are positioned to learn the input-output functions of the body and environment. Once such a neural system has learned (to some degree of accuracy) this mapping, the brain can then use this *emulator* in various ways: it can run the emulator off-line (meaning that the real body and environment are not involved in the control loop) in order to produce imagery; or it can run the emulator in parallel with the body and environment and use the output of the emulator to anticipate, fill in, or otherwise enhance the real sensory signal.

The emulation framework begins with neither traditional symbolic computation nor connectionism, but with *control theory* – there being a clear sense in which the brain is usefully conceived of as, and evolved to be, the controller of the body. The two best-known kinds of control scheme are open-loop and closed-loop (as shown in Figure 5.1).

[9] The best version currently available is Grush (2004). See also Grush 1997 and 2002 for treatments.

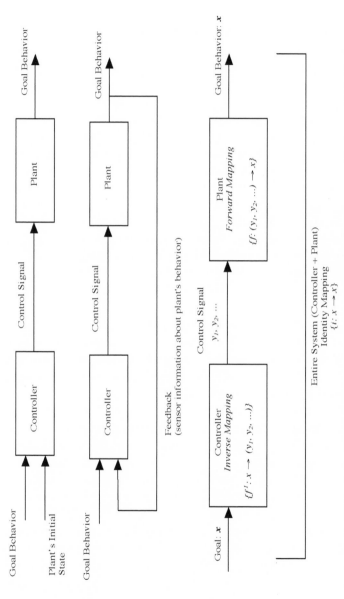

FIGURE 5.1. Basic control theory concepts and terminology. The scheme at the top is an open-loop scheme, where the controller gets no feedback from the plant. The middle diagram is feedback control, in which the controller exploits feedback from the plant. The bottom diagram explains the notions of inverse and forward mappings. *Source*: Grush, 2002.

Open-loop schemes take as input a goal state and produce a control sequence such that if the plant executes that sequence, it will end up in the goal state. The controller determines this control sequence in the absence of any information about the progress of the plant as it acts on the control sequence. A standard toaster is an example: A goal state (light, medium, dark) is input, and the controller determines how long to keep the heating elements on in order to achieve this goal state without any information about the actual state of the bread as it toasts.

A *closed-loop* controller also gets as input a desired goal state and produces a control sequence as output – a control sequence such that if the plant executes it, the desired goal will be achieved. Closed-loop controllers, though, have the benefit of feedback from the plant as the control episode progresses. A thermostat is an example of a closed-loop controller: Once the desired temperature is set, the controller need not determine the entire control sequence. It need not try to determine at once that it needs to turn the heater on for 44 minutes in order to achieve the desired room temperature. Rather, at each moment it merely compares the current temperature (the feedback in the form of a current temperature reading) with the desired temperature and takes one of a small number of actions on that basis, for example, turn heaters on, keep heaters on, turn air conditioner on, and so on.

Another, less well-known scheme is pseudo-closed-loop control (also called model reference control; see Figure 5.2).[10] In such schemes, the controlling system includes not only a controller per se but also a model or *emulator* of the plant – a system that implements the same input-output mapping as the plant, or close to it. In such a scheme, it becomes possible for the controller to operate as if it is in closed-loop contact with the plant even though it is not. A copy of the motor command is sent to the emulator, which, because it implements the same input-output mapping as the plant, produces as output a signal that is the same as the one the plant is producing. The controller can exploit the emulator's feedback in exactly the same way it would exploit feedback from the real system.

There are many potential uses for such a system. I will mention only two. First, the plant can be disengaged completely and the emulator run off-line by the controller in order to produce imagery and more

---

[10] The qualitative idea behind this sort of framework was articulated by Kenneth Craik (1943). To my knowledge, it was first put to use in theoretical neurophysiology by Masao Ito (1970, 1984) as a theory of one aspect of cerebellar function.

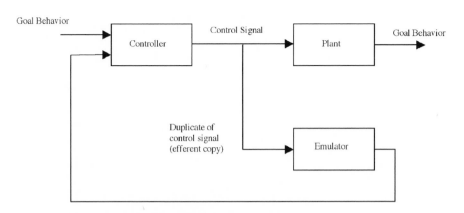

FIGURE 5.2. Pseudo-closed-loop control.

generally evaluate counterfactuals (if command sequence C were issued, what would the result be?). Second, in systems where real feedback is compromised or delayed, the emulator can be run in parallel with the plant to provide mock feedback. As a historical note, this was the use proposed by Masao Ito (1970, 1984) as a solution to the problem of delayed proprioceptive feedback: He argued that the cerebellum implements an emulator (not his term) of the musculoskeletal system that provides immediate feedback in order to overcome peripheral feedback delays for quick movements.

In such a scheme, an emulator is useful. But in another scheme, it can be even more useful. The next step to this more useful scheme is the *Kalman filter*. (Kalman 1960; Kalman and Bucy 1961; see also Bryson and Ho, 1969; Gelb, 1974; Haykin, 2001). The Kalman filter (henceforth KF) is an information-processing technique for filtering the noise from a certain kind of signal. The KF and the problem it solves are shown in Figure 5.3.

The problem is represented on the upper third of Figure 5.3. A process r(t) develops over time under three influences. The first is the manner in which the process's states lead to its own future states. This can be expressed as $r(t) = Vr(t-1)$, where the matrix V represents the function that determines future states of r on the basis of its past states (this is a Markov process). The second influence is random perturbations, or what is called process noise, represented as a small zero-mean Gaussian "noise" vector $n(t)$. Thus $r(t) = Vr(t-1) + n(t)$; this would be a Gauss-Markov process. The third influence is an external

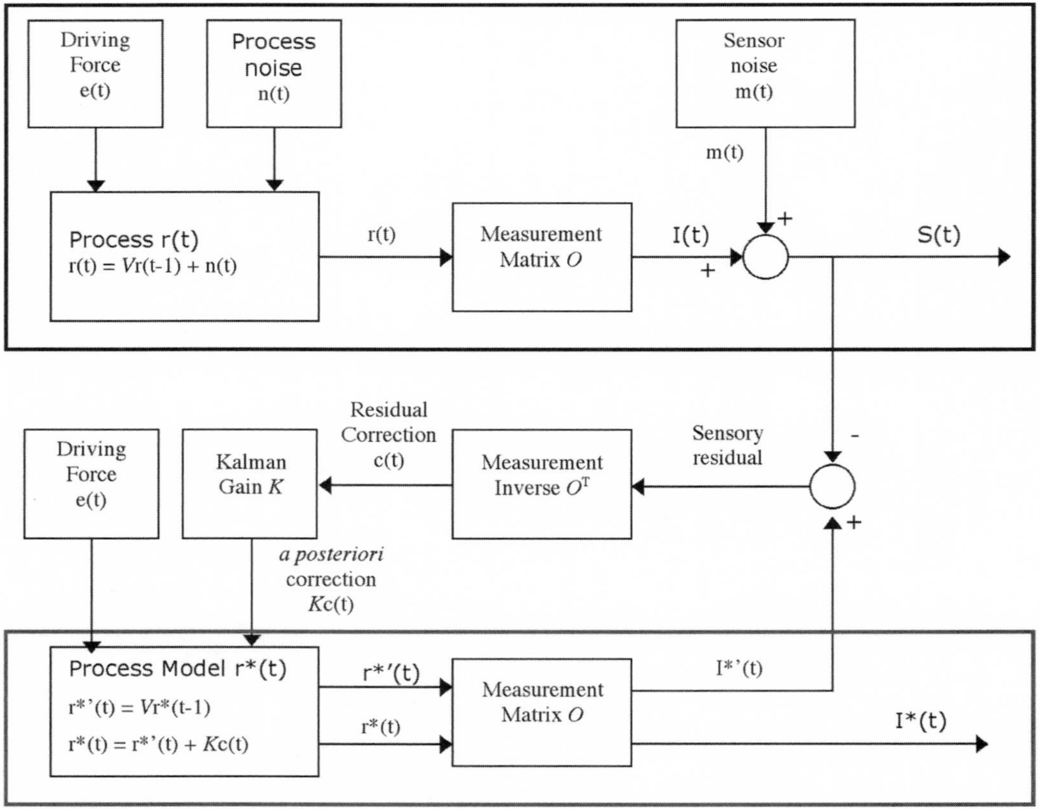

FIGURE 5.3. A standard Kalman filter.

driving force $e(t)$ that influences the process: $r(t) = Vr(t - 1) + n(t) + e(t)$; a *driven* Gauss-Markov process. An example would be a ship at sea. Its states are its position and velocity (speed and direction). At each time step, its position and velocity are determined by three things: (i) the position and velocity at the previous time; (ii) noise in the form of unpredictable winds or ocean currents, and anything else unpredictable; and (iii) driving forces such as the engine speed and rudder angle. Ships at sea can be faithfully described as driven Gauss-Markov processes, as can many other things.

A signal $I(t)$ is produced at each time t via a measurement that depends on the state of the process at t. To this signal is added some small *sensor noise* $m(t)$ (again, zero-mean Gaussian), and the result is the *observed signal* $S(t)$. In the ship example, the measurement might include bearings to known landmarks and stars (the values produced depend upon the ship's position and orientation). Such bearing measurements are not perfect, and the imperfection can be represented as sensor noise that is added to what "ideal" perfect bearing measurements would be. Everything just described is represented on the upper-third region of Figure 5.3.

The KF's job is to determine what the real signal $I(t)$ is, that is, to "filter" the sensor noise $m(t)$ from the observed signal $S(t)$. In the ship example, this would amount to determining what the ship's actual position is, or equivalently, what the perfect ideal bearing measurements would be. The KF does this in an ingenious way. It maintains an estimate of the current state of the process and then measures this *process model* with the same measurement function that produces the real measurement from the real process. Of course, if this state estimate is very accurate, then the KF's estimate of the real noise-free signal will be very accurate.

So the trick is to keep the state estimate embodied in the process model as accurate as possible. This is done in the following two-phase manner. In the first phase, the KF produces an a priori estimate of the real signal, $I^{*\prime}(t)$, by subjecting its a priori estimate of the state of the process $r^{*\prime}(t)$ to measurement O. The asterisk indicates that the value is an estimate produced by the KF, and the prime indicates that it is the a priori estimate (there will be another, a posteriori, estimate produced in the second phase). The KF compares this a priori signal estimate to the observed signal. The difference between them is called the *sensory residual*. It is the difference between what the KF *expected* to see and what it *actually* saw. From this sensory residual, the KF can determine how much it would have to modify its own a priori state estimate $r^{*\prime}(t)$ in order to eliminate the residual. This is the *residual correction* (and it is determined by pushing the sensory residual through an inverse of the measurement process). Qualitatively, the KF says: "I thought the process's state was X, and this led me to predict that the signal would be Y. But the observed signal was actually $Y + y$. If my state estimate had been $X + x$, rather than X, then my prediction of the signal would have matched the real signal exactly." Here y is the sensory residual, and x is the residual correction.

The key point, though, is that the KF does not (typically) apply the entire residual correction to its a priori estimate. Why not? Because the sensory residual – the mismatch between the a priori prediction of what will be observed and what is actually observed – has two sources. *One* is the difference between the KF's estimate of the process's state and the actual state of the process. This is the inaccuracy of the process estimate. The *second* is the sensor noise. That is, the KF's a priori state estimate might be entirely accurate, but there would still be a sensory residual because of the sensor noise. And even in the normal case where the a priori estimate is not entirely accurate, this inaccuracy is responsible only for part of the sensory residual. To return to the ship example, when the navigation team predicts that they should be at location L, they can also predict what the bearing measurements from the current fix cycle should be if their estimate is accurate. When the actual bearing measurements come in, they will typically not exactly match the navigation team's predictions. One source of the mismatch is that their prediction of where the ship actually is is probably not exactly right. Another source is the imperfection of the bearing-taking process itself.

The KF makes optimal[11] use of the residual correction in the following way. It makes an estimate of its relative confidence in its own estimates versus what the observed signal says, which is largely a matter of the relative size of the process noise and the sensor noise (this is the point of the KF process that I am glossing over. For more detail, see, e.g., Bryson and Ho 1969; Gelb 1974). Upon determining this relative confidence, the KF applies, to put it roughly, some fraction of the residual correction to its state estimate. *Very* roughly put, the KF determines a gain matrix, the *Kalman gain*, that determines what fraction of the residual correction to apply.

---

[11] The use is in fact optimal in the sense that, given the details of the various sources of noise and the way in which the KF determines the Kalman gain (explained shortly), it is provably the case that it is not possible to get more information from the signal than the KF gets.

If the KF knows that the sensor noise is large compared to the process noise (there are few unpredictable winds or ocean currents, for example, but the bearing measurements are known to be unreliable), it knows that, probably, most of the sensory residual will be due to inaccurate sensor readings and not to inaccuracies in its own a priori state estimate.[12] It therefore has reason to believe that most of the residual correction should be ignored (it is due to sensor noise). It applies only a small fraction of the residual correction in this case. On the other hand, if the KF knows that the process noise is large compared to the sensor noise (the instruments are good and those operating them are quite skilled, but there are large unpredictable winds and ocean currents), then the KF knows that in all likelihood, its own a priori prediction will not be very accurate, at least not in comparison to what the sensor signal says. So it will apply a large fraction of the residual correction.

Then begins the second phase of the KF's operation. However much of the residual correction gets applied, the result is a new updated estimate of the state of the process that has taken the observed signal into account. This new estimate is the a posteriori estimate. This estimate represents an optimal combination of two factors: the (fallible) a priori expectation of the process's state and the (fallible) observed signal. And this is how the KF keeps its estimate of the state of the process as accurate as possible. This a posteriori estimate is then subjected to a measurement, and the result of this measurement is the KF's final estimate of the real signal – its estimate of what one would get if one removed the sensor noise from the observed signal.

The process then repeats. The current a posteriori estimate is the basis for the next a priori estimate.

My introduction to KFs has been extremely incomplete. Suffice it to say that not only is it a provably optimal filter in a broad range of cases, but it is also one of the most widely used information-processing constructs in many areas of signal processing and related areas. The next task will be to integrate the KF into control contexts.

Figure 5.4 shows a schematic of what I have called the *emulation framework*. It is a combined control and signal-processing framework that blends pseudo-closed-loop control and the KF. It can be described as a

[12] This is because process noise is the biggest factor compromising the accuracy of a priori estimates. As process noise goes to zero, $r^{*\prime}(t)$ goes to $r(t)$, so long as the previous state estimate was accurate. As process noise increases, $r^{*\prime}(t)$ will deviate more and more from $r(t)$, even if the previous state estimate was accurate, because of the unpredictable influence on the process.

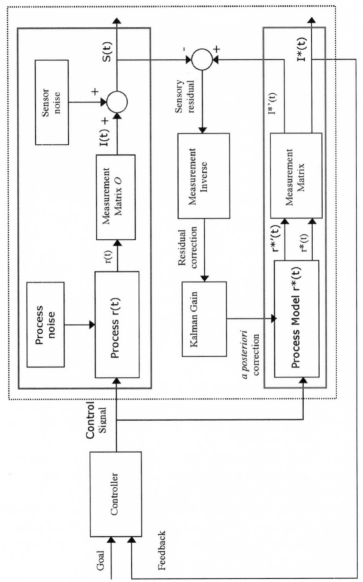

FIGURE 5.4. The emulation framework.

modified closed-loop control scheme that uses a KF to process the feedback signal from the plant. Or it can be described as a KF such that the process (together with the measurement process and sources of noise) corresponds to the plant and the driving force is the command signal from the controller. Note that the three control schemes described earlier in the section are all degenerate cases of this scheme: First, if the Kalman gain is set so that the entire residual correction is always applied, the scheme becomes functionally equivalent to closed-loop control. In such a case, the feedback the controller sees will always be identical to the signal sent from the plant, the observed signal. Second, if the Kalman gain is set so that none of the residual correction is applied, the scheme becomes functionally equivalent to the pseudo-closed-loop scheme of Figure 5.2. In such a case, the feedback sent to the controller will always be exactly what the a priori estimate says it should be, without any influence from the observed signal. Without correction, the state estimate continues to evolve only on the basis of its own estimates, just as in pseudo-closed-loop control. Third, if the feedback signal is suppressed completely, then the scheme becomes functionally equivalent to open-loop control.

The emulation framework is extremely useful for shedding light on phenomena in motor control, motor imagery, visual imagery, the relationship between visual imagery and motor processes, the nature of top-down processes in perception, and a host of others. In motor control, researchers have long known that peripheral proprioceptive feedback is too slow to be of direct use during fast goal-directed movements, and some have argued that the brain uses emulators (my expression, not theirs), in KF-control-type schemes to maintain an estimate of the state of the body during movement, since the sensory signal is delayed.

I have shown elsewhere how the same mechanisms can explain motor imagery, the imagined feelings of movement and exertion that are the proprioceptive and kinaesthetic counterparts of visual imagery. This is done, simply enough, by driving the process model/emulator with efference copies and suppressing the real motor command to the body. The emulator thus supplies off-line mock proprioceptive/kinaesthetic signals. Visual imagery is accounted for in an analogous manner, as the off-line operation of an emulator of the motor visual loop. The emulation framework also posits a relationship between imagery and perception (very analogous to the relation posited by Kosslyn (Kosslyn 1994; Kosslyn and Sussman 1995)), in that the same entity, the emulator, is what provides for imagery when run completely off-line and what provides anticipations and filling-in during perception by constructing the a priori estimate.

Striking confirmation of this view of the relation between motor-initiated anticipations (also called imagery) and perception comes from a phenomenon first hypothesized by H. von Helmholtz (1910) and discussed and verified experimentally by Ernst Mach (1897). Subjects whose eyes are prevented from moving and who are presented with a stimulus that would normally trigger a saccade (such as a flash of light in the periphery of the visual field) report seeing the entire visual scene momentarily shift in the direction opposite of the stimulus. Such cases are very plausibly described as those in which the perceptual system is producing a prediction – an a priori estimate – of what the next visual scene will be on the basis of the current visual scene and the current motor command. Normally, such a prediction would provide to specific areas of the visual system a head start for processing incoming information by priming them for the likely locations of edges, surfaces, and so on. Just less than 100 years after Mach published his experimental result, J.-R. Duhamel, C. Colby and M. E. Goldberg (Duhamel et al. 1992) found neurons in the parietal cortex of the monkey that remap their retinal receptive fields in such a way as to anticipate imminent stimulation as a function of saccade efference copies. That is, given the current retinal image and the current saccade motor command, these neurons in the parietal cortex anticipate what the incoming retinal input will be. In particular, a cell that *will* be stimulated because the result of the saccade will bring a stimulus into its receptive field begins firing in anticipation of the input, presumably on the basis of an efference copy of the motor command. (This is a neural implementation of the construction of an a priori estimate of the current visual scene.)

The applications of this framework could be multiplied, and I have done so elsewhere (see Grush 2002, 2004). Next, the project is to develop an extension of this framework: *moving window emulation.*

5 Temporal Representation: Moving Window Emulation

The KF, at any given point in time, is actively concerned with only one process state estimate, the current one. It uses the previous state estimate as one of the sources of information when developing the current state estimate, but it does not actively alter or update it once it has been arrived at. And as soon as the previous state estimate is used to construct the current state estimate, it is forgotten – its only trace being its influence on the current state estimate, which will itself be exploited and forgotten when used to construct the subsequent state estimate. The KF has no temporal *depth*. By the same token, the KF is not in the business of predicting *future* states: its current state estimate is used to determine a prediction of what the sensory signal will be, but it is not a prediction of what will be happening in the future with the process. It is a prediction of what sensory information will be coming in about the process as it is *right now*. At each t_i, the KF constructs and maintains an estimate of the process's state at t_i. It only had an estimate of t_{i-1} at t_{i-1}, and it will only get around to constructing an estimate for t_{i+1} at t_{i+1}. The KF is therefore temporally punctate in that it represents the states of the process only at one point in time at a time. And because of this, it does not need to represent time explicitly at all. It lets time be its own representation (as Carver Mead is widely reported to have said in a different context) in that the time of the KF's representing just is the time represented.[13] In this section, I will show how to generalize the temporally punctate KF to a system that maintains, at each time, an estimate of the behavior of the process over a temporal interval.

Information-processing systems that develop state estimates of past states using information up to the present are known as *smoothers*. There are different kinds of smoothers, depending on the point in the past that is being estimated. *Fixed point smoothers* keep an estimate of the process at some fixed temporal point and update that estimate as time progresses. Such a system would construct, at each time t_i, ($i > = 0$), an estimate of the process's state at some fixed time t_a. Presumably, as more future information is collected, the estimate of the process's state at that point in time can be improved.

A *fixed-lag smoother* constructs, at each t_i, an estimate of the process's state some fixed amount of time (the lag l) before t_i in the past: t_{i-1}. (The proposal by R. P. N. Rao, D. M. Eagleman and T. J. Sejnowski (Rao et al. 2001) is that visual awareness is the result of a fixed-lag smoother, with a lag of about 100 msec.)

Information-processing systems that produce estimates of *future* states of the process are *predictors*. Where the distance into the future for which a state estimate is produced is k, the predictor is a *k-step-ahead predictor*.

---

[13] This idea expressed by Mead recapitulates Hume's position, as expressed in the quotation at the beginning of this chapter. The same error was made by Helmholtz (von Helmholtz, 1910), and was known to be an error by James.

We can now phrase the general form of the problem addressed by filtering:

> Given the entire set of observed signals S_1, S_2, ... S_i (from the initial time t_1 to time t_i), find, for each t_j, the optimal estimate of the process state r_j.
>
> If $i = j$, this is filtering (the Kalman filter, as described, is thus a *filter* as defined here; later I will distinguish another kind of filter, and will specify the filter as defined here a *moving point filter*);
>
> If $j > i$, this is *prediction*; if j is some fixed point in the future, then this is fixed-point prediction; if $j = i + k$ for fixed k, then this is a k-step-ahead predictor.
>
> If $j < i$, this is *smoothing*; if j is fixed, then this is fixed-point smoothing; if $j = i - 1$ for some fixed lag l, this is fixed-lag smoothing.

With these definitions in hand, I can define a generalization of the emulation framework: what I will call a moving window filter. This is a scheme that combines smoothing, filtering and prediction to maintain, at all t_i, estimates of the process's state for all times within the interval $[t_{i-l}, t_{i+k}]$ for some constant lag l into the past, and some constant reach k into the future. A *moving window emulator* is an emulation framework information-processing structure that is like the normal emulation framework except that rather than exploiting a moving point filter (like the bare Kalman filter), it exploits a moving window filter to represent the behavior of the emulated system over some interval of time.

One way to express the distinction between a fixed-lag smoother, a bare moving point filter as defined earlier and a moving window filter is that at each point in time, the moving point filter constructs an optimal estimate of the process *at that time,* while the moving window filter constructs an optimal estimate of the behavior of the process over a temporal interval centered[14] on the current time.

Let's look first at the column in Table 5.1 under the moving point filter. The vectors r_4, r_5 etc., are the optimal estimates of the process at t_4, t_5, etc., respectively. But note that there is an ambiguity in the characterization given in the previous sentence. It could mean that it is the estimate, generated at t_4, t_5, of the process's state. Or it could mean that it is the estimate of what the state of the process is at t_4, t_5. With the moving point filter, this ambiguity does not make a difference, since the estimate

---

[14] It need not be literally *centered* – k need not equal l.

TABLE 5.1. *Fixed-Lag Smoothing, Moving Point Filtering and Moving Window Filtering*

	Fixed-Lag Smoother (l = 2)	Moving Point Filter	Moving Window Filter
t_4	r_2	r_4	$(r^4_2, r^4_3, r^4_4, r^4_5, r^4_6)$
t_5	r_3	r_5	$(r^5_3, r^5_4, r^5_5, r^5_6, r^5_7)$
t_6	r_4	r_6	$(r^6_4, r^6_5, r^6_6, r^6_7, r^6_8)$
t_7	r_5	r_7	$(r^7_5, r^7_6, r^7_7, r^7_8, r^7_9)$

generated at each time step *is* the estimate of the process's state at *that* time. Because the time of the estimating is always the same as the time estimated, we can get by with only one index.

Look now at the column for the fixed-lag smoother. Here, the ambiguity surfaces $-r_4$ is an estimate of the state that the process was in at t_4. But the time that this estimate is generated is t_6. For the fixed-lag smoother, the time represented always lags behind the time of representing by some constant amount. But exactly because this delay is fixed, we don't need separate indices for the time of representing and the time represented. The one index still uniquely determines both.

This is not true for the moving window filter. Because at each time step the filter maintains estimates of the process's state at multiple times, the time of estimating is not always the same as the time estimated, nor is it some fixed temporal distance from the time estimated. Thus, the need for two indices on each estimate: one for when the estimate is constructed, another to indicate the time index of the state being estimated. With r^4_2 for example, the superscript index "4" refers to the time of representing, the time at which that estimate is constructed. This is why at t_4, all of the states for the moving window filter have a 4 superscript. The subscript refers to the time represented. So r^4_2 is the estimate generated by the moving window filter at t_4 of what the state of the process was at t_2. The window $(r^4_2, r^4_3, r^4_4, r^4_5, r^4_6)$ is the filter's estimate, at t_4, of the trajectory of the process from t_2 (in the past) to t_6 (in the future).

One crucial thing to note is that the moving window emulator can, and often will, alter its estimates of what the process's state will be/is/was as time progresses. That is, r^4_4 need not equal r^5_4, even though they are both estimates of what the process's state is at t_4. The MWE's estimate at t_4 of what the process's state *is* at t_4 might not be the same as its estimate at t_5 of what the process's state *was* at t_4. The MWE might decide, on the basis of incoming information at t_5, that it had made an error – that while at t_4 it had thought that the state of the process was $r^4{}_4$, it now, at t_5, estimates that this was probably not correct. And so it now, at t_5, retroactively alters its estimate of what the process's state was at t_4.

For a qualitative example, imagine that we have a ship navigation team that employs the moving window filter technique. At each time, the team maintains an estimate of the ship's trajectory from two fix-cycles ago to two fix-cycles in the future. The estimates of the previous and current ship states (position, speed, bearing, etc.) are made on the basis of sensor information that has been received. There are three kinds of situation that would lead the team to retroactively modify its estimates of past states. First, suppose that at t_3, the best information available had the ship's state as being $r^3{}_3$. However, at t_4, new information comes in that strongly suggests that the ship's current state is $r^4{}_4$. But suppose that because of the nature of the process, if the ship's state at t_4 is r_4, then the ship could not have been in r_3 at t_3. The team can thus use this information concerning the ship's state at t_4 to correct its estimate of where it *was* at t_3. In such a case, $r^3{}_3$ (the estimate, at t_3, of the ship's state at t_3) would differ from $r^4{}_3$ (the estimate, at t_4, of the ship's state at t_3). This sort of correction of past estimates is based on consistency constraints.

A second kind of situation is as follows. Suppose that the ship's measurement systems – the people who take bearings and report the numbers to the navigation team – are spread throughout the ship and that their measurements are carried by hand to the navigation team. While all or most of the information eventually gets to the navigation team, some of it is significantly delayed because the ship is large and the couriers are slow. At t_3, the navigation team estimates that the ship's current state is $r^3{}_3$. At the next fix-cycle, t_4, a courier comes in with some bearing measurements that were taken at t_3. This information can then be used to modify the team's estimate of the ship's state at t_3. Again, $r^3{}_3$ will differ from $r^4{}_3$, but this time because of the incorporation of delayed information, not because of consistency constraints between present and past state estimates.

A third sort of case that can lead to changes of estimate are changes in *intentions*. Suppose that the captain has made it clear that the plan is to sail straight ahead at constant speed until further notice. On this basis, at t_3, the navigation team develops predictive state estimates $r^3{}_4$ and $r^3{}_5$. But the captain then changes the plan and communicates the intention to change speed at t_4. With this new information, the navigation team will, at t_4, construct a different predictive estimate of the ship's state at t_5. Thus, $r^3{}_5$ will differ from $r^4{}_5$, even though they concern a state that is in the future with respect to both times of representing. (This sort of case becomes important in section 7 when we turn to the issue of whether or not it is possible to "perceive the future.")

6 Temporal Import

Information-processing systems of the sort just described provide examples of how a physical system can have states that, at a given time, can carry information about states of other systems at times other than the present time. But as we saw in the case of spatial representation, a person's brain having states that carry X-ish information is not sufficient for that person having experience with X-ish import. In the case of space, the proposal was that having spatial import was a matter of the information-bearing states being appropriately mainlined into the organism's sensorimotor repertoire. To recap: the kind of spatial content that we wanted to explain was in terms of the behavioral space, and the axes and magnitudes of the behavioral space are defined in terms of the organism's behavioral capacities. The proposal was that the PPC, exactly because it is in the business of putting sensory information (together with postural information) into a format that cues and guides motor behavior, imbues those sensory episodes with behavioral spatial import.

The proposal for the case of temporal import will be parallel. What makes a given information-carrying state support experience with content specifiable in terms of behavioral time will be the fact that that state plays a role in the right sensorimotor dispositions, in this case, the *temporal* features of such dispositions.

In the case of the behavioral space, the point was made with the aid of the sonic guide: a device that produced a signal that carried spatial information, but was not such as to provide the subject with spatial information. We can imagine a similar device in the case of time. Suppose that an experimental condition is set up such that a small red point of light is quickly flashed on a subject's retina exactly 700 msec before an irritating itch, which will last a few seconds, occurs on her forearm. The flash thus carries temporally conditioned information, but its initial import will simply be of a quick flash of red light, lacking any interesting temporal import (except the trivial import to the effect that a red light is flashing *now*). Now I suppose that after a while, the subject might come to associate the red flash with the impending itch, and perhaps such an association would be enough to give the flash some kind of temporal significance.

Rather than pursue that thin hope, though, I want to upgrade the example so that the light has specific behavioral ramifications. (This in the spirit of spatial devices, for which it has long been known that having subjects use the devices in order to guide behavior is crucial for learning to use them as perceptual prostheses.) The experimental setup is such that if the subject scratches a specific spot on her forearm not earlier than 600 msec after the red light onset and not later than 650 msec after onset, then the itch will be prevented. Given these constraints, the timing will be quite tricky since it has to occur within, and only within, a 50 msec interval. We can suppose that when informed of the conditions, the subject will initially take the flash of light as a cue to try to time her motor response correctly. This would be parallel to your or my donning the sonic guide and, being informed that middle C at 35 dB carries information to the effect that a bowling ball is on the floor 2 m ahead, then trying to hit it with a dart by aiming at a spot we take to be about 2 m ahead whenever we hear middle C at 35 dB.

But as we saw in the case of the sonic guide, appropriately skilled subjects have a different phenomenology than this. To them, that same auditory input results in a perceptual state that is best described as "perceiving the ball just ahead" or something like that. Their phenomenology is different from ours, though the sensory inputs are the same. My *claim* is that our subjects will, after sufficient experience, undergo a similar phenomenological change. The red light will stop having the import of "red light in my visual field now right there" (where the "there" refers to a spot in their visual field), but will come to have the import "scratch *there* just *then.*" There "there" will have as its import the location on the forearm, and the "then" will have as its import a specific temporal location in the subjects' behavioral time, an anticipated point in the immediate future: a point in time exactly as phenomenally real as the point in the future at which you anticipate your hand catching the pencil as it rolls off the edge of the desk. (The proposal, of course, is that as the pencil rolls off the edge of the table, the MWE in your brain constructs states that carry information about the trajectory of the pencil over the next few hundred milliseconds. If combined with motor intentions, it constructs an estimate of the event over the next few hundred milliseconds that has your hand catching the pencil. One part or aspect of this state will be the one that carries information about your hand contacting the pencil. What gives this the phenomenal import of the anticipatory awareness of your hand contacting the pencil is the fact that this state and its various components have for a long time been appropriately connected to behavior.)

That, anyway, is the claim. Aside from the sonic guide analogy, I have not defended it, and I won't defend it here since that would take some time, and I'm not sure I could do a good job anyway. And I'm not an experimentalist, so I won't be running the experiment.

But although it is a speculation, I should point out that to my knowledge, it is the only speculation going (except for one by Timothy van Gelder, which will be discussed in section 7). There has to be *some* account of temporal phenomenology. And that account can't be one that just lets time be its own representation and then hopes that the phenomenology comes along with the representational content. A phenomenology that flows through time is not the same as temporal phenomenology (experimental results confirm this, some of which have been discussed, and others will be discussed briefly), and so the notion that we don't need to account for temporal phenomenology, but can instead just rely on the assumption that time is its own representation, is untenable. While the speculation I have offered might not ultimately be correct, the problem it is an attempt to address won't go away by ignoring it.

7 Discussion

7.1 Comparison with Two Recent Proposals

There are three psychological phenomena that bring out clearly the need for some sort of theory of temporal representation. The first is F. A. Geldard and C. E. Sherrick's cutaneous rabbit (Geldard and Sherrick 1972). These researchers placed tactile stimulators that deliver a small thump on subject's arms: one near the wrist, one near the elbow, and one near the shoulder. A series of taps is delivered to the subject with very brief interstimulus intervals, on the order of 50–200 msec. A series of 5 or 10 taps might take anywhere from a fraction of a section to a couple of seconds. If, say, 4 taps are delivered to the wrist, and nothing else, the subject will feel and report a series of taps all located on the wrist. But if the series delivered is, say 4 taps on the wrist, 3 at the elbow and 3 at the shoulder, the subject will feel, and report, a series of *evenly spaced taps*, like the footsteps of a small animal running up the arm from the wrist to the shoulder. The puzzle begins with the 2nd tap on the wrist. If *in the future* there will be some taps, say taps 5 through 7, located near the elbow, and maybe a few more after that near the shoulder, then the second tap will be felt a few inches up from the wrist in the direction of the elbow. If on the other hand, the taps will not continue beyond the 4th tap at the wrist, then the 2nd tap will be felt at the wrist. But how does the brain know, at

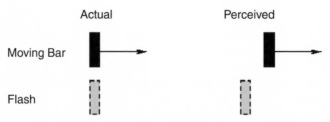

FIGURE 5.5. The flash-lag effect. *Source*: Rao, Eagleman and Sejnowski 2001.

the time the 2 tap is felt, whether or not taps will be delivered near the elbow later on? Obviously, it doesn't. Hence, the mystery.

The second is the finding concerns the flash-lag effect (MacKay 1958). The phenomenon (see Figure 5.5) is that subjects are shown a scene in which a moving object and small flash occur at the same place at the same time, but the moving object appears to be ahead of the flash. The finding of interest is that the magnitude and even direction of the flash-lag effect can alter, depending on what the moving object does *after* the flash (see Eagleman and Sejnowski 2000; Rao, Eagleman and Sejnowski 2001).

The third, found very recently by Vilayanur Ramachandran, is like the flash-lag effect but somewhat cleaner (see Williams, Hubbard and Ramachandran 2005). The basic phenomenon (see Figure 5.6) is that the kind of illusory motion that subjects perceive between pairs of flashing dots is influenced by what, if anything, happens after the flashes of the dots in question.

With phenomena like these in mind, Daniel Dennett (Dennett 1991; Dennett and Kinsbourne 1992) has proposed what he calls the "multiple drafts" model. According to this model, the brain is not a passive mirror of sensory input but is rather an active interpreter, and in particular, an interpreter that is not bashful about reinterpreting (hence, the multiple drafts metaphor) its previous interpretations of perceptual states on the basis of subsequent information. So when the 5th and subsequent taps occur near the elbow, the draft that said "4 taps at the wrist" is sent back to the editing table, and rewritten to say "4 equally spaced taps between the wrist and elbow." I will return to Dennett and Kinsbourne shortly.

The model proposed by Rao, Eagleman and Sejnowski in order to account for the flash-lag effect (and the same idea could be extended to Ramachandran's illusory motion retrodiction) is that the visual system is a fixed-lag smoother, as described in section 5. As Rao et al. put it: "In the smoothing model, perception of an event is not online but rather is delayed, so that the visual system can take into account information

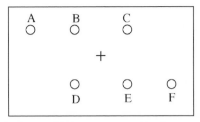

FIGURE 5.6. Ramachandran's illusory motion retrodiction. A screen is blank except for a central fixation point represented by the cross and by lights that might flash at any of the locations marked A–F. (i) When two lights, say B and C, are flashed in quick succession, subjects report seeing a single dot move from B to C. (ii) When a pattern of lights flashes, such as B + E, then D + C, subjects report seeing two moving dots. They report seeing either a) a dot moving from B to D and another moving from E to C or alternately b) a dot moving from B to C and another moving from E to F. If the pattern shown is C + D, then B + E, subjects will report seeing one of two movements: c) C to B and D to E, or d) C to E and D to B. In either case, which of the two patterns is seen appears random. (iii) If, however, A + F is flashed just before the sequence B + E, D + C, subjects will always see the dots as moving rectilinearly: one dot moving from A to B to C, and a second moving from F to E to D. They are less likely to see a dot moving from A to B to D, and a second from F to E to C. In this case, the first illusory motion pair (A to B and F to D) biases the visual system for the next chunk of illusory motion. (iv) Most interestingly, if the pattern C + D then B + E is shown, as in (ii) followed by A + F, subjects are less likely to see the (iii)(d) pattern of C to E and D to B; usually they only see the (iii)(c) pattern of C to B and D to E, continued by B to A and E to F, for an illusory straight-line motion of C to B to A and D to E to F. Given that in cases where A + F is not shown after the initial pattern, subjects see pattern (iii)(d) 50% of the time, the later appearance of the A + F dots apparently retroactively interprets all the motion as (iii)(c), and so the illusory motion is more likely to be rectilinear.

from the immediate future before committing to an interpretation of the event" (2001, p. 1245). As Eagleman has elsewhere put it, we are living in the past, by around 100 msec or so. (See Rao, Eagleman and Sejnowski, 2001 for an explanation of how it is that a smoothing model can postdict the perceptual episode in the conditions of the flash-lag effect to come up with an estimate to the effect that the flash lagged behind the moving stimulus.)

Although Eagleman and Sejnowski cite Dennett as a theoretical ally, their model is closer to the Cartesian Theatre that Dennett criticizes than to a Dennettian multiple drafts model. On their model, there is one definitive percept – one draft, so to speak, not multiple drafts. The departure from the Cartesian Theatre model as Dennett presents it is that this single definitive percept is delayed by a hundred milliseconds or so, and what is shown on the Cartesian screen as having occurred at *t* depends in part on the nature of the sensory input that comes in after that sensory input that is most directly implicated in carrying information about what happened at *t*. But a theater performance that starts 100 msec late in order to accommodate a tardy script writer is still a theater performance, even if the reason that the script writer is tardy is that he wants his script to be consistent with things that take place after the time of the performance.

The moving window emulation theory is similar in some respects to Rao, Eagleman and Sejnowski's fixed-lag smoother model (it incorporates a smoother, for example). It has all the advantages and lacks the shortcomings. First, to the advantages it shares. Rao et al.'s fixed-lag smoother applies its interpretive energies at the endpoint of the relevant temporal window, the fixed lag. This allows such a system to exploit, in its construction of percepts, information that has arrived after the event the percept concerns. This is why they proposed the model, after all. Obviously, the moving window emulator does the same. The trailing edge of the window is a fixed lag, and at this trailing edge and beyond, the two models are functionally equivalent. For an event that occurs at *t*, both models predict that from $t + lag$ onward, the percept associated with that event is influenced by what happened after *t*, up to and including $t + lag$. So long as subjects' perceptions are probed after $t + lag$, the two models make the same predictions.

Now to four disadvantages that the moving window emulator does not have. First, the problem with fixed-lag smoothers is that they delay their estimates, and in sensorimotor contexts, this can be costly. A fixed-lag smoother must balance these costs against the gains made by having a more accurate (though delayed) percept. The moving window emulator does not need to delay its processing. In fact, it even engages in anticipatory estimates before the event. While this has the theoretical advantage that it avoids the disadvantages of delaying the percept, it also seems to accord with empirical results that appear to indicate that anticipatory percepts are in fact constructed (e.g., the Helmholtz phenomenon, but more generally, visual perception appears to exploit a wide range of anticipatory mechanisms). This is the second disadvantage that the moving window emulation account does not have. If anything is clear about the nervous system, it is that a great deal of its evolution was devoted to prediction, not postdiction, for obvious reasons. The fixed-lag smoother model, positing a *delay* in order to get a better estimate of what happened in the *past*, seems to cut directly against the grain of what the nervous system ought to be doing. The moving window emulation framework incorporates predictive mechanisms as part of the same framework that has the retrodictive capacity.

The third avoided disadvantage is connected with the first. On the fixed-lag model, our phenomenology is necessarily epiphenomenal. Our brain and its mechanisms do the needed sensorimotor work, and then after the fact, our phenomenology shows up and provides us with the delayed illusion that we were in control. While the mere fact that this is counterintuitive is no decisive argument against it – cognitive neuroscience has demonstrated that many counterintuitive theses are true – it should at least motivate us to take seriously any contenders that have all the advantages but lack the counterintuitive consequences. The moving window emulation theory does not entail that our phenomenal experience is epiphenomenal, nor does it entail that it is not. Because on the model there are state estimates (percepts) concerning what is happening at *t* available at *t* (and maybe in some cases even a bit before *t*, as with anticipatory percepts), it remains possible that our percepts are genuine players in our sensorimotor engagements.

Finally (and this may or may not ultimately be an advantage), the model here is much more consistent with Dennett and Kinsbourne's multiple drafts model. In the first place, there really are *multiple* drafts. The moving window emulator constructs estimates for the entire trajectory at each time, and it is in the business of rewriting many of these estimates as a function of anticipations, surprising input, consistency constraints or changes in intention.

The primary advantage of the moving window emulation model over the multiple drafts model is clarity. It allows us to cash in the loose copy-editing metaphor for some mathematically clear apparatus from control theory and signal processing. A second advantage is that unlike the multiple drafts model per se, there is a quite clear connection on the moving window emulation framework between mechanisms that maintain and modify multiple drafts and the exigencies of biological sensorimotor control systems. (Such connections will be further explored shortly.)

Although this section has been, to some extent, critical of Dennett and Kinsbourne and also Rao, Eagleman and Sejnowski, I hope it is recognized that this is friendly criticism. By the lights of the model I have articulated, these researchers are, more than anyone else, on the right track. It is because their views are so close to my own that it is useful to clarify my own proposal by highlighting the points of divergence. I am hopeful that these researchers will see my proposal not as an adversarial competitor but as a friendly successor. It captures everything that

Dennett and Kinsbourne wanted to capture with their multiple drafts model, at least for phenomena within the window of the behavioral now, but replaces the metaphor with a clear mathematical apparatus. And it captures everything that Rao et al. wanted: a filtering mechanism that accommodates the phenomenon of perceptual postdiction. But the model here is simply more general in that it is not limited to postdiction but is capable of addressing a wider range of phenomena.

7.2 *How Big Is the Window?*

The information-processing structure just introduced shows how a system can maintain, at a given time, an estimate of what is occurring with some target system over an interval. But if this is in fact what the brain is doing, the questions arise: Why is there a window? How big is the window? What is the lag of the trailing edge of the window, and what is the reach of the leading edge? Eagleman and Sejnowski's data suggest that the trailing edge is about 100 msec or so. They have an explanation for why there is this lag and why it is the magnitude it is. The reason, they suggest, is that it is beneficial to have more accurate percepts, and because the smoother uses information that comes in after the event to interpret the event, the percept will typically be more accurate. The reason that the lag has the magnitude it has is that this particular length of lag represents an optimal (or at least very good) trade-off between the benefit of having more accurate estimates and having those estimates delayed. As I have already pointed out, this reasoning highlights a theoretical advantage of the moving window filter over the fixed-lag smoother – the fixed lag smoother must delay coming to a decision about what happened, and this has clear disutility, as Eagleman and Sejnowski's analysis recognizes.

But the proponent of the moving window filter cannot make use of this form of reasoning because the moving window filter does not have this cost. So why, according to the moving window account, is there a window, and why does it have the magnitude that it has? There are facts that provide a possible explanation for why there should be not only a lag but a reach of some particular length: feedback delays. First, to lag. At a given time t, the sensory receptors of an organism are in some state of stimulation that comprises the sensory snapshot of the organism's state. But due to the fact that these signals are carried by channels with finite information transmission velocity, some parts of that snapshot will not be centrally available until after some time has passed. Call the typically maximum sensory delay d_s. In the case of humans, this is probably proprioceptive information from the extremities, and its maximum delay is probably on the order of one to two hundred milliseconds at most. If the central nervous system wants to have an accurate central representation of what is happening or has happened at any time, then it needs to have its representation of what is happening at a given time t remain open to revision for a time at least as long as d_s, for the obvious reason that information relevant to that state estimate might be delayed as long as d_s. Clearly, then, this is a theoretically motivated reason for an organism's CNS to adopt d_s as the minimum lag of the trailing edge of its MWE. But since d is the typical *maximum* delay of the sensory information's arrival at the CNS, there won't be any strong reason to make the lag longer than d_s, especially since increasing the lag is likely to be of nontrivial cost in terms of neural computational resources.

What about the reach of the leading edge of the MWE? Again, perhaps surprisingly, feedback delay provides a possible answer. According to the moving window emulation account, one of the things the CNS does is to continually generate predictions of what is going to happen. How far into the future should the CNS generate predictions? The major class of predictive phenomena are those based on the CNS's own motor commands (this is what the Helmholtz phenomenon is: the current eye movement motor command generates an expectation as to what the retinal image will change over the next few hundred milliseconds). But motor commands are subject to a delay in execution, just as sensory states are subject to a delay. Call the maximum typical motor command delay d_m. Given this, the CNS at t has the information it needs to generate predictions about what will be happening (as a function of its own motor commands) up to $t + d_m$. This is, so to speak, how far into the future the CNS's causal influence reaches, and thus it sets a minimum desirable reach for the leading edge of the prediction side of the MWE.

While these considerations are obviously not watertight, they do provide some prima facie motivation for the CNS to have evolved to exploit a MWE information-processing structure, and why the window involved might have some specific magnitude.

At the same time, these considerations, together with the moving window filter model, the details of which they are meant to help explain, place certain conceptions of the "specious present" in a new light. The immediate specious present, what I have called the behavioral now, is indeed a window spanning a hundred or so milliseconds into both the past and future. The extent into the past is the region within which the nature of what is/was experienced can be overwritten without the change being phenomenally registered *as a change*, but rather as what was experienced all along. These are Husserl's primary retentions: the present awareness of the immediate past, but not as a memory, rather as the trailing edge of the present. And Husserl's primary protentions, the hundred milliseconds or so within which anticipations of experience are constructed, find an analogue in the moving window filter as well. They are the leading edge, within which predictions are constructed of what will be experienced – predictions that are so tightly woven into one's current experience that unless something goes very wrong (as it does in the Helmholtz/Mach phenomenon), they are not experienced as separate predictions at all, but become silently absorbed into the perceptual event along with incoming sensory information, in a way that erases all record of their status as anticipations.[15]

7.3 *The Specious Present*

From its inception, the notion of the specious present has been beset with conceptual problems. Nevertheless, I think there is something fundamentally correct behind the intuition that our conscious awareness is not confined to a moving point (that what we are aware of at an instant is more than what is happening at just that instant).

The central problems with the standard notion of the specious present are three. First, it appears to be too big, or of an unacceptably variable size. Second, even if confined to a respectable interval, it seems to go flatly against facts of our phenomenal experience. Third, it can seem to involve a sort of spooky backward causation if it is taken to include a glimpse into the future (something like Husserlian protentions). I will address each in turn.

William James' specious present is rather large – up to 12 seconds – and variable between subjects, and within subjects as a function of, for example, fatigue. If this temporal interval is taken to be what is phenomenally present to one as the present, then odd consequences can follow, some of which are exposed with delight by Gilbert Plumer (1985):

Suppose you are an earnest and energetic entry in a footrace; you want to begin when you can *first* hear the starting gun, and the length of the sensory present

---

[15] C. D. Broad (1923) and C. W. K. Mundle (1954), while advocating a brief specious present, confine it to the objective present and a brief *past* interval, unlike Husserl, who seems to treat the future reach of protentions as parallel to the lag of retentions. Their reasons have to do with what they seem to feel is the oddity of saying that we *perceive* the future. Indeed, van Gelder (1996) chides Husserl on just this point.

for you is at maximum. Suppose the gun will go off at time F, a full 30 seconds in the objective future. It follows that you will start running immediately after *now* (E), because that's when you will first *hear* the gunshot; it is something of which you are 'sensible', however indistinctly. Of course, at the finish line you have an earnest and energetic timekeeper, a friend, who *sees* you hit that line, however indistinctly, a half-minute before you do so objectively. If the sensory present for him had been at minimum ('a few seconds') when he first heard the starting gun (suppose he had no sleep but snorted cocaine during the race), you could break the world's record for the mile by almost a minute. (Plumer 1985, p. 22)

... On the other hand, some contestant might be a nonfatigued but unsportsmanlike lackey who would want to do nothing else while enjoying the dawning, then maximal intensity, then fading of the gunshot. He could do this for 'probably' a minute – beginning his trot about 30 seconds (though maybe 'hours') after the gun objectively fires, still able to claim that he *hears* it. (Plumer 1985, p. 23)

If one's notion of the specious present is of a temporal interval such that at any moment one is phenomenally aware of this entire interval, then an interval of 12 seconds seems to be way too large. (James is largely concerned with what we would now call short-term memory phenomena.)

This leads to the second issue, which is that it seems to falsify facts of phenomenology. Hearing a series of notes as a melody is one of the central phenomena discussed by the proponents of the specious present, but if we are aware of the entire interval at each moment, then what we should hear is a single chord or set of co-present tones, not a melody. And seeing a white patch moving from left to right should look like a white line segment. And thus, one should not be able to tell the difference between looking at a white patch moving left to right, or right to left, or of a long, static white stick stretching from left to right. But clearly we experience such things quite differently. Hence, the specious present so conceived cannot correctly capture the phenomenological facts.

Proponents of the specious present must be more careful in formulating the doctrine. In particular, they must take care in distinguishing the time of representing from the temporal content represented. A better formulation would be that at each moment, we are aware of a temporal interval in the sense that the *time of representing* is a single moment, but the temporal content represented is an interval. So at t, what I am phenomenally aware of is the interval $[t - 2, \ t + 2]$, say. But crucially, I am aware of this interval *as a temporally extended interval*, not as a superimposed mash of states onto one static scene like a time-lapse photograph where the headlights of moving cars look like long, static streaks. Only a confusion of the time of representing with the time represented generates the odd consequences. For only under such a confusion can it seem plausible to move from the claim that I am aware, at an instant, of the trajectory of the white patch to the conclusion that what I am aware of is the content one would get if one projected this trajectory into a single time-slice (like a time-lapse photograph). And many proponents and opponents of the specious present have been guilty of this conflation.

Although one is aware of an interval at each moment, the content of what one is aware of is a temporally extended interval, with a future pole and a past pole, and a now in between. Far from being collapsed onto an undifferentiated superposition, it is within this interval that our brain's capacity to make fine temporal discriminations is at its highest. But those discriminations are made on the basis of the different perceptual and behavioral affordances these states provide, not on the basis of the states' temporal features (they are all present now).

Once we disarm this confusion, Plumer's foot race example is easily handled. What such runners need to understand is that they start when the gun sounds, not when they are aware of the gun's sounding (though these two events are often close to simultaneous). Even if they are very good at anticipating the time of the gun sounding (perhaps one of them is the one firing the gun, and hence has a vivid phenomenal anticipation of its going off in 200 msec), they should still be aware of *that* sound – the vivid anticipation – as about-to-happen, not as happening-now, in a way analogous to how I am aware of the point of contact between my hand and the falling pencil as about-to-happen, not as happening-now. Similar remarks hold for the person timing the finish.

While a fuller discussion of these matters is called for, I hope to have shown in outline form how the proponent of the specious present can handle a few of the more common objections by carefully formulating the doctrine so as to distinguish the time of representing from the times phenomenally represented. The usual talk of "fading" or "sinking" to capture present awareness of just-past events (among Husserl's favorite metaphors) invites exactly the confusions that it is crucial to avoid.

Now to the third prima facie problem with the idea of the specious present. On some formulations, the specious present interval includes only the present and past. On others, it includes a bit of the future as well as present and past. Husserl, for example, took protentions into the future to be part of the three-part structure of time-consciousness (along with retentions and the "primal impression" of the now). Husserl treats these protentions as "perception" of the immediate future, a claim for which he is often chided (see Brough 1989; van Gelder 1996), and over which he himself was troubled – the future, after all, does not yet exist, and so how can it be perceived? (The past no longer exists, but it at least did exist, and was experienced, and continues to have causal influence, and so the mysteriousness of perceiving the past seems less pronounced.)

Even if we take causation to somehow be necessary for perception, clearly we cannot be so restrictive in our use of "perceive" so as to limit it only to those entities with which we have direct causal interaction. Surely I can perceive the entire barn even though only its outer surface – and indeed only a fraction of that – is causally responsible for my sensations.

One obvious response in the case of the barn is that I am in causal contact with the barn, not just the facing parts of its outer surface. Quite so. But notice that in just the same way, I can be in causal contact with a temporally extended process, not all parts of which are in direct causal contact with my sensory apparatus. The cases are more parallel than it might at first seem. My perceiving the barn as a barn depends on my perceptual systems having enough experience with human buildings and structures, even barns themselves, so that when the facing parts of the surface become manifest, they interpret that stimulus as being a barnlike structure (and not, e.g., a wafer-thin red surface suspended in the air, or a facade). Similarly, to perceive a process correctly, our perceptual systems need a certain amount of experience with processes of that type, so that sensory contact with parts of that process can be taken as cues to the presence of the process as a whole. Seeing the pencil roll to the edge of the table and drop down an inch is enough for my perceptual systems to perceive the process of the pencil dropping, perhaps all the way to the ground. Things can go awry in both cases. The pencil might, in fact, not fall all the way. I might be surprised to see the pencil caught in the jaws of a dog speeding by. My perception, even of the future, need not always be veridical. I might equally be surprised to learn that what I thought was a barn is, in fact, a wafer-thin red surface suspended in the air.

On the theory I am pushing here, the specious present extends only 100 or 200 milliseconds into the future and past. On these short time-scales, the way that many common processes will unfold can be predicted fairly well, well enough to trigger the normal, fallible, interpretive perceptual mechanisms, at least. Processes that are initiated and controlled by my own intentions are such as to be extremely likely candidates for such anticipatory perception.

I will close this section on the specious present with a few remarks on a proposal by Timothy van Gelder (van Gelder 1996). His goal is similar to that of the present chapter: an outline of an account of the physical implementation of the mechanisms supporting temporal phenomenology. And there appears to be broad agreement on the nature of the phenomenology. Where the disagreement occurs is over the correct account of implementation. Van Gelder, best known for championing dynamical systems models of cognitive processing, draws a number of analogies between Husserlian phenomenology of time consciousness and properties of dynamical systems. The key points about dynamical systems are that they are dynamic: They are defined as a set of possible states, a temporal ordering, and a *dynamic,* or a set of rules that specifies how a system's state evolves over time. Van Gelder then makes the connections:

When we understand how dynamical models...work, we are already (without realizing it) understanding what retention and protention are. When we look at schematic diagrams of the behavior of these models, we are (without realizing it) already looking at schematic depictions of retention and protention. The key insight is the realization that the state of the system at any given time models awareness...at that moment, and that state builds the past and the future into the present, just as Husserl saw was required. (van Gelder 1999, p. 260)

How is the past built in? By virtue of the fact that the current position of the system is the culmination of a trajectory which is determined by the [system's state] up to that point. In other words, retention is a geometric property of dynamical systems: the particular *location* the system occupies in the space of possible states....It is that location, in its difference with other locations, which "stores" in the system the exact way in which the [system's state] unfolded in the past. It is how the system "remembers" where it came from. (van Gelder 1999, p. 260)

How is the future built in? Well, note that a dynamical system, by its very nature, continues on a behavioral trajectory from any point. This is true even when there is no longer any external influence. The particular path it follows is determined by its current state in conjunction with its intrinsic behavioral tendencies. Thus, there is a real sense in which the system automatically builds in a future for every state that it happens to occupy....Protention is the location of the current state of the system in the space of possible states, though this time what is relevant about the current state is the way in which, given the system's intrinsic behavioral tendencies, it shapes the future of the system. It is the current location of the state of the system which stores the system's "sense" of where it is going. (van Gelder 1999, pp. 260–261)

The essentials of this proposal were entertained by William James (1890, pp. 606–607) and correctly diagnosed as insufficient. In discussing a thought-experiment in which a mind had the sequences of its ideas fixed by rules (also known as dynamic governing the evolution of its states) that produced them in orderly ways, James observes that

> [w]e might, nevertheless, under these circumstances, *act* in a rational way, provided the mechanism which produced our trains of images produced them in a rational order. We should make appropriate speeches, though unaware of any word except the one just on our lips; we should decide upon the right policy without ever a glimpse of the total grounds of our choice. Our consciousness would be like a glow-worm spark, illuminating the point it immediately covered, but leaving all beyond in total darkness. (1890, p. 607)

Van Gelder's proposal will have influence only on those victimized by a certain kind of content-vehicle confusion. In this case, the confusion takes the form of sliding from the observation that the brain states (or perhaps a more inclusive set of physical states, including aspects of the body and environment) that are in some sense responsible for our phenomenal consciousness are momentary slices of temporally extended trajectories through a state space, to the conclusion that the phenomenal awareness supported by such states is an awareness whose intentional content includes temporal protentions and retentions of future and past states of affairs. (The fact that this proposal has no obvious resources to address the kind of temporal illusions discussed here, such as the flash-lag effect, is a symptom of this assimilation of content to vehicle. Dynamical systems don't illusorily go into states they aren't really in.)

The account I have been outlining makes no such illicit assimilation of properties of the underlying mechanisms to the phenomenal properties supported by those mechanism. Those aspects of phenomenal consciousness at time t that concern the immediate future are present at time t. What makes their content concern time $t + 1$, even though they are occurring at time t, are two related conditions: The state carries information about what will be happening at time $t + 1$, and the state is appropriately related to the organism's perceptual and behavioral repertoire in such a way as to imbue it with temporal import (see section 6). And because the content of the phenomenology is not taken to be just determined by properties of the underlying dynamical system, room is made for illusions, corrections, and so on.

7.4 Toward a Kantian Theoretical Cognitive Neuroscience

Those who are involved in cognitive neuroscience and who have an opinion on Kant are typically not fans. And Kant certainly thought that the project of trying to understand the mind by studying the brain was an ill-conceived fool's errand.[16] This latter fact is no doubt part of the explanation of the former. Such postures notwithstanding, cognitive neuroscience stands to gain much by exploring various Kantian lines of thought.[17] While both Hume and Kant conceived of the mind as quite active in its processing of sensory information, Hume conceived of this activity as essentially reconfiguring and recombining, in perhaps novel ways, materials that have been gained exclusively through sensation. Kant, on the other hand, argued that at least some of the materials with which the mind worked were not provided to it through sensation but were supplied by the mind itself.

The basic idea is represented quite clearly in the two quotations with which I headed this article. Hume (1739/1909, p. 340) takes it that the mind's ideas of time and duration are no more than the temporal features of what is provided to it in sensation: our impression of succession comes from a succession of impressions. For Kant (1783/1950, p. 31), though, the time (and space) that we experience are constructed by the mind, not given in to it in sensory experience. Space and time, as forms of intuition, are the structures within which the mind interprets its experience, and hence, time as experienced is not read off of mind-independent objects or sensation but is added by the mind in its interpretation of what it perceives.

The theoretical moves by Dennett and Kinsbourne, and by Eagleman and Sejnowski, are nods in Kant's direction here. Whether the correct metaphysics of time is tensed (one privileged present moment moving forward as the only existing time) or detensed (all times are equally existent, ordered by before, after and simultaneity relations), it is clear that

---

[16] See, e.g., Kant's preface to neuroanatomist Thomas Sommerring's *Über das Organ der Seele* (1795).

[17] I am not the first to push this point. Patricia Kitcher's *Kant's Transcendental Psychology* (1990), and Andrew Brook's *Kant and the Mind* (1994) are recent examples from the philosophical end. I should mention that though I share the large-scale view of these authors, their interpretation of Kant and their views on how to best understand a Kantian contribution to cognitive science differ from my own on many specifics.

temporal phenomenology is not simply a reflection of, or determined by, the temporal facts. The only way that we could experience things at times or in orders other than those in which they objectively occur is if our experience of time is not determined by the temporal features of the things themselves, but is at least in part a product of the brain's own interpretation.

And time is not the only such element: for Kant, the mind-supplied contents included space, objects, and the knowledge of causal relations.[18] And as in the case of time, various results and theories in cognitive neuroscience are beginning to suggest that these elements are, as Kant maintained, interpretive elaborations supplied by the mind/brain, and not contents merely received from without. The present theory is best viewed as an attempt to contribute to this larger project of helping to move cognitive neuroscience from its Humean phase to a Kantian one.[19]

The challenge, of course, is to explain where these mind-supplied elements come from if not the world itself, and how they have the content that they do. It's not magic, and so the accounting has to be done somehow. This is where cognitive neuroscience should part company with Kant. We cannot let these puzzles force us to recognize another realm beyond the one science studies, either as Kant himself did or as some of the scientists who have worked on temporal experience have been seduced into doing (e.g., Benjamin Libet). This is not an easy undertaking.

Acknowledgments

I would like to thank the McDonnell Project in Philosophy and the Neurosciences and Project Director Kathleen Akins for financial support during which this research was conducted, and audience at the Carleton University Philosophy and Neuroscience Conference and UC San Diego's Experimental Philosophy Lab, where an early progenitor of this chapter was presented in October of 2002, for helpful discussion and feedback. David Eagleman also provided very helpful comments on a near-final version.

---

[18] I am, of course, advocating a particular reading of Kant that has him engaged in a sort of transcendental psychology, and not just spelling out nonpsychological "epistemic conditions." And it will be clear momentarily that I am also attributing to Kant a metaphysically robust and healthy transcendental idealism, not some wimpy form of a two-aspect view. Though many contemporary interpreters try to make Kant relatively innocuous (and to my mind, less interesting), I think that the Kant that Peter Strawson (1966) objected to is relatively close to the real Kant, even if I don't agree with all his objections.

[19] I am here echoing Wilfrid Sellars' suggestion that his own work was an attempt to move analytic philosophy from its Humean phase to a Kantian one.

References

Bach-y-Rita, P. (1972). *Brain Mechanisms in Sensory Substitution*. New York and London: Academic Press.

Broad, C. D. (1923). *Scientific Thought*. London: Kegan Paul.

Brook, Andrew (1994). *Kant and the Mind*. Cambridge: Cambridge University Press.

Brough, J. (1989). Husserl's phenomenology of time-consciousness. In J. N. Mohanty and W. R. McKenna (eds.), *Husserl's Phenomenology: A Textbook*. Lanham, MD: University Press of America.

Bryson, A., and Ho, Y.-C. (1969) *Applied Optimal Control: Optimization, Estimation, and Control*. Waltham, MA: Blaisdell.

Craik, Kenneth (1943). *The Nature of Explanation*. Cambridge: Cambridge University Press.

Dennett, Daniel (1991). *Consciousness Explained*. Boston: Little, Brown.

Dennett, Daniel, and Kinsbourne, Marcel (1992). Time and the observer. *Behavioral and Brain Sciences* 15(2):183–247.

Duhamel, J.-R., Colby, C., and Goldberg, M. E. (1992). The updating of the representation of visual space in parietal cortex by intended eye movements. *Science* 255(5040):90–92.

Eagleman, David, and Sejnowski, Terrence (2000). Motion integration and post-diction in visual awareness. *Science* 287 (5460):2036–2038.

Evans, Gareth (1981). Understanding demonstratives. In H. Parret and J. Bouveresse (eds.), *Meaning and Understanding*. Berlin: W. de Gruyter.

Evans, Gareth (1985). Molyneux's question. In Gareth Evans, *Collected Papers*. Oxford: Oxford University Press.

Gelb, A. (1974). *Applied Optimal Estimation*. Cambridge, MA: MIT Press.

Geldard, F. A., and Sherrick, C. E. (1972). The Cutaneous 'rabbit': A perceptual illusion. *Science* 178(4057):178–9.

Grush, Rick (1997). The architecture of representation. *Philosophical Psychology* 10(1):5–25.

Grush, Rick (1998). Skill and spatial content. *Electronic Journal of Analytic Philosophy* 6(6). (http://ejap.louisiana.edu/EJAP/1998/grusharticle98.html).

Grush, Rick (2000). Self, world and space: On the meaning and mechanisms of ego- and allo-centric spatial representation. *Brain and Mind* 1(1):59–92.

Grush, Rick (2002). An introduction to the main principles of emulation: Motor control, imagery, and perception. UCSD Philosophy Tech Report.

Grush, Rick (2004). The emulation theory of representation: Motor control, imagery and perception. *Behavioral and Brain Sciences* 27(3):377–442.

Haykin, S. (2001). *Kalman filtering and neural networks*. New York: Wiley.

Heil, John (1987). The Molyneux question. *Journal for the Theory of Social Behavior*. 17:227–241.

Hoerl, Christoph (1998). The perception of time and the notion of a point of view. *European Journal of Philosophy* 6(2):156–171.

Hume, David (1739/1909). *A Treatise of Human Nature*. (T. H. Green and T. H. Grose, eds.) London: Longmans, Green, and Co.

Husserl, Edmund (1905). *The Phenomenology of Internal Time-Consciousness*. (M. Heidegger, ed.; J. Churchill, trans.) Bloomington: Indiana University Press.

Ito, M. (1970). Neurophysiological aspects of the cerebellar motor control system. *International Journal of Neurology* 7:162–176.

Ito, Masao (1984). *The cerebellum and neural control.* New York: Raven Press.

James, William (1890). *The Principles of Psychology.* New York: Henry Holt.

Kalman, R. E. (1960). A new approach to linear filtering and prediction problems. *Journal of Basic Engineering* 82(d):35–45.

Kalman, R., and Bucy, R. S. (1961). New results in linear filtering and prediction theory. *Journal of Basic Engineering* 83(d):95–108.

Kant, Immanuel (1783/1950). *Prolegomena to Any Future Metaphysics.* Indianapolis: Bobbs-Merrill.

Kitcher, Patricia (1990). *Kant's Transcendental Psychology.* Oxford: Oxford University Press.

Kosslyn, S. M. (1994). *Image and Brain.* Cambridge, MA: MIT Press.

Kosslyn, S. M., and Sussman, A. L. (1995). Roles of imagery in perception: Or, there is no such thing as immaculate perception. In M. S. Gazzaniga (ed.), *The Cognitive Neurosciences.* Cambridge, MA: MIT Press.

Le Poidevin, Robin (1999). Egocentric and objective time. *Proceedings of the Aristotelian Society* XCIX, 19–36.

Mabbot, J. D. (1951). Our direct experience of time. *Mind* 60:153–167.

Mach, Ernst (1897). *Contributions to the analysis of sensations* (C. M. Williams, trans.). Chicago: Open Court Publishing.

MacKay, D. M. (1958). Perceptual stability of a stroboscopically lit visual field containing self-luminous objects. *Nature* 181:507.

Mundle, C. W. K. (1954). How specious is the 'specious present'? *Mind* 63:26–48.

Plumer, Gilbert (1985). The myth of the specious present. *Mind* 94:19–35.

Pouget, Alexandre, and Sejnowski, Terrence (1997). Spatial transformations in the parietal cortex using basis functions. *Journal of Cognitive Neuroscience* 9(2):222–237.

Rao, R. P. N., Eagleman, David, and Sejnowski, Terrence (2001). Optimal smoothing in visual motion perception. *Neural Computation* 13:1243–1253.

Stern, L. W. (1897). Psychische Präsenzzeit. *Zeitschrift für Psychologie und Physiologie der Sinnesorgane* 13:325–349.

Strawson, Peter F. (1966). *The Bounds of Sense.* London: Methuen.

van Gelder, T. (1999). Wooden iron? Husserlian phenomenology meets cognitive science. In Jean Pettitot et al. (eds.), *Naturalizing Phenomenology.* Stanford, CA: Stanford University Press.

von Helmholtz, H. (1910). *Handbuch der Physiologischen Optik,* vol. 3, 3d ed. (A. Gullstrand, J. von Kries and W. Nagel, eds.) Leipzig: Voss.

Williams, L. E., Hubbard, E. M., and Ramachandran, V. S. (2005, April). Retrodiction in Apparent Motion. Poster presented at the Annual Meeting of the Cognitive Neuroscience Society, New York, NY.

Zipser, David, and Andersen, Richard A. (1988). A back-propagation programmed network that simulates response properties of a subset of posterior parietal neurons. *Nature* 331:679–684.

6

The Puzzle of Temporal Experience

Sean D. Kelly

1 Introduction

There you are at the opera house. The soprano has just hit her high note –
a glass-shattering high C that fills the hall – and she holds it. She holds
it. She holds it. She holds it. She holds it. She holds the note for such a
long time that after a while a funny thing happens: You no longer seem
only to hear it, the note as it is currently sounding, that glass-shattering
high C that is loud and high and pure. In addition, you also seem to hear
something more. It is difficult to express precisely what this extra feature
is. One is tempted to say, however, that the note now sounds as though
it has been going on for a very long time. Perhaps it even sounds like
a note that has been going on for *too* long. In any event, what you hear
no longer seems to be limited to the pitch, timbre, loudness, and other
strictly *audible* qualities of the note. You seem in addition to experience,
even to *hear*, something about its temporal extent.

This is a puzzling experience. For surely, it seems, you never actually
hear anything but the note as it is currently sounding. You never actually
hear anything, one might have thought, except for the *current audible*
qualities of the note. But how long the note has been going on – its
temporal extent – does not seem to be an audible feature of the note that
is on a par with its pitch, timbre, and loudness. So how is it possible for
one to experience the note as *having gone on for a very long time?* How is
possible, indeed, for one *now* to hear something as *having been* a certain
way? This is a particular example of the kind of problem I am interested
in. It is a very specific kind of problem that concerns our capacity to
experience time.

When I first began thinking about our experience of time, I expected to find a vast quantity of literature devoted to the problem. I expected this literature to span the disciplines of philosophy, psychology, and neuroscience, and I therefore expected the area to be ripe for interdisciplinary research. In some sense I was not disappointed. There is, for instance, a vast literature in neuroscience on the neural basis of temporal illusions like the flash-lag illusion;[1] likewise, there is in psychology a considerable industry devoted to such issues as the temporality of gestalt switching, the temporal limits on attention, and the psychophysics of visual motion;[2] and in philosophy, of course, there is a growing movement concerned with questions about the metaphysics of time.[3]

To my surprise, however, I discovered that there is, in fact, very little written in any of these areas nowadays about the particular problem I am interested in. There is very little written, in other words, about our experience of the passage of time, at least if the issue is construed in a certain way. This is partly because, I believe, the problem I am interested in is rather difficult to articulate. The goal of my chapter, therefore, is quite modest: I hope to show you there is a problem here, one that is being overlooked in all of these fields. I will have succeeded if I manage to make you puzzled about something that you hadn't already known was puzzling.

2 Preliminary Statement of the Puzzle of Temporal Experience

For the purposes of this chapter, I will approach the puzzle through the examination of a cluster of phenomena that share the same general form. These phenomena include 1) the perception of objects as persisting through time; 2) the perception of objects as moving across space; 3) the perception of events as occurring in a unified but temporally ordered manner (as, for instance, in a melody); and, generally speaking, 4) our experience of the passage of time.

Any careful and sensitive treatment of these phenomena would, of course, distinguish them in a variety of ways. In this chapter, however, I will treat all of these phenomena as manifestations of the same general philosophical problem. At the general level I am interested in, I do

---

[1] Eagleman and Sejnowski 2000, 2001, 2002; Krekelberg and Lappe 2000; Whitney et al. 2000; see, e.g., Nijhawan 2002. See also footnote 34.

[2] See, e.g., Pöppel 1988; Palmer 1999; and references therein.

[3] See, e.g., Le Poidevin 1998; Mellor 1998; Sider 2001.

not believe that the differences among these phenomena are as relevant as their similarities. As a result, I will be cavalier about my use of these examples, blithely moving among them as if they were identical in all the relevant respects. This is clearly false, but I hope my strategy is not therefore illegitimate. For my main goal is to isolate a philosophical problem the various sides of which are seen more or less clearly through the various phenomena listed here.

The puzzle I am interested in can be stated, in preliminary form, as follows:

The Puzzle of Temporal Experience: How is it possible for us to have experiences as of continuous, dynamic, temporally structured, unified events given that we start with (what at least seems to be) a sequence of independent and static snapshots of the world at a time?

This philosophical puzzle of time and experience has largely been ignored in philosophy for the past 50 years. Even so, it is possible to isolate in the history of philosophy two traditional ways of resolving the puzzle.

The first approach centers on what has been called the theory of the Specious Present. According to this theory, we are wrong to think of our experience as providing us with static snapshots of the world. Rather, we are in direct perceptual contact with an ordered, temporally extended window on the world. The strategy of the Specious Present Theorist, therefore, is to reject the puzzle by rejecting the model of experience it presupposes. The principal advocates of the Specious Present Theory, in its various forms, are William James, C. D. Broad, and more recently, the British philosopher Barry Dainton.[4]

The second approach to the philosophical puzzle of temporal experience depends upon what I will call the Retention Theory. According to the Retention Theorist, our experience is fundamentally a presentation of the world at a time. The Retention Theorist, in other words, accepts the presupposition that experience presents us with a snapshot of the world. But this is not *all* that experience gives us, according to the Retention Theory. Rather, in experience, the snapshot that we get of the world is always supplemented with memories or "retentions" from the past and anticipations or "protentions" of the future. These retentions and protentions are the key to our experience of the passage of time.

---

[4] See James 1890; Broad 1923, 1938; and Dainton 2000.

The principal advocates of the Retention Theory are Immanuel Kant, Edmund Husserl, and possibly Maurice Merleau-Ponty.[5]

Although these are the only going ways of resolving the puzzle, I do not believe that either of them works. The Specious Present Theory, I will argue, simply makes no sense. It is committed to claims about experience that have no sensible interpretation. The Retention Theory, by contrast, though it is correct as far as it goes, ends up raising more questions than it answers. Although one can make sense of the theory, it does not give an explanatory account of the puzzle. This is what makes the philosophical puzzle of temporal experience so puzzling: The only natural ways to resolve it are either senseless or insufficient. My strategy, therefore, is to make you puzzled about temporal experience by showing how difficult it is to deal with it adequately.

Before I get to this stage, however, I want to begin by distinguishing the puzzle I have proposed from two other, more familiar problems that also deal with time and experience. I call these problems, instead of puzzles, since they seem to admit of various more or less tractable solutions. The puzzle of temporal experience, by contrast, seems to require a reconsideration of some of its motivating assumptions.

3 Two Problems of Time and Experience: The Time Stamp Problem and the Simultaneity Problem

3.1 The Time Stamp Problem

One problem that philosophers and psychologists have worried about is what I will call the Time Stamp Problem. The leading question behind the Time Stamp Problem is this:

Time Stamp Problem: How do we come to represent events as occurring at a particular time, and therefore to represent some events as occurring before others?

The Time Stamp Problem is familiar from recent work in philosophy, but it has a long history in psychology as well. Major contributors to the

---

[5] See Kant 1781/1787; Husserl 1893–1917; and Merleau-Ponty 1945. It may seem odd to Husserl scholars that I distinguish the Specious Present Theory from the Retention Theory since, at least on a cursory reading of Husserl's time-consciousness lectures, he appears to advocate both. It is crucial, however, to recognize that these two views are opposed to one another. To the extent that Husserl seems to advocate the Specious Present Theory, it is because, I believe, he understands this theory differently from the way I will be presenting it, and indeed differently from the way its proponents intend it.

discussion of this problem include Daniel Dennett, Hugh Mellor, and the gestalt psychologist Wolfgang Köhler.[6] Although the Time Stamp Problem lies at the intersection of issues concerning experience and temporality, it is not the problem I am interested in here. Let me say a word or two about this problem, however, in order to distinguish it from my own.

There are two main approaches to the Time Stamp Problem, each presented with typical clarity by Dennett in his book *Consciousness Explained*.[7] According to the first approach, the time at which an event is represented to occur is determined by the time at which the relevant brain process occurs. The relevant brain process is the one that, so to speak, correlates with the experience. So, for example, if the brain process correlating with my experience of a green flash of light occurs before the brain process correlating with my experience of a red flash of light, then I will experience first a green flash and then a red. On such an account, one can say that time is its own representation. The time at which an event is represented to occur is itself represented only by the already existing temporal order of the brain processes involved. There is an isomorphism, in other words, between the temporal order of the brain processes and the represented temporal order of the events in the world. Köhler was perhaps the main proponent of this isomorphism view, though it was typical of gestalt psychologists generally. As Köhler writes in his classic book *Gestalt Psychology*, "Experienced order in time is always structurally identical with a functional order in the sequence of correlated brain processes."[8]

The second approach to the Time Stamp Problem rejects Köhler's simple isomorphism view. According to this second strategy, the time at which the correlated brain process occurs is completely irrelevant, since representations can have time stamps – independent indicators of the time at which an event is represented to occur. On such a view, time is represented by something other than itself. This idea is very familiar to us from the way we date our letters. If we are interested in the time at which the events in a letter are represented to occur, we should look at the date on the letter, rather than pay attention to the time at which the letter arrives. (Think of the time of the arrival of the letter here as analogous to the time of the occurrence of the brain process.) Indeed, just because letter B arrives in my mailbox on Monday and letter A arrives on Tuesday, there is no guarantee at all that the events in letter B are also *represented to occur* before the events in letter A. If letter A has come from farther away,

[6] Köhler 1947; Dennett 1991; Mellor 1998.   [7] Dennett 1991.   [8] Köhler 1947, p. 62.

for example, and therefore took a longer time to arrive, or if it simply got lost in the mail, then it could very well happen that letter A tells me about events that are represented as occurring before the events in letter B, even though letter A arrives later. As Dennett writes, in discussing brain processes instead of letters: "What matters is that the brain can proceed to control events 'under the assumption that A happened before B' whether or not the information that A has happened enters the relevant system of the brain and gets recognized as such before or after the information that B has happened."[9]

Dennett's own Multiple Drafts Theory of consciousness depends upon the plausibility of this second approach to the Time Stamp Problem. I certainly believe the problem is an interesting one, and that much work remains to be done in this area.[10] But I do not intend to say anything more about it here. My main goal at this juncture is simply to emphasize that the Time Stamp Problem is not the problem with which I am concerned.

3.2 The Simultaneity Problem

A second problem in the general area of time and experience is what I call the Simultaneity Problem. This problem is deceptively simple to state:

Simultaneity Problem: Which events do we experience as simultaneous?

The Simultaneity Problem is not much discussed in the literature today, but it was important for Bertrand Russell during a certain period of his development,[11] and in any case, it is quite easy to generate. Let us assume a simple, nonrelativistic notion of simultaneity for the purposes of discussion. Obviously, then, there are events in the world that are not as a matter of fact simultaneous, which we nevertheless experience as simultaneous.

---

[9] Dennett 1991, p. 149.

[10] One principal problem is to determine what a neural time stamp could look like and how we should go about finding one. It is one thing to look for correlations between the firing rate of a neuron and the presence within a receptive field of, for instance, a horizontal line. But if the stimulus is not a horizontal line but instead the time at which the horizontal line is represented to appear, then it seems the methodology for varying the stimulus will be trickier. The problem is even more complicated when you realize that in the simple cases, like neurons in V1, Köhler's isomorphism story is likely to be right. More complicated cases of temporal illusion are clearly the place to focus here, and the burgeoning literature on the flash-lag illusion is perhaps a case in point. (See references in footnote 1.) But these models do not go unequivocally in Dennett's direction, since they tend to rely upon a combination of low-level temporal isomorphism and higher-level nonisomorphic time stamping.

[11] See chapter 6 of Russell's unpublished 1913 manuscript *Theory of Knowledge* (Russell 1913).

(This follows trivially from the fact that the temporal acuity of experience is not infinitely fine-grained.) The Simultaneity Problem, then, is to determine which events these are.

One might wonder why experienced simultaneity is a problem. After all, one might think, can't we simply say that two events are experienced to be simultaneous if and only if they seem to the subject to be simultaneous when they are experienced? That would certainly be the simplest approach. But a problem quickly arises if we proceed this way. For if we define experienced simultaneity in this obvious manner, then the relation turns out to be nontransitive. Since it is meant to be an equivalence relation, this would be a devastating blow.[12] Let us see in a bit more detail how this works.

Consider any two events A_1 and A_2 that seem to me to occur simultaneously. For example, imagine that A_1 is a flash of light and A_2 is a tone. Then take some third event A_3, which actually occurs later than both A_1 and A_2 but is experienced as simultaneous with at least the second. Perhaps A_3 in this case is another flash. By this method, we can generate a series of events $A_1, \ldots, A_n$ such that each is experienced as simultaneous with the next when considered pairwise, but such that A_1 and A_n, when considered pairwise, are experienced as nonsimultaneous. In this way, we can show that the relation of experienced simultaneity is nontransitive.

The Simultaneity Problem is formally equivalent to Nelson Goodman's problem about the nontransitivity of appearance properties.[13] Recall that Goodman considers a series of color chips $R_1, \ldots, R_n$ such that R_j appears identical to R_{j+1} for all j from 1 to $n - 1$, but it is not the case that R_1 appears identical to R_n. Just as with the Simultaneity Problem, if we define the "appears identical" relation in terms of the way the colors look to the subject in pairwise comparisons, then we can show that this relation is nontransitive.

There is an obvious way to avoid this deficiency, and it is perhaps not surprising that Russell seems to have discovered the same solution

---

[12] One possibility, of course, is simply to deny that experienced simultaneity is an equivalence relation. Few have been tempted by this approach, however, perhaps because it has the effect of divorcing the world so radically from our perceptual representation of it. This divide is more uncomfortable for some perceptual theorists than others. Disjunctivists about perception, for example, are typically committed to the idea that in perception, we are presented with facts about the way the world is. (McDowell is a good example of a view like this.) On such a view, it would be extremely awkward to let simultaneity in the world be an equivalence relation but to deny this for our perceptual representation of it.

[13] See Goodman 1951.

to the problem as Goodman. The trick is to deny that two events are experienced to be simultaneous if and only if they seem to the subject to be simultaneous when they are experienced. Rather, two events A and B are experienced to be simultaneous if and only if they meet the following two conditions:

1. A and B seem to be simultaneous.
2. There is no third event C such that one of A or B seems to be simultaneous with C and the other does not.

If experienced simultaneity is defined in this more restrictive way, then the problem of nontransitivity does not arise. That is because the second condition effectively excludes, by fiat as it were, all of the cases that could give rise to a problem. But the solution has a surprising consequence as well. It turns out, on this approach, that just because two events seem to me at the time to be simultaneous, I cannot thereby conclude that I am experiencing them to be simultaneous. I will only know that I have experienced them to be simultaneous once I can show in addition that there is no possible third event such that it seems to be simultaneous with one but not with the other. Experienced simultaneity, in other words, is more fine-grained than momentary reflection would indicate. As a phenomenologist, I find this result congenial. That is because phenomenologists believe, in general, that experiences are richer and more complicated than one can know simply by having them. But again, my goal here is not to discuss the Simultaneity Problem in any great detail; it is only to emphasize that this is not the puzzle with which I am concerned.

4 Development of the Puzzle of Temporal Experience

Both the Time Stamp Problem and the Simultaneity Problem are focused on *when* we experience events to occur. The Time Stamp Problem is concerned with whether we experience A to be before or after B. The Simultaneity Problem is concerned with how to define the equivalence class of events that are experienced to be simultaneous. But the question of time and experience in which I am interested is not the question *When* do we experience events to occur? but, rather, the question How do we come to experience events as occurring *through* time at all? This is a question about experiencing the *passage* of time, not just a question about *at what time* we experience an event to occur.

The distinction between experiencing events to occur *at a time* and experiencing them to occur *over or through time* has been discussed since at least the writings of Saint Augustine.[14] By the period of the Renaissance, this distinction had worked its way into the culture more generally. Consider, for instance, Shakespeare's sonnet number 104, in which he writes of a gracefully aging friend:

> Ah! yet doth beauty, like a dial-hand,
> Steal from his figure and no pace perceived,
> So your sweet hue, which methinks still doth stand,
> Hath motion and mine eye may be deceived.

In this stanza, Shakespeare compares the graceful decline of his friend's beauty with the slow movement of the hand on a clock: Although one never sees any change occur ("no pace perceived"), it is nevertheless clear after some time that there has been a change ("Hath motion and mine eye may be deceived").[15]

In general, it is a common observation that some movements or changes happen too slowly for us to experience them *as movements or changes occurring through time.* For example, John Locke considers, in the *Essay,* cases in which

[t]he Body, though it really moves ... [nevertheless] seems to stand still, as is evident in the Hands of Clocks, and Shadows of Sun-dials, and other constant, but slow Motions, where though after certain Intervals, we perceive by the change of distance, that it hath moved, yet the Motion it self we perceive not.[16]

By contrast, successive events sometimes occur too swiftly for us to perceive them as part of a movement or change through time. Locke makes this point in a colorful way:

Let a Cannon-Bullet pass through a Room, and in its way take with it any Limb, or fleshy Parts of a Man; 'tis as clear as any Demonstration can be, that it must strike successively the two sides of the Room: 'Tis also evident, that it must touch one part of the Flesh first, and another after; and so in Succession: And yet I believe, no Body, who ever felt the pain of such a shot, or heard the blow against the two distant Walls, could perceive any Succession, either in the pain, or sound of so swift a stroke.[17]

But there is a wholly distinct phenomenon that seems to lie somewhere between perceived precedence and perceived simultaneity, and that is the

---

[14] See Augustine, *Confessions,* Book 11.

[15] Shakespeare does not specify which dial-hand he is talking about, but naturally the metaphor would not work if it were the second hand on a clock. There were no second hands in Shakespeare's day.

[16] Locke 1690, Book 2, chapter 14, §11.   [17] Ibid. §10.

phenomenon of perceived movement or change. Perceived movement does not seem to be the kind of thing in principle that can be explained in terms of perceived precedence or perceived simultaneity. It seems, in other words, to be a basic kind of perception. As C. D. Broad writes:

It is a notorious fact that we do not merely notice that something *has* moved or otherwise changed; we also often see something *moving* or *changing*. This happens if we look at the second-hand of a watch or look at a flickering flame. These are experiences of a quite unique kind; we could no more describe what we sense in them to a man who had never had such experiences than we could describe a red color to a man born blind. It is also clear that to see a second-hand *moving* is a quite different thing from "seeing" that an hour-hand *has* moved.[18]

If perceived movement is a basic kind of perception, then a certain question naturally arises: What must perceptual experience be like if it is to allow for the possibility of the perception of motion or change? Since all motion takes place not only across space but also through time, this question is closely related to the question with which we are principally concerned: What account of perceptual experience explains how it is possible for us to experience the passage of time? Two theories are prominent in historical accounts, but neither seems to do the job.

5 The Specious Present Theory

The Specious Present Theory found its first major proponent in William James, who developed the theory at length in *The Principles of Psychology*. The central idea behind the theory is that instead of giving us a snapshot of the world at a time, perception presents us with a temporally extended window of events. In particular, the Specious Present Theory proposes that a subject is in direct perceptual contact with an ordered, temporally extended, unified expanse. As James writes, in a famous passage from *Principles:*

The practically cognized present is no knife-edge, but a saddle-back, with a certain breadth of its own on which we sit perched, and from which we look in two directions into time. The unit of composition of our perception of time is a *duration*, with a bow and a stern, as it were – a rearward- and a forward-looking end.... We do not first feel one end and then feel the other after it, and from the perception of the succession infer an interval of time between, but we seem to feel the interval of time as a whole, with its two ends embedded in it.[19]

[18] Broad 1923, p. 351.   [19] James 1890, pp. 609–610.

According to this theory of the nature of perceptual experience, we are in direct perceptual contact not only with what is *now* occurring but also with what *has recently occurred* and indeed with what is *about to occur* as well.[20]

Although I believe there is something important about the perceptual phenomenon that this theory is intended to explain – and indeed I believe that no theory of perceptual experience would be complete without an explanation of the temporal effects in question – it nevertheless seems to me that there are at least three devastating objections to the Specious Present Theory itself. These objections coalesce around the following three questions:

1. How can I be directly aware of something that is no longer taking place?
2. How can I be directly aware of a duration?
3. How can I be directly aware of the future?

I will take these questions in order.

5.1 Problem 1: The Experience of the Recent Past

The Specious Present Theory proposes that I am in direct perceptual contact with events that occurred in the recent past. This is at best an odd suggestion. After all, the events in the recent past are no longer occurring, and one might naturally wonder how I can be directly aware of something that is no longer taking place. I can think of two obvious ways to interpret the theory so that it gives an answer to this question. Unfortunately, neither of them does the work that the Specious Present Theory needs. I conclude that the theory has no obvious interpretation on which this central claim makes sense.

So, then, how can I be directly aware of something that is no longer taking place? One suggestion is to emphasize the time lag that always exists between the occurrence of an event and the subject's experience of the event's occurring. In extreme cases, this time lag can be millions

---

[20] James's version of the Specious Present Theory is one of the few I know of that advocates both a past and a future aspect to the specious present. Most accounts limit the window of the Specious Present to a short period of time that stretches back from now. James's proposal follows (as always) from his acute observations of the perceptual phenomena. For it is certainly true that our current experience is directed toward both the recent past and the near future; theories of the Specious Present that deny this fact are certainly less attuned to the phenomena they hope to explain. As we will see in a moment, though, the claim that we are in *direct perceptual contact* with the future is itself specious at best.

of years. When you look into the night sky and see the explosion of a star, it may be that the event you are now witnessing actually took place many millions of years ago (if the supernova occurred millions of light-years away). Of course, not every time lag between event and experience is this extreme. Sometimes the experience of the event occurs only a very short time after the event itself. Even so, the experience always occurs at least some time after the event, since it always takes time for the waves carrying information about the event to reach the sensory organs, and some further time for the brain to process this information in such a way that it gives rise to an experience. The result, therefore, is that I am *always* aware of events that are no longer taking place, since experience lags behind the event experienced. Perhaps, then, this time lag between an event and the experience of it is a way of making sense of the claim that I can experience what is past.

The time lag suggestion, however, is irrelevant to the Specious Present Theory for at least two reasons. First, this suggestion implies, against the Specious Present Theory itself, that I am never aware of the present, never mind aware of the future. The Specious Present Theory proposes, you will recall, that I am aware of a temporal window of events that reaches into the past but includes at least the present and possibly also the future. But the time-lag argument emphasizes that the events I am aware of are always events that occurred in the past. Second, the Specious Present Theory says not just that I am perceptually aware *of* the past but that I am aware of it *as* the past. Without this, the theory cannot make any sense at all of my experience of the passage of time. The time-lag phenomenon, however, shows only that the events I am aware of occurred in the past, not that I am aware of them as occurring in the past. Indeed, these two phenomena usually come apart: It is always a surprise for children to learn, for example, that looking into the night sky is the same as looking back in time.

A second suggestion makes more headway, but in the end is insufficient as well. This suggestion derives from what Russell calls, in *The Analysis of Mind,* "akoluthic sensations." The akolouthoi, in ancient Greece, were the so-called camp followers who trailed behind the soldiers on their way to battle, providing services for them and collecting valuables they left behind. The akoluthic sensations, in Russell's analysis, are the sensations one has of events as just having occurred. Take, for example, the acoustic case. When I clap my hands together, the primary sensation is as of a sharp sound that lasts for an instant. But after this primary sensation runs its course, it is natural to say that one continues to experience the clap of the hands, but now one experiences it as recently past. This is an akoluthic sensation of the clap. Russell's classically empiricist analysis of this phenomenon proposes that hearing a sound *as* past in this way is hearing the same sound but with less "force and vivacity." He writes:

> Succession can occur within the specious present, of which we can distinguish some parts as earlier and others as later. It is to be supposed that the earliest parts are those that have faded most from their original force, while the latest parts are those that retain their full sensational character.... Sensations while they are fading are called "akoluthic" sensations.[21]

Perhaps Russell's notion of akoluthic sensations gives us a clue as to how we should develop our theory of perception so that it allows us to be directly aware of the recent past.

I am not sanguine about this approach. The problem is that Russell's account fails to make sense of one of the main problem cases we are hoping to solve – that is, the case of perceived motion. Consider the experience of watching Derek Jeter's throw travel through the air from shortstop to first. On Russell's account, the earlier phases of the moving ball are now seen, but are seen less forcefully than the current phase. If that were right, then seeing a moving object would be a matter of seeing it vividly at its current position, but with a continually fading trail. Needless to say, this does not capture the experience of watching Jeter's throw, unless your television is significantly worse than mine.

5.2 *Problem 2: The Experience of Duration*

A second challenge for the Specious Present Theory is to explain how I can be directly aware of a duration. The default position seems to be that I am directly aware of what is presented to me now. As with awareness of the past, however, awareness of a duration of time requires that some part of what I am directly aware of is not occurring now. So how can I be perceptually aware of something that is not now occurring?

Once again, the most natural approach to this problem turns out to be flawed in a crucial way. The most natural approach emphasizes that we always experience a duration, but we experience it as a moment in time. Recall Locke's example of the Cannon-Bullet. The point of that example was to show that our perceptual experience is relatively coarse-grained – it groups together events that happen successively into a single undifferentiated experience. Even though the bullet hits first one wall and then the other, for example, we are apt to experience only a single

---

[21] Russell 1921, p. 145.

sound. In this sense, therefore, we always experience events that happen across a span or duration of time.

It is certainly true that our experience is relatively coarse-grained in this way. It is, however, irrelevant to the challenge raised for the Specious Present Theory. That is because there is a difference between the claim made by the Specious Present Theory and the fact explained by the suggestion here. The suggestion proposes rightly that we experience temporally distinct events as simultaneous. But the Specious Present Theory claims more than this. It claims that we experience temporal extension itself. This strange claim is not in any way clarified by the correct but trivial observation that experience groups together successive events. Perhaps there is some other way to explain the claim that we experience temporal extension directly. If so, it will require a broader defense of the theory of perception that lies behind the story of the specious present.

5.3 Problem 3: The Experience of the Near Future

A third challenge for the Specious Present Theory is to explain how I can be directly aware of the near future. I have already mentioned in passing that not every version of the theory insists that the specious present has a future dimension. But perhaps the most famous version of it, manifest in the passage previously quoted from James, has experience looking in two directions into time. It is odd enough to think that we could be directly aware of something that was occurring but no longer is. But it is odder still to think that we can be directly aware of something that has not yet occurred. This view of perception seems to smack of extrasensory powers. What could the Specious Present Theorist mean?

As with the previous two problems for the theory, there is a natural way to answer this third question. But once again it turns out not to do the necessary work. The natural suggestion is to give examples in which it really does seem to make sense to say that I can be aware that something is *about to occur*. Here's one. Sometimes if you are standing in the batter's box watching the pitch come in, it becomes immediately clear to you that you are about to be hit by the pitch. It is not ESP. It is just that you can see the pitch heading for your shoulder and you are immediately inclined to duck out of the way. This kind of thing seems to happen all of the time. You can see that the light is about to turn red, that the pedestrian is about to walk out into the street, that the professor is about to trip over the extension cord at the front of the lecture hall, and so on. Perhaps it sounds strange in the abstract to say that we are directly aware of these events before they occur, but after all, isn't that the phenomenon?

I do not intend to deny the phenomenon. As with the experience of the recent past and the experience of duration, there is no doubt that there is a sense in which our experience has a futural dimension to it. The question is whether the best explanation of this phenomenon is to say that we have direct perceptual awareness of the future. Without a more robust and independently motivated theory of perception to back up a claim like this, it seems a very large pill to swallow. After all, nobody can deny that there are some intentional states by means of which we can be directed toward the future. Anticipation, expectation, hope, desire, and so forth are all quite good ways of doing this. But that we have a direct perceptual awareness of the future is an unsubstantiated claim at best.

5.4 A Final Problem

Let us suppose for the sake of argument that the Specious Present Theorist can get around the previous three objections. Let us suppose, in other words, that we can make sense of the central claims of the theory. There is still an important final problem for the theory – namely, that it fails to explain the central problem case it was invented to explain. That is the problem, as Shakespeare puts it, of pace perceived.

To see that the perception of motion remains unexplained, we need only to notice that the Specious Present, by nearly all accounts, lasts only a relatively limited time. Recent estimates generally agree that it is in the area of three seconds or so.[22] But we often experience things to be moving for periods that are longer than this. If you watch an airplane taking off from the runway, you can follow its continuous motion for several minutes before it disappears. Even on the Specious Present Theory, therefore, we must *keep track* of the earlier phases of long movements in some way other than by perceiving them directly. That we have some relation to the past and the future other than direct perception of it, however, is the main point of the Retention Theory. I turn to this alternative theory now.

6 The Retention Theory

The Retention Theory has its origins in the empiricism of Locke and Hume. Recall that one central tenet of their empiricism is that in the most basic cases, our ideas of properties and objects in the world originate in

---

[22] These estimates are ascertained empirically by measuring the amount of time between, for instance, gestalt shifts in the perception of a Necker cube. See Pöppel 1988. William James is again a renegade here, since his estimates for the length of the Specious Present have it lasting somewhere between 3/4 of a second and 12 minutes. The principle of charity requires that we leave this aspect of his position behind.

our perceptions or impressions of them. We have the idea of squareness for example, on such a view, because we have had a perception or impression of a square. The question naturally arises, then, what the origin is of our ideas of succession and duration, and generally of our idea of time. Locke denies outright that we have any perception of duration by means of which we can get our idea of it.[23] That is to say, Locke denies the central tenet of the Specious Present Theory. Hume expands upon this idea in the *Treatise*:

> The idea of time is not deriv'd from a particular impression mix'd up with others, and plainly distinguishable from them; but arises altogether from the manner, in which impressions appear to the mind, without making one of the number. Five notes play'd on a flute give us the impression and idea of time; tho' time be not a sixth impression, which presents itself to the hearing or any other of the senses.[24]

If we do not get our ideas of succession and duration from perceptions of time, according to the empiricists, then what precisely is the origin of these ideas? Locke and Hume do not give a particularly detailed story about this. They do suggest, however, without much further explanation, that the succession and duration of our ideas themselves are central in the development of our ideas of succession and duration. As Hume says, "From the succession of ideas and impressions we form the idea of time."[25]

If the suggestion here is that merely by *having* a succession of ideas we can thereby get an idea of one thing's following another, then the suggestion is clearly wrong. For imagine a creature who has a succession of experiences, but at each moment forgets all the previous ones. Such a creature, though it has a succession of experiences, does not seem thereby to have the resources for forming the idea that each experience follows the previous one. For without any memory of the previous experience, such an idea is impossible to form. That is why William James (following Kant) says, rightly, "A succession of feelings, in and of itself, is not a feeling of succession."[26]

If a mere succession of experiences is not sufficient for the idea or experience of succession, what must be added? Kant famously develops a story about this in the A-Deduction of the first *Critique*. It is the story of the so-called threefold synthesis. The key to Kant's account is that at every moment, I not only must have an experience of the thing before me now but also must "reproduce in imagination" the things I experienced

---

[23] "We have no perception of duration," Locke 1690, Book 2, chapter 14, § 4.
[24] Hume 1739, Book 1, chapter 2, § 3, pp. 34–5.   [25] Ibid. p. 35.
[26] James 1890, p. 628.

in the recent past. Notice that I don't *perceive* these past events, according to Kant, but rather *reproduce* them in *imagination*. As he writes:

When I seek to draw a line in thought, or to think of the time from one noon to another, or even to represent to myself some particular number, obviously the various manifold representations that are involved must be apprehended by me in thought one after the other. But if I were always to drop out of thought the preceding representations (the first parts of the line, the antecedent parts of the time period, or the units in the order represented) and did not reproduce them while advancing to those that follow, a complete representation would never be obtained.[27]

But what kind of intentional attitude, we might naturally ask, is the attitude of reproduction in imagination? There are lots of views among Kant interpreters, and I certainly do not intend to wade into this thorny literature here. But one prominent view, which is suggested at least implicitly by Robert Paul Wolff, is that *reproducing* an event in imagination is like *remembering* it. As Wolff explains:

What I must do . . . as I proceed from one moment to the next, is to reproduce the representation which has just been apprehended, carrying it along in memory while I apprehend the next. In looking at a forest, I must say to myself, "There is a birch; and there is an elm, plus the birch which I remember, etc.[28]

Whatever the virtues of this as an interpretive claim, I do not believe that the subject's intentional relation to recently past events is anything like a memory of them. In this I follow Husserl, whose notion of retention is meant to do the work that Wolff believes something like memory can do. Before I present Husserl's approach, therefore, let me say why memory is not sufficient.

Memory comes in a variety of different forms, and so it is important to specify which kind we have in mind. I will consider two. First, there is the kind of memory in which one is reminded of something in a flash. It occasionally happens to me, for example, that as I walk out the front door of my apartment, and hear it closing locked behind me, I am reminded instantly that my house keys are on the kitchen table inside. This kind of memory brings something to mind in a flash and apparently without any effort on the subject's part at all. It has the feeling of a kind of instantaneous revelation, in which all of a sudden something completely new is brought to mind.

---

[27] Kant 1781, p. A102. [28] Wolff 1963, p. 128.

This kind of instantaneous remembering cannot be the relation we have to the recent past. This becomes clear when we think again of Hume's example of hearing a melody. Hearing the fifth note of the melody does not involve being instantaneously reminded of the earlier ones at all. It is not as if when I hear the fifth note, the earlier notes jump immediately to mind like the image of the keys when the door clicks shut. As Husserl says, "A present tone can indeed 'remind' one of a past tone, exemplify it, pictorialize it; but . . . the intuition of the past cannot itself be a pictorialization."[29]

A second kind of memory falls naturally under the heading of *entertaining* a memory. One entertains a memory when one goes over in one's head, often from a first-person point of view, events in which one was previously involved. I might, for example, entertain the happy memory of my wedding ceremony. When I do so, I go back over the events of the ceremony in my mind, experiencing them again as if they were occurring now. Some people are extremely good at this kind of remembering and can even produce autonomic responses, such as increased heart rate, by entertaining memories in this way.

The problem with this kind of remembering is that in putting myself again in the situation, I experience the events as occurring *now*. If we model the experience of the melody this way, then the earlier notes in the melody are experienced as present together with the current one. It would be as if, while hearing the fifth note, I simultaneously entertain the memory of the notes before. But this would give me the experience of a chord instead of the experience of an extended event. It would be a very odd type of chord, of course, since one of the notes would now be *heard* while the others would simultaneously be *entertained in memory*, but it would be a type of chord nevertheless.

Husserl concludes from considerations like these that our relation to recently past events cannot be any kind of memory. He coins the term "retention" to characterize our actual relation to the recent past and claims that retention is a unique kind of intentional act that is unlike any kind of reproduction or memory. Retention is, by definition, a way of being directed toward objects and events *as just-having-been*. About perceived motion, for example, Husserl says:

During the time that a motion is being perceived, a grasping-as-now takes place moment by moment; and in this grasping, the actually present phase of the motion

[29] Husserl 1893–1917, §12.

itself becomes constituted. But this now-apprehension is, as it were, the head attached to the comet's tail of retentions relating to the earlier now-points of the motion.[30]

The problem with Husserl's account, as I see it, is that it seems only to name the phenomenon instead of explain it. We have no interesting account of what it is now to experience something *as just-having-been*, except to say that it is the phenomenon involved in the experience of the passage of time. But this is the phenomenon we are trying to *explain*. It does no good just to give a name to its various parts. What, after all, is it now to experience something *as just-having-been?* I know what it is now to think of George W. Bush *as the president of the United States*. Our whole theory of intentionality is built around cases of linguistic predication such as this. But what kind of perceptual phenomena can we point to that will help us unravel the *temporal aspect* of perceptual experience? Husserl does not even attempt to say.

I conclude from this that Husserl's theory is some kind of advance, since at least it does not have the disadvantages of Kant's view or of the Specious Present. But Husserl's account seems to raise more questions than it answers. What we would like is a standard set of examples that gives us the feel for what it is to experience something now *as just-having-been*. It would be even nicer to formulate these examples in such a way that we could pose empirical hypotheses about their features. In short, our project should be to give examples like the kind that Russell uses to expound his theory of akoluthic sensations, but to give examples that are phenomenologically apt instead of obtuse. I believe there are good examples of this sort, and I would like to think more about how to discuss them in the future. But for the moment, I will be content if I have managed to convey how puzzling the phenomenon of temporal experience really is.

7 Appendix: Temporal Experience and the Science of the Brain

The goal of this chapter was to make you puzzled. I have discovered, however, that it is very difficult to make psychologists and neuroscientists (as well as some philosophers) puzzled about this topic in the appropriate way. Usually these unflappable types have one or the other of two different responses. The first is to assert that we are already busy filling in the

[30] Ibid. §11.

neuroscientific details that, when complete, will constitute a satisfying answer to the problem I have posed. There is no particularly gnawing *puzzle* of temporal experience, on this account, just an empirical *problem* that we are busy solving. The second is to admit that there is a puzzle, but to identify it as a version of the so-called hard problem of consciousness. (The hard problem is really a puzzle in my terminology, since we don't have the conceptual resources yet to understand what the form of a satisfying answer would be.) On this account, there is a genuine puzzle about how the brain could give rise to anything like experience at all, but the question how it could give rise to temporal experience is just a special version of that familiar puzzle.

I believe that neither of these responses is to the point. The puzzle of temporal experience will not be resolved by empirical research *of the type now being done.* Even so, the issue is not hopeless: Temporal experience is a much more tractable puzzle than that posed by the hard problem of consciousness. Indeed, the really nice thing about the puzzle of temporal experience is that it seems to be just the kind of puzzle that would benefit most directly from the interaction of conceptual, philosophical work and empirical psychological and neuroscientific progress. The goal of this appendix is to say how this interaction might work. I will pursue this goal by showing precisely why each of the two usual responses to the puzzle misses the point.

7.1 Temporal Experience Really Is a Puzzle

I begin by explaining why current neuroscientific research, though interesting and important, is not the right kind of thing to resolve the puzzle I have identified. I don't know about all the empirical research there is, of course, and it may be that some neuroscientist somewhere really is making progress. But I doubt it. All of the research I am familiar with is either directed at some version of the Time Stamp and/or Simultaneity Problems, or else depends implicitly on either the Specious Present or the Retention Theory. For this reason, the current empirical research seems either to miss the puzzling issue of temporal experience or to fail to appreciate its puzzling nature. Let me give one example in a bit of detail.

Some of the most interesting neuroscientific work on time and experience focuses on temporal illusions like the flash-lag illusion.[31] In the flash-lag illusion, a flash occurs at a certain location precisely when a

---

[31] See footnote 1 for references.

moving object reaches that location. The illusion is that the moving object and the flash appear to be offset. On one version of the illusion, for example, the moving object is a bar of constant velocity traveling in a straight line from left to right. When the bar reaches a certain point, a second bar flashes directly above it; the first bar continues moving in the same direction. After being presented with this stimulus, the subject is asked to say where the flash seemed to be, relative to the moving bar at the instant the flash occurred. In the case just described, subjects systematically report that the flash seemed to be behind (to the left of) the moving bar. The flash, in other words, seemed to lag behind the bar; hence, the name of the illusion. It is interesting to note, however, that if the bar stops moving at the moment the flash occurs, then subjects report the flash and the bar to be in the same location; and if the bar reverses direction when the flash occurs, then subjects report the bar to lag behind the flash instead of the other way around.[32]

The principal neuroscientific problem here is to provide a computational model that accounts for the psychophysical data. Since the judgment about the relative locations of the flash and the bar is affected by what happens after the flash occurs, one model proposes that the brain is taking these later events into account in determining what the percept represents to be happening at the instant of the flash. This is the "postdiction" account preferred by D. M. Eagleman and T. J. Sejnowski.[33] There are also prediction and online accounts of some of these phenomena.[34]

The question of what model the brain is using in order to give rise to this range of psychophysical results is an enormously interesting and important one. No doubt the final answer will be established not just by the computational neuroscientists but also by their brethren in the cell-recording industry. What I wish to emphasize here, however, is that this problem, interesting though it is, has no bearing on the puzzle of temporal experience that I have described. The reason is that the flash-lag illusion is a straightforward hybrid of (admittedly empirical versions of) the Time Stamp and Simultaneity Problems. Insofar as it combines elements of these two problems, it is more complicated than either on its own, but it is not for that reason a problem of a radically different kind.

[32] See Eagleman and Sejnowski (2000) for details. These authors used a moving ring and a flash tha occurs inside the ring, but illusion is the same.

[33] Ibid.

[34] For the prediction account, see Nijhawan 1994. For the online account, see Baldo and Klein 1995; Purushothamen et al. 1998; and Whitney and Murakami and 1998; expanded upon in Whitney, Murakami, and Cavanaugh 2000.

To see this, recall that the Time Stamp Problem is the problem of determining how we come to represent events as occurring at a particular time and, therefore, to represent some events as occurring before others. One issue in the neuroscientific accounts of the flash-lag illusion is just that: what model is the brain using to represent the time at which the flash occurred and to represent the position of the moving bar at that time? The models currently under discussion do not so far take a stand on whether the brain uses time as its own representation in implementing these models, but they have turned this into an empirical question that they can begin to address.

In addition to the question of how we represent events occurring at a time, there is the question of which events we experience as happening together. This is the Simultaneity Problem, and it is the second problem under discussion in the flash-lag illusion. The psychophysical data that define the illusion just are data about which events the subject experiences to be simultaneous. Russell's discussion of the Simultaneity Problem, of course, should make us wonder whether we really can take the subject's report about perceived simultaneity as a good indicator of what the experience actually represents. To the extent that the psychologists assume the subject's report to be a precise indicator of his percept, they have failed to grapple with the philosophical issues behind the Simultaneity Problem. But the question of which events we experience as simultaneous is nevertheless a central question for these experimenters.

Insofar as the Puzzle of Temporal Experience is different from either the Time Stamp or the Simultaneity Problem, as I have argued it is, it should be clear that the neuroscientific accounts of the flash-lag illusion fail to impinge on the puzzle I have tried to describe. They do not have any bearing at all, in other words, on the question of how we are to conceptualize perceptual experience from the point of view of the subject so as to allow for the possibility of experiences as of continuous, dynamic, temporally structured, unified events. The flash-lag illusion, of course, is not the only temporal phenomenon now being studied. There are many other neuroscientific projects that cover the general territory of time and experience. These include, for example, discussions of the echoic buffer, of apparent motion and the motion detection neurons in area MT, of the neural basis of circadian rhythms, of the temporal structure of gestalt switching phenomena, and so on. Many of these are fascinating problems, and in the case some of them, important neuroscientific research is being done. But in no case that I know of is the work of the sort to address the Puzzle of Temporal Experience that I have characterized.

There is a simple reason why no empirical work is being done on this puzzle. The reason is that we're not yet clear enough on what kind of work would make progress. The puzzle, in other words, has not yet been turned into a problem. We would be clear enough about how to approach the issue if either the Specious Present Theory or the Retention Theory provided the right kind of background model for temporal experience. Some empirical work presupposes one or the other of these theories, and thus gives the appearance of making progress.[35] But if my criticisms of these theories have been on target, then what we need is a new background model on the basis of which we can begin to pursue empirical research. It is not my goal to provide such a model here – that will have to wait for another day. But what I can do is say something about why I think the prospects for such a model are better in the case of temporal experience than they are in the more general case of consciousness. I turn to this final topic now.

7.2 *The Puzzle of Temporal Experience Is Not a Version of the Hard Problem (Puzzle) of Consciousness*

The hard problem of consciousness, as David Chalmers calls it, is the problem of determining how the brain – a mass of physical stuff – can be the kind of thing that causes conscious experience. The hard problem is different from various easy problems of consciousness. These concern the question of which part of the brain causes this or that kind of experience. The easy problems are the kind that are well studied using, for instance, fMRI. There are many details and niceties to this kind of work, and it can be very challenging empirically, but the basic idea is very easy to understand. Induce a certain kind of experience in the subject and see which part of the brain lights up when he or she is having it. Having discovered the relevant kind of correlation, we have some evidence about which part of the brain causes the kind of experience in question.

The various easy problems of consciousness are vulnerable to just the kind of empirical attack that science is very good at making, and we do seem to be making progress on some of them. The hard problem, however, the question precisely *how* activity in the relevant part of the brain causes anything like *experience*, is left completely untouched by this technique. This is unfortunate, of course, but it is no great embarrassment

---

[35] Work on the temporal structure of gestalt-switching phenomena, for example, is sometimes taken to be telling us about the temporal extent of the Specious Present. See Pöppel 1988.

for science. Scientists generally have accepted that the hard problem is, for the time being anyhow, outside their domain.

Sometimes when psychologists and neuroscientists, and even some philosophers, have heard me present the ideas in this chapter, they have accused me of trying to saddle them with a particular version of the hard problem. Their reasoning is something like this: You say that the experience of motion is one of the phenomena you're trying to explain. This is, after all, one of the central examples under discussion, from Locke through Broad and up to your case of Derek Jeter. (Often they pause at this point either to disparage or to applaud the Yankees more generally.) But the problem of motion perception is a good old empirical problem that we are making progress on. After all, we have known since S. Zeki's work in the early 1970s that a good number of the neurons in cortical area MT are direction-selective: They respond selectively to the movement of objects or fields of dots when they are traveling in one direction but are silent when they are traveling in a different direction.[36] There are lots of further details to be learned, of course, but in this work we have the foundation for an account of the neural basis of motion perception. Any further question about *how* the neurons in MT cause experiences of motion is nothing but another version of the hard problem. We admit that we have nothing to say about that, but after all, that was not our task anyhow. The claim that there is a further issue about temporal experience that we should be addressing but are not is simply unfair. So they say.

I think the scientists are being too modest here. There are further issues in the area, and if properly conceptualized, they can be pursued empirically. Let me try to say something about what these issues are.

An important clue comes from the so-called miracle cures. These are cases in which vision is restored to a patient who has been congenitally blind or blind from a very early age. The first and most well known patient of this sort is S. B., who was initially studied by Richard Gregory and Jean Wallace in the early 1960s.[37] S. B. had lost effective sight in both eyes as an infant due to corneal disease. At the age of 52, he received a corneal graft that restored the optical mechanisms in both eyes. There was naturally some start-up time in learning to deal with his new kind of

[36] See, e.g., Dubner and Zeki 1971 and Zeki 1974.
[37] See Gregory and Wallace 1963. The material from this monograph, plus some additional material, is reprinted in Gregory 1974. An overview discussion of this case and others like it can be found in Wandell 1995, pp. 388–390.

sensory input, but S. B. learned in remarkably short order to recognize various visual forms, such as uppercase letters and the face of a clock. (His success in these areas, by the way, may shed some light on the famous Molyneux Problem, at least if the usual claim about S. B. is correct. The usual claim is that he learns to recognize shapes as quickly as he does because he is transferring his tactile understanding of shapes into the new visual domain.) Perhaps even more surprising than his ability to recognize shapes, S. B. learned to identify colors by their names without much difficulty, though it should be noted that even when he was blind, he had the ability to detect the difference between light and dark.

Despite these great successes, however, it is the limitations of S. B.'s visual understanding that are important for us here. The central limitation was his difficulty, even after a long period of recovery, in dealing with visually presented moving objects.[38] He had difficulty both when the object was moving and he was standing still, and when the object remained stationary as he moved around it. As to the first of these, crossing the street became a terrifying affair for him since he could not track the motion of the cars. As to the second, it was very difficult for S. B., as he walked around an object, to integrate the various perspectival presentations of it into an experience of a unified thing. As Gregory writes:

Quite recently he had been struck by how objects changed their shape when he walked round them. He would look at a lamp post, walk round it, stand studying it from a different aspect, and wonder why it looked different and yet the same.[39]

The problem of seeing an object as unified throughout various perspectival presentations of it is precisely the kind of problem Kant was concerned with in his discussion of the threefold synthesis. As we saw, Kant believed that a certain kind of memory was needed in order to have this unified experience. Another miracle cure patient speaks to just this point. A. Valvo's patient H. S. described his difficulty in learning to read after his vision was restored:

My first attempts at reading were painful. I could make out single letters, but it was impossible for me to make out whole words; I managed to do so only after weeks of exhausting attempts. In fact, it was impossible for me to remember all the letters together, after having read them one by one. Nor was it possible for

---

[38] Also notable, and perhaps related, was his difficulty in recognizing depth.

[39] Gregory 1974, p. 111, quoted in Wandell 1995, p. 389. The question in what sense it *looked* the same for S. B. is an important one. The lamp post looks *to be* a single, unified object to us as we walk around it. But presumably this is just what is lacking is S. B.'s experience. It is very possible that the claim that it looked the same to S. B. is misleading.

me, during the first weeks to count my own five fingers: I had the feeling that they were all there, but . . . it was not possible for me to pass from one to the other while counting.[40]

Difficulties like these have led some researchers to posit a kind of short-term visual memory on the basis of which normal subjects are able to have experiences of moving objects of the sort that is extremely difficult after the miracle cure. This short-term visual memory is reminiscent of the kind of memory proposed by Kant and the kind of retention proposed by Husserl. As Brian Wandell writes:

To perceive motion, the visual system must be able to integrate information over space and time. To perform this integration, one needs a means of short-term visual storage that can be used to represent recent information and visual inferences. If this visual storage fails, perhaps because it did not develop normally during early blindness, motion perception will be particularly vulnerable.[41]

The kind of visual experience that requires this type of short-term visual storage seems to me to be quite distinct from the pure visual experience of motion. This is an empirical claim, but one that bears strongly on the Puzzle of Temporal Experience. I do not know of any empirical work that has been done in this area, and so it seems to me quite likely that the conceptual distinction between the pure visual experience of motion, on the one hand, and the kind of visual experience of a moving object that requires short-term visual storage, on the other, has not been made clear enough in the neuroscientific literature. Let me say a bit more about this distinction.

There is a kind of pure visual experience of motion that can be isolated in the so-called motion aftereffect (sometimes known more popularly as the waterfall illusion). To induce the motion aftereffect, the subject habituates to a constantly moving scene, like a waterfall, and then focuses his gaze on a stationary object. One has the very strong impression of motion in the direction opposite to habituation, although it does not look as though the stationary object *is moving*. We might think of this aftereffect as the pure visual experience of motion. The aftereffect shows that it is possible to have the pure visual experience of motion in the absence of the experience of an object as moving. Recent fMRI studies show that the neural basis of the aftereffect is principally to be found in area MT.[42]

---

[40] Valvo 197, quoted in Wandell 1995, p. 390.   [41] Wandell 1995, p. 390.
[42] Niedeggen and Wist 1998.

The pure visual experience of motion seems to me not to depend on anything like short-term visual memory. There is no question of needing to remember which unified object has the property of moving, since the experience is not as of *an object* moving at all. It would be very interesting to know, therefore, whether miracle cure patients are susceptible to the motion aftereffect. As long as there is no cortical damage to MT, it seems reasonable to think they will be. If that's right, then their inability to perceive objects as moving, or to perceive an object as identical throughout perspectival variations, would not be due to an inability to experience motion per se. Rather, it would be due to an inability to keep track, in some kind of short-term visual storage, of the thing that is moving and thereby to experience it as identical over time.

If this hypothesis is right, then the miracle cure patients are likely to be very different from patients who suffer from akinetopsia. Akinetopsia is the deficit, sometimes called motion blindness, in which a subject is unable to perceive objects as moving despite stationary objects remaining more or less visible. In the standard example, a patient suffering from akinetopsia is unable to pour a cup of coffee because the level appears frozen even while the cup is filling up. Similarly, a car moving toward such a patient will appear to be first far away and then all of a sudden much closer, without any experience of its moving in between. It is interesting that akinetopsia seems to be tied closely to a deficit in cortical area V5 (MT). It can be induced temporarily, for instance, by disabling V5 through transcranial magnetic stimulation (TMS).[43]

Such a deficit seems quite distinct from what is described by the miracle cure patients. The akinetopsic patient experiences a series of snapshots in each of which a unified object appears to be located successively in different places; the miracle cure patients, by contrast, have difficulty keeping track of the moving object as the same throughout variations. These two deficits seem to be dissociable: The akinetopsic patient has no trouble counting his fingers, for instance. If in addition the miracle cure patients are susceptible to the motion aftereffect, then we will be on the way to a double dissociation.

Admittedly, what I have offered is a set of empirical hypotheses. These are the kinds of hypotheses one can make from the armchair but can prove only with good empirical science. But they are hypotheses that are made on the basis of a conceptual distinction, between the pure visual experience of motion and the visual experience of an object as

---

[43] Beckers and Homberg 1992.

moving, that seems not to have been emphasized in the empirical literature. To be sure, the motion aftereffect itself highlights this distinction.
But until one thinks about the miracle cure cases in addition, it is all to
easy to think that activity in MT is all one needs to experience Derek
Jeter's throw as moving from shortstop to first. If that's what you think,
then the Puzzle of Temporal Experience really will seem like a version
of the hard problem of consciousness. But it's much more interesting
than that.

If it is right, as Kant and Husserl and Wandell all think, that some kind
of memory or retention or storage is required for the experience as of a
unified, temporally persisting object, then we can ask all sorts of further
empirical questions about the psychophysics of this kind of short-term
visual memory. Because of the considerations adduced in section 6, it
seems to me that this short-term visual storage is unlikely to be assimilable
to any familiar kind of memory. But what kind of experience is it, then?
Is it, for instance, a version of iconic memory?[44] It seems to me very
different from this as well. In iconic memory, a subject can retain for
short amounts of time a tachistoscopically presented visual image and
can read off some of its details after the fact. But the phenomenology of
short-term visual storage seems not to be like this at all. Rather, what the
miracle cure patients seem to be missing is tied much more directly to
our visual concept of an object as a unified entity persisting through time.
Can we isolate this kind of short-term visual memory and distinguish it
from other kinds of visual phenomena? That is an important empirical
question, and one which, if answered in the affirmative, would do an
enormous amount to turn the Puzzle of Temporal Experience into a
genuine scientific problem.

Acknowledgments

I have benefited, in writing this paper, from the feedback of many people. I
would like to thank audiences at MIT, Wake Forest, Georgetown, and Bryn Mawr,
as well as what seemed like the entire Psychology Department at Princeton. All
these audiences listened and responded, sometimes vociferously, to early drafts
of the chapter. I would like especially to thank Andy Brook and Kathleen Akins
for the wonderful conference they organized on Philosophy and Neuroscience
at Carleton University, at which I presented the original draft. In developing
the ideas in the chapter, and other related ideas that I hope will appear later, I
remember especially interesting discussions with Adrian Bardon, Michael Berry,

---

[44] Sperling 1960.

Dave Chalmers, Cheryl Kelly Chen, Bert Dreyfus, Adam Elga, Brian McLaughlin, Frank Tong, Ann Treisman, and Michael Tye. In addition to several very helpful discussions on this topic, I would like to thank my wife, Cheryl Kelly Chen, for everything else, not least of which includes all her help and support.

References

Augustine, *Confessions*. See esp. Book 11.

Baldo, M. V., and Klein, S. A. (1995) Extrapolation or attention shift? *Nature*, 378: 565–566.

Beckers, G., and Homberg, V. (1992) Cerebral visual motion blindness: Transitory akinetopsia induced by transcranial magnetic stimulation of human area V5. *Proc R Soc Lond B Biol Sci.*, 249(1325): 173–178.

Broad, C. D. (1923) *Scientific Thought*. Bristol: Thoemmes Press, 1993. Originally published in 1923. See esp. chapter 10, "Date and Duration."

Broad, C. D. (1938) *Examination of McTaggert's Philosophy*, Vol. II Part I. London: Cambridge University Press. See esp. chapter 35, "Ostensible Temporality."

Dainton, Barry. (2000) *Stream of Consciousness: Unity and Continuity in Conscious Experience*. London: Routledge.

Dennett, Daniel. (1991) *Consciousness Explained*. Boston: Little, Brown. See esp. the chapter on the perception of time.

Dubner, R., and Zeki, S. (1971) Response properties and receptive fields of cells in an anatomically defined region of the superior temporal sulcus in the monkey. *Brain Research*, 35: 528–532.

Eagleman, D. M., and Sejnowski, T. J. (2000) Motion integration and postdiction in visual awareness. *Science*, 287(5460): 2036–2038.

Eagleman, D. M., and Sejnowski, T. J. (2001) The flash-lag illusion: Distinguishing a spatial from a temporal effect, and why that matters for interpreting visual physiology. *Journal of Vision*, 1(3): 16a.

Eagleman, D. M., and Sejnowski, T. J. (2002) Untangling spatial from temporal illusions. *Trends in Neurosciences*, 25(6): 293.

Goodman, Nelson. (1951) *The Structure of Appearance*. Cambridge, MA: Harvard University Press.

Gregory, Richard L. (1974) *Concepts and Mechanisms of Perception*. London: Duckworth.

Gregory, Richard L., and Wallace, Jean G. (1963) *Recovery from Early Blindness: A case Study*. Experimental Psychology Society Monographs, no. 2.

Hume, David. (1739) *A Treatise of Human Nature*. L. A. Selby-Bigge (ed.). Oxford: Clarendon Press, 1978. Originally published in 1739. See esp. Book I, section III.

Husserl, Edmund. (1893–1917) *On the Phenomenology of the Consciousness of Internal Time*. John Barnett Brough (trans.). The Netherlands: Kluwer Academic Publishers, 1980. Lectures given between 1893 and 1917.

James, William. (1890) *Principles of Psychology*, Vol. I. New York: Dover Books, 1950. Originally published in 1890. See esp. chapter 15, "The perception of time."

Kant, Immanuel. (1781/1787) *Critique of Pure Reason*. Paul Guyer and Allen Wood (trans.). Cambridge: Cambridge University Press, 1998. Originally published in 1781/1787. See especially the A-deduction.

Köhler, Wolfgang. (1947) *Gestalt Psychology: An Introduction to New Concepts in Modern Psychology*. New York: Liveright.

Krekelberg, B., and Lappe, M. (2000) A model of the perceived relative positions of moving objects based upon a slow averaging process. *Vision Research*, 40: 201–215.

Le Poidevin, Robin (ed.). (1998) *Questions of Time and Tense*. Oxford: Clarendon Press.

Locke, John. (1690) *An Essay Concerning Human Understanding*. Peter H. Nidditch (ed.). Oxford: Oxford University Press, 1975. Originally published in 1690. See esp. Book 2, chapter 14.

Mellor, D. H. (1998) *Real Time II*. London: Routledge.

Merleau-Ponty, Maurice. (1945) *Phénoménologie de la Perception*. Paris: Gallimard.

Niedeggen, Martin, and Wist, Eugene R. (1998) The Physiologic Substrate of Motion Aftereffects. In *The Motion Aftereffect: A Modern Perspective*. George Mather, Frans Verstraten, and Stuart Antis (eds.). Cambridge, MA: MIT Press, pp. 125–155.

Nijhawan, Romi. (2002) Neural delays, visual motion and the flash-lag effect. *Trends in Cognitive Sciences*, 6(9): 387–393.

Nijhawan, Romi. (1994) Motion extrapolation in catching. *Nature*, 370: 256–257.

Palmer, Stephen E. (1999) *Vision: From Photons to Phenomenology*. Cambridge, MA: MIT Press.

Pöppel, Ernst. (1988) *Mindworks: Time and Conscious Experience*. Tom Artin (trans.). Boston: Harcourt, Brace, Jovanovich.

Purushothamen, G., et al. (1998) Moving ahead through differential visual latency. *Nature*, 396: 424.

Russell, Bertrand. (1913) *Theory of Knowledge: 1913 Manuscript*. New York: Routledge, 1999. Written in 1913. See esp. chapter 6, "On the experience of time."

Russell, Bertrand. (1921) *The Analysis of Mind*. London: G. Allen and Unwin. See esp. chapter 9, "Memory."

Sider, Theodore. (2001) *Four-Dimensionalism: An Ontology of Persistence and Time*. Oxford: Clarendon Press.

Sperling, G. (1960) The information available in brief visual presentations. *Psychological Monographs*, 74: 1–29.

Valvo, A. (1971) *Sight Restoration After Long-Term Blindness: The Problems and Behavior Patterns of Visual Rehabilitation*. New York: American Foundation for the Blind.

Wandell, Brian A. (1995) *Foundations of Vision*. Massachusetts: Sinauer Associates.

Whitney, David, and Murakami, I. (1998) Latency difference not spatial extrapolation, *Nature Neuroscience*, 1: 656.

Whitney, D., Murakami, I., and Cavanaugh, P. (2000) Illusory spatial offset of a flash relative to a moving stimulus is caused by differential latencies for moving and flashed stimuli. *Vision Research*, 40: 137–149.

Wolff, Robert Paul. (1963) *Kant's Theory of Mental Activity.* Cambridge, MA: Harvard University Press.

Zeki, S. (1974) Functional organisation of a visual area in the posterior bank of the superior temporal sulcus of the rhesus monkey. *Journal of Physiology,* 236: 549–573.

VISUOMOTOR TRANSFORMATION

7

Grasping and Perceiving Objects

Pierre Jacob

Introduction

There is today, or so I will argue in this chapter, a vast array of empirical evidence in favor of the 'two visual systems' model of human vision. Human beings are so visually endowed that they can see a wide variety of things. Some of the things that they can see are objects that they can also reach, grasp and manipulate with their hands. Many of the things that they can see, however, are not objects that they can reach and grasp.[1] As J. Austin (1962) pointed out, humans can see, for example, mountains, lakes, liquids, gases, clouds, flames, movies, shadows, holes, stars, planets, comets and events. Among events, humans can see behaviours or actions, some of which are performed by conspecifics. Some visible human actions are directed towards inanimate objects (e.g., actions of reaching and grasping an object or a tool). Others are directed towards animate objects, including animals and conspecifics. In P. Jacob and M. Jeannerod (2003), we argue that there is evidence for the view that the human brain contains two complementary networks that respond to the perception of respectively object-oriented actions and actions directed towards conspecifics.

In this chapter, I will restrict myself to the visual processing of objects that can be both perceived and grasped with a human hand. My goal will be to try to formulate an adequate version of the two visual systems model of human vision, which ought, I think, to be properly restricted to seeing objects that can be reached and grasped. At the heart of my version of

---

[1] Nor are actions that humans can perform reducible to manual actions of prehension.

the two visual systems model is the claim that the human visual system can process one and the same stimulus in two fundamentally different ways. It can give rise to a visual percept (or to perceptual awareness), and to a visuomotor representation, of one and the same object, according to the task. The challenge is to provide an explicit characterization of the differences between visual perception and visually guided actions or between perceptual and visuomotor representations of a given visual stimulus.

Now, in healthy human adults, the two systems of visual processing beat to a single drum. So only via carefully designed experimental tasks can one system be dissociated from the other. The two visual systems model of human vision is first and foremost a view about the anatomy of the human visual system. In the second section of the chapter, I will briefly recount the history of the two visual systems model of human vision. I believe that the two visual systems model of human vision runs counter to well-entrenched assumptions in the philosophy of perception, according to which the main purpose of human vision is visual perception. Thus, in the first section of the chapter, I will lay out my five main assumptions, some of which will be supported by empirical evidence that I will selectively review in subsequent sections, some of which is inconsistent with what I take to be orthodox views in the philosophy of perception. In sections 3, 4 and 5, I will review experimental evidence, respectively, from electrophysiological studies in the monkey, neuropsychological studies of human patients and psychophysical studies in healthy subjects. Finally, in section 6, I will argue that the perceptual awareness of the visual attributes of objects asymmetrically depends upon the representation of the spatial relations among two or more such objects.

1 Five Background Assumptions[2]

I am presently sitting on a chair in front of a table, facing an open window. On the table, I can see a tea-pot and a tea-cup next to it. I can see the shapes, contours, sizes, orientations, colors, textures of both the tea-pot and the tea-cup, and the level, the color and texture of the tea in the cup. The perception of the visual array containing the pair of objects on the table has a distinctive phenomenology: my visual percept makes me visually aware of a pair of objects with specific visual properties. The peculiar

---

[2] In the course of stating my 'assumptions', I shall offer some justification for them.

combination of visual properties bound in my percept is distinctive of the phenomenology of visual perception (by contrast with the phenomenology of experiences in other sensory modalities). Now, I can extend my right arm, reach and grasp the handle of the tea-cup with precision grip between my right thumb and index finger, bring the cup to my lips and drink a sip of tea from it. It may seem as if my visually guided action of reaching for, and grasping, the handle of the tea-cup is under the control of my fully detailed visual percept representing the shapes, contours, sizes, orientations, colors and textures of both the tea-pot and the tea-cup in front of me. It may seem so, but it is not, or so I will argue.

1.1 First Assumption

My first assumption (shared by many but not all philosophers) is that the detailed content of my visual percept outstrips my conceptual repertoire. I lack the detailed concepts and words in either French or English whose content could match the content of my visual experiences of the shapes, contours, sizes, orientations, colors and textures of the pair of objects in front of me. I will not rehearse but merely accept the arguments for the view that the conceptual content of my linguistically expressible thoughts and judgements fails to capture the fine-grainedness and the informational richness of the nonconceptual content of my visual percept.[3]

Conversely, there are things that one can think about but that one cannot see. For example, my visual percept can represent the cup as being to the right of the pot. It can represent the cup, the pot and their spatial relation. Now, I can form the *thought* that the cup is to the right of the pot *from my present perspective*, that is, facing the open window. I can also form the thought that the cup is to the left (not to the right) of the pot for an observer sitting across the table from me with his or her back turned against the window. I can think that were I to sit across the table, I would see the cup to the left of the pot. But strictly speaking, I cannot

---

[3] See, e.g., Dretske (1981), Evans (1982), Peacocke (1992, 1998, 2001), Tye (1995) and Bermudez (1998). For criticism, see McDowell (1994). Here, I merely state my view that demonstrative concepts fail to jointly satisfy requirements on concepts and to match the fine-grainedness and informational richness of visual percepts. If they match the fine-grainedness and informational richness of visual percepts, then they fail to satisfy conceptual requirements. If they satisfy conceptual requirements, then they fail to match the fine-grainedness and informational richness of visual percepts. See Jacob and Jeannerod (2003, ch. 1) for details.

see the cup as being to the right of the pot from my current perspective, let alone see the cup as being to the left of the pot from the opposite perspective. I cannot see the perspective from which (or by means of which) I can see the cup as being to the right of the pot while I occupy that very position. Of course, someone else can see the perspective in question. But I cannot. Granted, at $t + 1$, I can see both the cup to the left of the pot and the position I occupied at t after I have switched position from t to $t + 1$. But at t, when I see the cup to the right of the pot, I can think but I cannot see the position by means of which I see the cup to the right of the pot.

1.2 Second Assumption

Since (by the first assumption) the conceptual content of thoughts and linguistic utterances cannot match the fine-grainedness and informational richness of the nonconceptual content of visual percepts, many philosophers (e.g., O'Shaughnessy, 1992, and Peacocke, 1992) have linked the nonconceptual content of visual percepts to the requirements of bodily actions. In particular, they have assumed that the distinctive fine-grainedness and informational richness of the nonconceptual content of visual percepts are attuned to the on-line guidance of an agent's bodily movements.

According to the assumption which A. Clark (2001: 496) labels the assumption of 'experience based-control' (or EBC), the function of the detailed content of visual percepts is to guide and monitor bodily movements:

> [C]onscious visual experience presents the world in a richly textured way . . . that presents fine detail . . . (that may exceed our conceptual or propositional grasp) and that is, in virtue of this richness, especially apt for, and typically utilized in, the control and guidance of fine-tuned, real-world activity.

I reject EBC because it is, I think, inconsistent with the version of the two visual systems model of human vision to which I subscribe. I also reject J. K. O'Regan and A. Noë's (2001) version of the enactive theory of perceptual content, according to which the detailed pictorial content of a creatures' visual experience is constituted by their implicit (or practical) knowledge (or mastery) of sensorimotor contingencies. As I shall argue (in section 4.3), the double dissociation between perceptual and visuomotor responses in brain-lesioned human patients, which is evidence for the two visual systems hypothesis, is also, on my view, evidence against the enactive theory.

Instead, I accept what Clark (2001) calls the assumption of 'experience-based selection' (EBS), according to which the fine-grainedness and informational richness of the nonconceptual content of visual percepts contribute to an agent's *selection* of the target of his or her motor activity. In accordance with EBS, my second assumption is, in M. Goodale and G. K. Humphrey's (1998: 185) terms, that the nonconceptual content of visual percepts is not 'linked directly to specific motor ouputs but [is] linked instead to cognitive systems involving memory, semantics, spatial reasoning, planning, and communication'. Thus, I assume that the nonconceptual content of visual percepts is poised for further conceptualization. In other words, visual percepts of objects provide the 'belief box' with visual information about objects. But a visual percept is not ipso facto a belief (or a judgement) with conceptual content.

1.3 Third Assumption

The computational requirements made by the perceptual system on human vision are clearly different from the requirements made by the visual monitoring of object-oriented actions. The task of the former is to encode pictorial visual information about the enduring properties of objects that will facilitate identification or recognition from many different perspectives on many different occasions in which such things as the lighting, the distance from the objects, the relative motions of objects and the observer's own movements may vary considerably.[4] Recognition of a perceived object requires that visual information match conceptual information stored in long-term memory. The task of the latter is both to encode the geometrical properties of the target relevant for prehension (such as size, shape and orientation) and to keep track of the spatial position of the target relative to the moving agent. As I will argue, it is of the essence of a visual percept that it represents objects in a visual array together with their spatial relations. Perceptual processing makes

---

[4] By contrast with the conceptual content of thoughts, the content of visual percepts is pictorial in the sense that several visual attributes are always bound together in a visual percept. Philosophers sometimes misleadingly assume that enduring properties cannot be part of the content of representations with pictorial content. This is a mistake. Consider shape, which is uncontroversially an enduring property of an object. What makes a visual perceptual representation of shape pictorial is that it cannot represent the shape of an object without representing other properties (e.g., size and color) of the object. You can think of a triangle without thinking of its color or size. But in good lighting conditions, healthy observers cannot visually perceive a triangular object unless they perceive its color, size and orientation. Nor can they visually image a triangular object without imaging its color, size and orientation.

me visually aware of the tea-cup, the tea-pot and their spatial relation, for example, the former's being to the right of the latter. By contrast, for me to grasp the handle of the tea-cup and lift the cup off the table, my visual system must have computed two landing sites on the handle where to apply my right thumb and index finger. But I am not visually aware of these points. Furthermore, in order to reach the cup and grasp its handle, I must represent its location relative to my body, not relative to the tea-pot, unless the latter is on my way to the former.

According to my second assumption, conscious visual experience is involved in the selection of the target of an object-oriented action, not in the monitoring of the bodily action itself. According to my third assumption, what is crucial to the visual awareness of objects provided by visual perception is the representation of the spatial relations among constituent objects in a visual array. As we shall see, there is much prima facie paradoxical evidence that accurate, visually guided actions do not require visual full awareness of the target.

1.4 Fourth Assumption

My fourth assumption is that much of human object-oriented behavior (such as pointing to, or reaching for and grasping, a target) is intentional. On the one hand, much of the visual behavior of humans and nonhuman animals (such as the pupillary light reflex, the synchronization of circadian rythms and the visual control of posture) is reflexive behavior, not intentional behavior. On the other hand, humans also perform many acts that are neither reflexive nor intentional, and which O'Shaughnessy (1980: 58–73) labels 'sub-intentional acts', such as the movement of one's tongue in one's mouth as one reads, the movements of one's fingers as one is attending to a philosophy lecture or the tapping of one's feet to the rhythm of a piece of music. Such subintentional acts are not reflexive behaviors: unlike the dilation of one's pupil, an agent can interrupt the tapping of feet to the rhythm of the music if requested to do so. They are not intentional since the movements are not intended by the agent, let alone produced by his or her beliefs and desires. By contrast, human movements of prehension are caused by the agent's intentions, and they are monitored by a visuomotor representation of the target.

1.5 Fifth Assumption

In accord with G. E. Anscombe (1957) and J. R. Searle (1983), I assume that perceptions and beliefs have a mind-to-world direction of fit. If their content matches a fact, then they are veridical. If it does not, then they are not. I assume that intentions and desires have a world-to-mind direction of fit. Their function is not to match a fact but to represent a goal, that is, a nonactual state of affairs. In addition, percepts have a world-to-mind direction of causation: A percept is caused by the fact that it represents. Intentions have a mind-to-world direction of causation: An intention causes the obtaining of the state of affairs that it represents. R. G. Millikan (1996) has interestingly argued for the existence of Janus-like representations with both a mind-to-world and a world-to-mind direction of fit, which she calls 'pushmi-pullyu' representations. She argues convincingly that such representations with a twofold direction of fit are evolutionarily more primitive than representations with either a purely descriptive or a purely prescriptive direction of fit. My fifth assumption is that visuomotor representations are such hybrid mental representations with a twofold direction of fit. They represent what J. J. Gibson (1979) called affordances for manual actions. Thus, they provide motor intentions with visual information about the properties of objects relevant for actions of prehension.

2 A Brief History of the Two Visual Systems Hypothesis

Early versions of the two visual systems hypothesis were first entertained by neurophysiologists working on the visual systems of nonhuman animals. In nonmammalian vertebrates, for example, in amphibians, it was demonstrated by D. J. Ingle (1973) that prey-catching behavior is mediated by retinal projections onto the optic tectum, while the visual control of barrier-avoidance is mediated by retinal projections onto pretectal nuclei. Similarly for mammalians, it was demonstrated by G. E. Schneider (1967) that a hamster with a lesioned superior colliculus could discriminate vertical from horizontal stripes but could not run a maze. Conversely, a hamster with a lesioned visual cortex could run a maze but not do pattern recognition. Since the earlier evidence came from the study of animals with little or no visual cortex, early versions of the two visual systems hypothesis emphasized the contrast between, on the one hand, 'ambient' vision controlled by peripheral retinal information, based on subcortical structures, and, on the other hand, 'focal' vision based on cortical structures.

A first major step was taken by L. Ungerleider and M. Mishkin (1982), who located the two visual systems within the primate visual cortex. More than 30 specialized areas with many reciprocal connections have been discovered in the visual cortex of primates. But the connections turn out to be segregated into two main streams of processing: the ventral stream and the dorsal stream. The former projects the primary visual cortex (area V1) onto the inferotemporal area. The latter projects the primary visual cortex onto the posterior parietal lobe, itself a relay on the way towards the premotor cortex and the motor cortex (see Figure 7.1).

Ungerleider and Mishkin (1982) examined the selective effects of lesions in the brains of macaque monkeys on two kinds of behavioral tasks: a landmark task and an object-discrimination task. In the former task, the monkey had to discriminate between two covered wells – one empty and one full – according to whether they were located far away or near a landmark. In the latter task, the monkey had to discriminate two objects of different shapes, colors and textures. Ungerleider and Mishkin found that a lesion in the inferotemporal cortex severely impaired the animal in the object-discrimination task, but not in the landmark task. Conversely, they found that a lesion in the posterior parietal cortex severely affected the animal's performance in the landmark task, but not in the object-discrimination task. On the basis of these experiments, they concluded that the ventral stream is the *What*-system (or 'object-channel') and the dorsal stream is the *Where*-system (or 'space-channel'). Importantly, the landmark task tested the animal's ability to *perceive* spatial relations, not to act on a target. Thus, on Ungerleider and Mishkin's version of the two visual systems in the primate brain, the predominant function of both the ventral and the dorsal pathways is to mediate visual *perception*: While the former is involved in the perception of so-called intrinsic attributes of an object, such as its shape, size, texture and color, the latter is involved in the perception of such 'extrinsic' properties of objects as their spatial relations. On this view, visuomotor functions involved in guiding reaching and/or grasping movements are relegated to subcortical structures.

A second major step was taken by M. A. Goodale and A. D. Milner (1992) and Milner and Goodale (1995) when they provided room for the visuomotor transformation within their amended version of the two visual systems model of human vision.[5] On their revised view, the ventral stream underlies what they call 'vision-for-perception' and the dorsal stream underlies what they call 'vision-for-action'. The crucial evidence on which Goodale and Milner based their revised interpretation of the two visual systems model of human vision is the neuropsychological double dissociation between two visual impairments produced by two selective lesions in the human visual system: Whereas apperceptive visual agnosia results

---

[5] The visuomotor transformation is the network in the human brain whereby visual information is converted into motor commands of arm and hand movement towards objects.

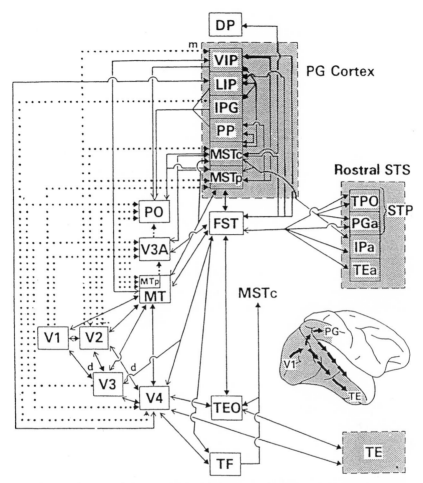

FIGURE 7.1. A view of the anatomical bifurcation between the ventral and the dorsal pathways in the primate visual system originating from the primary visual cortex (areas V1 and V2). The ventral pathway projects towards the inferotemporal cortex (areas TE and TEO). The dorsal pathway projects towards the parietal cortex. Many of the connections are reciprocal. *Source*: Reprinted with permission of Wiley-Liss, Inc., a subsidiary of John Wiley & Sons, Inc., from D. Boussaoud, L. Ungerleider, and R. Desimone (1990), Pathways for motion analysis: Cortical connections of the medial superior temporal sulcus and fundus of the superior temporal visual areas in the macaque monkey, *Journal of Comparative Neurology*, 296, 462–495, copyright © 1990 C. Wystar Institute of Anatomy and Biology, Wiley Interscience.

from lesions in the inferotemporal area, optic ataxia results from lesions in the posterior parietal cortex. Apperceptive visual agnosic patients are deeply impaired in the visual recognition of the size, shape and orientation of objects. But they can reach and grasp objects whose shapes, sizes and orientations they cannot visually recognize. Conversely, optic ataxic patients cannot reach and grasp objects whose shapes, sizes and orientations they can visually recognize.[6]

Thus, in the mid-1990s, the two major versions of the two visual systems model of human vision disagreed on the functional significance of the dorsal pathway and the role of the posterior parietal lobe. Ungerleider and Mishkin's (1982) model subscribes to the assumption that the major function of the primate visual system is to allow visual perception: all cortico-cortical connections in the primate visual brain underly perceptual awareness. By contrast, according to Milner and Goodale's (1995) model, neither is perceptual awareness the exclusive (or the main) function of vision in primates nor are cortico-cortical pathways in the primate and the human brains limited to visual perception. Unlike Ungerleider and Mishkin (1982), Milner and Goodale (1995) locate the visuomotor transformation firmly within the visual cortex of nonhuman primates and humans.

3 Evidence for Dual Visual Processing in the Monkey

I will briefly review a sample of electrophysiological recordings of single cells in the brain of either anaesthetized or awake monkeys that fall under two basic categories: cells in the ventral pathway of the monkey visual system that are involved in the discrimination of basic visual attributes of complex stimuli such as their colors, textures, shapes and contours; and cells in the posterior parietal cortex that respond to the geometric properties of objects that are relevant for tasks of prehension.

3.1 Perceptual Cells in the Inferotemporal Area
K. Tanaka et al. (1991) recorded cells in the posterior part of the inferotemporal cortex of macaque monkeys that respond preferentially to relatively simple visual stimuli, such as elongated bars or small colored disks. Other cells in this area fire in response to the detection of some favored shape (e.g., an eight-arm star as opposed to a circle or a diamond). Still others fire in response to a complex combination of shape, orientation and texture (e.g., a triangle whose apex is directed towards the

---

[6] See section 4 for detailed examination of the double dissociation between optic ataxia and apperceptive visual form agnosia.

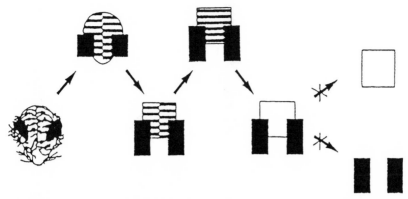

FIGURE 7.2. Single cell recorded in inferotemporal area of the brain of a macaque monkey. The neuron responds to the sight of a tiger's head viewed from above (as represented in lower left). The ordered sequence of arrows shows the progressive simplification of the stimulus that keeps triggering the discharge of the neuron. It still responds to a white square flanked by two black rectangles. Further reduction fails to trigger cell response. *Source*: Reprinted with permission from K. Tanaka (1993), Neuronal mechanisms of object recognition, *Science*, 262, 685–688, Fig. 1, copyright © 1993 AAAS.

right with vertical stripes). In a more anterior part of the inferotemporal cortex, Tanaka (1993) found more elaborate cells that are sensitive to the contours of complex stimuli, such as brushes, leaves and other biological objects. One such cell responded preferentially to the head of a tiger seen from the top. Tanaka found that the cell keeps responding to a series of simplifications of the stimulus until it is reduced to a combination of a pair of symmetrical black rectangles flanking a white square. It stops responding to either the pair of black rectangles or the white square alone (see Figure 7.2).

3.2 *Visuomotor Cells in the Posterior Parietal Lobe*

In the 1970's, V. B. Mountcastle et al. (1975) recorded cells in the posterior parietal cortex of awake macaque monkeys. They showed that unlike inferotemporal cells, cells in the posterior parietal cortex covary with many aspects of the animal's visual behavior, such as saccadic eye movements, head movements and the monitoring of reaching towards, and manipulating, objects. H. Sakata et al. (1995) and A. Murata et al. (2000) recorded neurons in the anterior intraparietal (AIP) area of the awake monkey. They showed three-dimensional black and white objects with well-demarcated geometrical properties: a plate, a ring, a cube, a cylinder, a cone and a sphere (see Figure 7.3). Four different conditions were

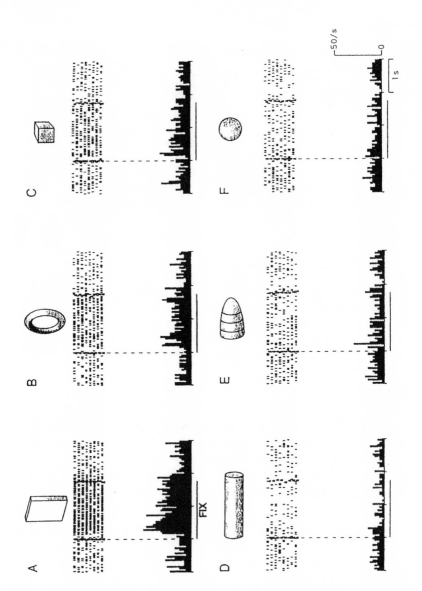

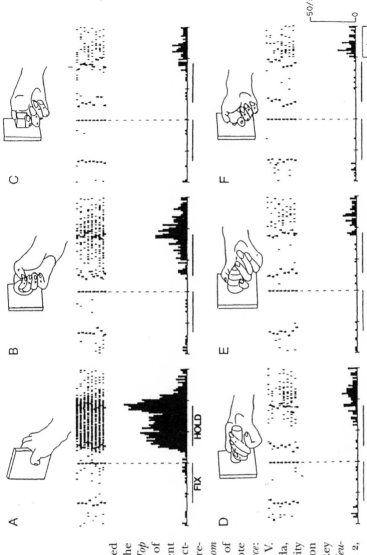

FIGURE 7.3. Single cell recorded in the intraparietal sulcus of the brain of a macaque monkey. *Top half* shows different responses of a cell to objects with different geometrical properties in object-fixation task. Note maximal response in A and lowest in D. *Bottom half* shows different responses of a cell in a visuomotor task. Note maximal response in A. *Source:* Reprinted from A. Murata, V. Gallese, G. Luppino, M. Kaseda, and H. Sakata (2000), Selectivity for the shape, size, and orientation of objects for grasping in monkey parietal area AIP, *Journal of Neurophysiology*, 79, 2580–2601, Fig. 2, used with permission.

253

compared, two of which involved an object-manipulation task (with or without a delay and with or without light), and two of which consisted of an object-fixation task (either in the presence or in the absence of light). AIP neurons turned out to divide into three groups: The visual dominant neurons fired preferentially when the animal manipulated the object in the light, not in the dark. The motor dominant neurons fired when the animal manipulated the object either in the light or in the dark. The visual-and-motor dominant neurons fired more when the animal manipulated the object in the light than in the dark. Thus, AIP neurons are sensitive primarily to the geometric affordances of objects relevant for grasping and manipulating them.

Several experiments in the monkey have further showed that reaching and grasping can be selectively impaired. For example, Gallese et al. (1994) managed to cause a reversible impairment of the control of grasping that did not affect reaching. This experiment confirmed Mountcastle et al.'s (1975) distinction between 'arm projection' neurons and 'hand manipulation' neurons.

4 Double Dissociations in Brain-Lesioned Human Patients

As indicated in section 2, much of Milner and Goodale's (1995) version of the two visual systems model of human vision relies on the double dissociation between apperceptive visual form agnosic patients and optic ataxic patients. Before examining this dissociation in more detail, it is, however, necessary to review the psychophysical properties of movements of prehension in healthy human subjects.

4.1 Reaching and Grasping in Healthy Human Subjects

The articulation of their upper limbs and the dexterity of their hands allow nonhuman primates and especially humans to execute actions of prehension and the manipulation of objects. The human hand allows two basic kinds of grip: precision grip and power grip (see Jeannerod, 1997). The former is involved in grasping the handle of a tea-cup. The latter is involved in grasping a hammer. In the early 1980s, it was found that an action of prehension involves three major components or phases: a selection phase, a reaching phase and a grasping phase. During the selection phase, which is the perceptual part of the process, the target of prehension must be sorted out from a number of possible competitors and distractors on perceptual grounds. The rest of the process is an automatic visuomotor process in which the visual information is exploited for the purpose of action.

The reaching part of the process of prehension involves ballistic movements of the different segments of the upper limbs connecting the hand to the shoulder. It was found that during transportation of the hand to the target, there is an automatic process of grip formation: The preshaping of the fingers is programmed much before the hand contacts the object. During transportation of the hand, first the finger-grip opens up and then closes down on the target. The opening reaches its peak – maximum grip aperture (or MGA) – at about 60% of the movement of transportation. It was found that the size of the finger-grip both is much larger than the size of the target and is linearly correlated with it. Calibration of the finger-grip does not require perceptual comparison of the hand with the target: it is made automatically on the basis of a visuomotor representation of the target without visual access to the hand (in so-called open loop condition).

4.2 Visuomotor Impairments Following Lesions in the Dorsal Stream

Optic ataxia (also called Balint syndrome) is a disturbance in visually guided actions of pointing towards, and prehension of, objects in patients who have otherwise no purely motor impairment. It is subsequent upon a lesion generally located in the superior parietal lobe that can be located in either hemisphere and that causes impairment in movements of the hand contralateral to the lesion. The disturbance can either affect transportation of the hand towards the target or formation of the finger-grip appropriate for grasping or affect both (see Figure 7.4).

Patient AT (examined by Jeannerod et al. 1994; see also Jeannerod, 1997) presented a bilateral optic ataxia. She was impaired in the accuracy of both her reaching and grasping movements. In fact, she was far more severely impaired in grasping than reaching movements. In particular, close examination revealed grossly exaggerated grip aperture. By contrast, her perceptual judgements were normal, as she could recognize and identify objects visually presented to her. Jeannerod et al. (1994) used the same effector in both a visuomotor task and a perceptual estimation task: the distance between her thumb and index finger. They found that while her grip aperture was severely inadequate in a task of grasping an object, the distance between her thumb and index finger was positively correlated with the size of objects in a task of perceptual estimation. Also

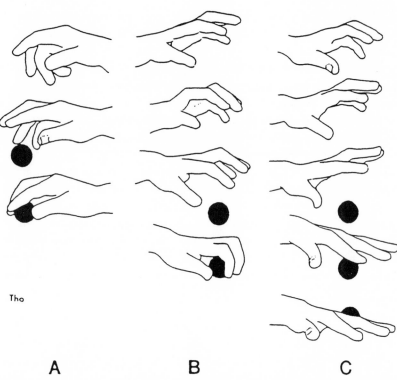

Tho

A   **B**   **C**

FIGURE 7.4. (A) shows unfolding of hand movement in grasping task in healthy subject. (B) shows abnormal pattern of opening and closure of fingers in grasping task in optic ataxic patient. (C) shows lack of both preshaping of finger grip and misreaching in reaching task in optic ataxic patient. *Source*: Reprinted from M. Jeannerod (1986), The formation of finger grip during prehension: A cortically mediated visuomotor pattern, *Behavioural Brain Research*, 19, 99–116, copyright 1986, with permission from Elsevier.

when asked by Milner et al. (1999) to locate a source of light by pointing her index finger, her performance improved significantly when tested after a 5 second delay between the occurrence of the signal and the initiation of the action.

4.3 Perceptual Impairments Following Lesions in the Ventral Stream

According to M. J. Farah (1990), there are two grades of visual form agnosia, both caused by lesions in the ventral pathway of the human visual system: associative agnosia and apperceptive agnosia. The former, caused at a later stage of visual processing within the ventral stream, results, in H. L. Teuber's (1960) phrase, in a 'normal percept stripped of its meaning'.[7] The patient can process the local shapes of objects but cannot identify and recognize them. The latter, caused at an earlier stage of visual processing within the ventral stream, deprives the patient of the ability to perceive and thus be visually aware of the shapes, sizes and orientations of objects altogether.

Possibly, the distinction between the two grades of perceptual impairment in visual form agnosia could be matched onto F. Dretske's (1969, 1978) distinction between two distinct stages of normal visual perception: nonepistemic visual perception and primary epistemic visual perception. Suppose, for example, that you are driving too fast to identify some obstacle lying on your trajectory, but you skillfully manage to avoid hitting it. Unless you saw it, you would have hit it. Since you avoided it, you did see it. But since you failed to identify it, your perception of what you avoided was nonepistemic.[8] Visual identification of an object yields perception of something as instantiating some property or other (or as falling under some concept or other). What Dretske (1969) calls 'primary epistemic perception' consists in seeing a fact involving a perceived object. For example, seeing that the cat is on the mat by seeing the cat is primary epistemic perception: Unless you possess the concepts of a cat and of a mat, you could not see that the cat is on the mat.[9] You could not form the belief that the cat is on the mat by visual means. Arguably, apperceptive visual form agnosics could not come to believe that the cat is on the mat by seeing a cat lying on a mat: Visually presented with a cat, they could not assemble its local shapes into the contour of a cat. In apperceptive visual form agnosic patients, earlier stages of nonepistemic perception of objects involving the assemblage of the elementary visual attributes into a percept of an object would be impaired. In associative visual form agnosic patients, later stages of primary epistemic perception of facts involving

---

[7] Hence, Jeannerod's (1994, 1997) contrast between the 'semantic' and the 'pragmatic' processing of visual information. In visual form agnosic patients, the semantic processing is impaired. In Jacob and Jeannerod (2003), we further distinguish a lower level and a higher level of pragmatic processing of objects. The former is involved in the visuomotor transformation and is impaired in optic ataxic patients. The latter is involved in the skilled manipulation of tools and is impaired in apraxic patients (with a lesion in the left inferior parietal lobe).

[8] Here I am assuming that a driver is forming a visual percept, not a visuomotor representation, of an obstacle lying on the road on which he or she is driving.

[9] In addition, what Dretske (1969, ch. 3) calls 'secondary' epistemic visual perception is visually based belief about a fact involving an unperceived object, for example, seeing that the tank in your car is full by seeing not the tank but the gas gauge.

the matching between visual percepts and concepts of objects would be impaired.

It was an important finding to discover the residual visuomotor capacities of an apperceptive agnosic patient DF, first examined by Milner et al. (1991). This finding is reminiscent of the earlier finding of residual visuomotor capacities in blindsight patients after a lesion in their primary visual cortex (see Weiskrantz, 1997). DF suffered a bilateral occipital lesion destroying a large part of her ventral pathway as a consequence of carbon monoxide poisoning. When asked to name the shapes of geometrical objects and when asked whether two shapes were the same or different, DF's ability to get the answer right was not greater than chance. When asked to match the size of an object by scaling the distance between the index finger and the thumb of her right hand, her answers were likewise no better than chance. When asked to report the orientation of a line (or a slot) or to match it by turning a hand-held card so as to match the perceived orientation of the line (or slot), her performance was very poor. DF's visual imagery was preserved: Although she could not copy a visually presented object, she could draw one from memory. Finally, when asked to locate a source of light without a delay, DF's performance was normal and it deteriorated after a 2 second delay (see Goodale et al., 1991, Goodale et al., 1994, Goodale, 1995, and Milner and Goodale, 1995: 136–137).

By contrast with her deep perceptual impairment, DF was surprisingly accurate when the shape, size and orientation of an object had to be processed in the context of a goal-directed hand action. During reaching and grasping of objects between her index finger and her thumb, DF was able to perform prehension movements with the very same objects whose shapes, sizes and orientations she could not visually recognize. While transporting a hand-held card in order to insert it into a slot at different orientations around the clock, she could normally turn her wrist and orient her hand (see Goodale et al., 1991, Goodale et al., 1994, Goodale, 1995, Milner and Goodale, 1995, and Carey et al., 1996) (see Figure 7.5). DF's visuomotor capacities, however, turn out to have interesting perceptual limitations. For example, when asked to insert a T-shaped object into a T-shaped aperture, she had trouble matching the orientation of the stem with the orientation of the top. As noted by Goodale (1995: 197), it suggests that a dedicated system for fast and accurate visuomotor behavior (such as grabbing a branch) may need 'to coopt more flexible perceptual systems' when required to process the several axes of symmetry of a complex object.

The double dissociation between apperceptive visual form agnosia and optic ataxia is hard to square with O'Regan and Noë's (2001) and Noë's

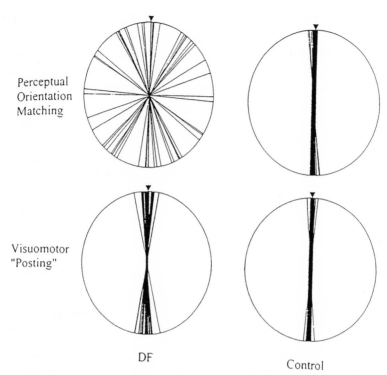

Perceptual Orientation Matching

Visuomotor "Posting"

DF

Control

FIGURE 7.5. The top two circles show the performances of an apperceptive agnosic patient (*left*) and a healthy subject (*right*) in a perceptual task in which the subject is asked to match the orientation of a slot by orienting her wrist over successive trials. The bottom circles show the performances of an apperceptive agnosic patient (*left*) and a healthy subject (*right*) in a visuomotor task in which the subject is asked to insert a hand-held card through a slot at different orientations. *Source*: Reprinted with permission from M. A. Goodale (1995), The cortical organization of visual perception and visuomotor control, in D. Osherson, *An Invitation to Cognitive Science: Visual Cognition*, vol. 2 (Cambridge, Mass.: MIT Press), copyright © 1995 Massachusetts Institute of Technology.

(2002) version of the enactive theory of perceptual content. On the one hand, optic ataxia shows that practical knowledge of some basic sensorimotor contingencies is not a necessary condition of perceptual identification and recognition of visually presented objects. Although we may not be too sure about the phenomenology of her visual experience,[10] patient AT is quite able to correctly identify and recognize objects presented

---

[10] As Kathleen Akins reminds me. In addition to her optic ataxia, patient AT also happens to have dorsal simultagnosia that impairs her recognition of more than one object at a time. But according to Milner et al. (1999), her simultagnosia is separable from her optic ataxia.

to her in the visual modality, but she is deeply impaired in visuomotor behaviours of reaching for, and grasping, objects. Impairment in reaching and grasping should presumably, according to the enactive theory, alter patient AT's ability to perceptually recognize visual objects. But it does not. On the other hand, apperceptive visual form agnosia shows that knowledge of some basic sensorimotor contingencies is not a sufficient condition for perceptual experience and recognition of visually presented objects. Patient DF has kept pretty remarkable visuomotor capacities. But she is nonetheless deprived of the perceptual experience of, and the ability to recognize the shapes, sizes and orientations of, visually presented objects. If reaching and grasping an object do not constitute the sensorimotor contingencies the knowledge of which should count for the perceptual experience of the object, I wonder which are the sensorimotor contingencies the knowledge of which could underlie perceptual experiences.[11]

5 Dissociations Between Perception and Action in Healthy Human Subjects

Milner and Goodale's (1995) version of the two visual systems model of human vision was elaborated mostly on the basis of the double dissociation between optic ataxia and apperceptive visual form agnosia. The question arises: Is the model consistent with what is known of the psychophysical responses of normal human subjects? This is the question addressed in the present section. I will start with dissociations between perceptual and visuomotor responses when the latter task consists in pointing towards a visible target with one's index finger. Pointing to a target involves the visual computation of the location of the target in 'egocentric coordinates', that is, relative to the agent's body. Then I shall examine dissociations between perceptual and visuomotor responses when the relevant task consists in grasping a target. Grasping an object involves processing the position of the object in egocentric coordinates and such attributes of the object as its shape, size and orientation.

---

[11] I am reminded of the molecular biologist Jacques Monod's reaction to a claim made by the Piagetian psycholinguist Bärbel Inhelder that 'sensorimotor intelligence' is crucial to language acquisition, at the 1975 Royaumont Conference with Noam Chomsky and Jean Piaget. Monod observed that this claim could be tested by examining the acquisition of language in children with serious motor disabilities (see Piattelli-Palmarini, 1980). In response, Inhelder so weakened her claim that it became untestable.

5.1 Pointing to a Target and Perceiving It

B. Bridgeman et al. (1979) and Goodale et al. (1986) have exploited the phenomenon of saccadic suppression. When the motion of a target coincides with a saccadic eye movement, a healthy human subject is not consciously aware of the motion of the target. Bridgeman et al. (1979) and Goodale et al. (1986) found that normal human subjects can point accurately to a target on the screen of a computer whose motion they could not consciously notice because it coincided with one of their saccadic eye movements (see Bridgeman, 2002). U. Castiello et al. found that subjects were able to correct the trajectory of their hand movement directed towards a moving target some 300 milliseconds before they became conscious of the target's change of location (see Jeannerod, 1997: 82).

L. Pisella et al. (2000) and Y. Rossetti and Pisella (2002) performed experiments involving a pointing movement towards a target. Subjects were presented with a green target towards which some of them were requested to *stop* their pointing movement when and only when they saw the target change location by jumping either to the left or to the right. Among the responses of the subjects whose task was to *stop* pointing, they found a significant percentage of very fast unwilled correction movements towards the target in response to a change of location to the left or to the right. Pisella et al. (2000) and Rossetti and Pisella (2002) called the mechanism responsible for these fast unwilled correction movements the 'automatic pilot' for hand movement. In a second experiment, Pisella et al. (2000) presented subjects simultaneously with pairs of a green and a red target. They were instructed to point to the *green* target, but the color of the two targets could be interchanged unexpectedly at movement onset. Unlike a change of target location, a change of color did not elicit fast unwilled corrective movements by the automatic pilot. Unlike motion, which is processed by brain area MT, which belongs to the dorsal pathway, color is processed in areas that are more ventrally located. On this basis, Pisella et al. (2000) drew a contrast between the fast visuomotor processing of the location of a target in egocentric coordinates and the slower visual processing of the color of an object. In an optic ataxic patient with a lesion in the superior parietal lobe, the fast automatic corrections for changes of target-location were missing. The authors conclude that the automatic pilot that produces fast corrections of hand movements by normal subjects in response to target-motion is located in the superior parietal cortex.

In an experiment by M. Gentilucci et al. (1996), subjects sat in front of a table on which they could see three possible stimuli: one of the two

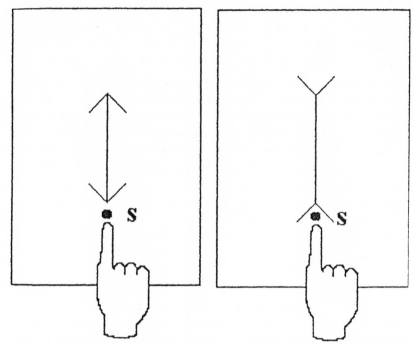

FIGURE 7.6. Subjects are asked to move their index finger from the closer to the more distant vertex of a Müller-Lyer segment. *Source*: Adapted from Gentilucci et al., 1996.

segments of the Müller-Lyer size-contrast illusion with opposing arrow configurations or a control segment without arrows (see Figure 7.6). Subjects positioned their right index finger towards the nearest vertex of the displayed segment. They were asked to point their index finger from its initial position towards the more distant vertex of the segment. Gentilucci et al. compared four conditions:

- in the 'full-vision' condition, subjects could see both their hand and the target during the action;
- in the 'nonvisual feedback condition', they could see the target, but not their hand;
- in the 'no-vision' condition, they could see neither their hand nor the target during the action;
- in the 'no-vision 5 second delay' condition, they could see neither their hand nor the target during the action, which started 5 seconds after the light went off.

Gentilucci et al. found an increasing effect of the Müller-Lyer illusion on the pointing movement from the first to the fourth condition. When subjects can see both their hand and the target, they code its location within a visuomotor representation of the target, in which the target's location is coded in egocentric coordinates, relative to either their body or their hand. When subjects build such a visuomotor representation of the target, they do not need to represent the full segment at all. The target is just a point in egocentric space at which to direct one's index finger. A fortiori, the arrow configurations surrounding the vertices of a Müller-Lyer segment are not part of the visuomotor representation of the target. By contrast, when subjects see neither their hand nor the target, they code the location of the target as the farthest vertex of a segment. Thus, they build a perceptual representation of the length of the segment and the orientations of the arrows become relevant.

5.2 *Grasping an Illusory Display*

In standard displays of the Titchener (or Ebbinghaus) size-contrast illusion, two circles of equal diameters are surrounded by two annuli of circles either smaller or larger than the central circle. The central circle surrounded by an annulus of circles smaller than it is looks larger than the central circle surrounded by an annulus of circles larger than it is (see Figure 7.7). Pairs of unequal circles can also be made to look equal. Aglioti et al. (1995) replaced the 2-D central circles by 3-D graspable disks. In a first row of experiments with pairs of unequal disks whose diameters ranged from 27 mm to 33 mm, they found that on average, the disk in the annulus of larger circles had to be 2.5 mm wider than the disk in the annulus of smaller circles in order for both to look equal. Aglioti et al. alternated presentations of physically unequal disks, which looked equal, and presentations of physically equal disks, which looked unequal. Both kinds of trials were presented randomly and so were the left versus right positions of either kind of stimuli. Subjects were instructed to pick up the disk on the left between the thumb and index finger of their right hand if they thought the two disks to be equal or to pick up the disk on the right if they judged them to be unequal (see Figure 7.8).

The sequence of subjects' choices of the disk on the right versus left provided a measure of the magnitude of the illusion prompted by the perceptual comparison between two disks surrounded by two distinct annuli. In the visuomotor task, the measure of grip size was based on the unfolding of the natural grasping movement performed by subjects while their hand approached the object. During a prehension movement,

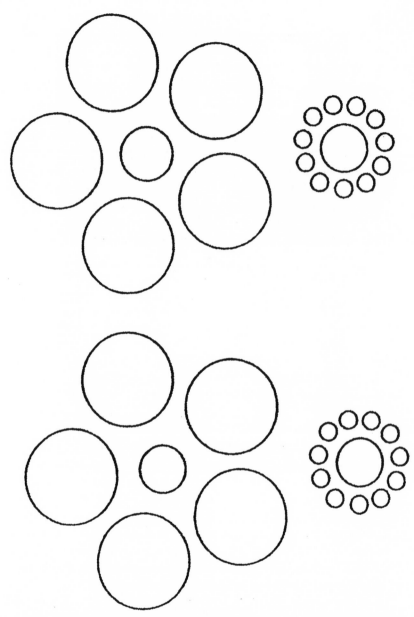

FIGURE 7.7. The Titchener circles (or Ebbinghaus) illusion. In the bottom cluster, the circle surrounded by an annulus of smaller circles looks larger than a circle of equal diameter surrounded by an annulus of larger circles. In the top cluster, the circle surrounded by the annulus of smaller circles is smaller than, but looks equal to, the circle surrounded by an annulus of larger circles.

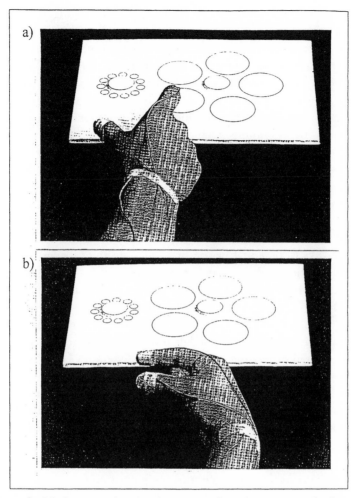

FIGURE 7.8. (a) shows maximum grip aperture in a visuomotor task of grasping a disk in a Titchener circles condition. (b) shows manual estimation of the size of a disk in the same condition. Note the sensors attached to thumb and index finger. *Source*: Reprinted with permission from Angela M. Haffenden and Melvyn A. Goodale (1998), The effect of pictorial illusion on prehension and perception, *Journal of Cognitive Neuroscience*, 10:1 (January), 122–136. © 1998 by the Massachusetts Institute of Technology.

fingers progressively stretch to a maximal grip aperture (MGA) before they close down until contact with the object (see section 4.1). Aglioti et al. (1995) measured MGA in flight using optoelectronic recording. They found that grip aperture was significantly less affected by the size-contrast illusion than comparative perceptual judgement as expressed by the sequence of choices of disks on the left versus right.

This experiment, however, raises a number of methodological problems. The main issue, raised by F. Pavani et al. (1999) and V. H. Franz et al. (2000), is the asymmetry between the two tasks. In the perceptual task, subjects are asked to compare two distinct disks surrounded by two different annuli. But in the grasping task, subjects focus on a single disk surrounded by an annulus. So the question arises whether, from the observation that the comparative perceptual judgement is more affected by the illusion than the grasping task, one may conclude that perception and action are based on two distinct representational systems.

Aware of this problem, A. M. Haffenden and Goodale (1998) performed the same experiment, but they designed one more task: In addition to instructing subjects to pick up the disk on the left if they judged the two disks to be equal in size or to pick up the disk on the right if they judged them to be unequal, they also required subjects to manually estimate between the thumb and index finger of their right hand the size of the disk on the left if they judged the disks to be equal in size, and to manually estimate the size of the disk on the right if they judged them to be unequal. Haffenden and Goodale found that the effect of the illusion on the manual estimation of the size of a disk (after comparison) was intermediary between comparative judgement and grasping.

Furthermore, Haffenden & Goodale (1998) found that the presence of an annulus had a selective effect on grasping. They contrasted the presentation of pairs of disks either against a blank background or surrounded by an annulus of circles of intermediate size, that is, of size intermediary between the size of the smaller circles and the size of the larger circles involved in the contrasting pair of illusory annuli. The circles of intermediate size in the annulus were slightly larger than the disks of equal size. When a pair of physically different disks were presented against either a blank background or a pair of annuli made of intermediate size circles, both grip scaling and manual estimates reflected the physical difference in size between the disks. When physically equal disks were displayed against either a blank background or a pair of annuli made of circles of intermediate size, no significant difference was found between the grasping and manual estimate. The following dissociation, however, turned up: when physically equal disks were presented with a middle-sized annulus, overall MGA was smaller than when physically equal disks were presented against a blank background. Thus, the presence of an annulus of middle-sized circles prompted a smaller MGA than a blank background. Conversely, overall manual estimate was larger when physically equal disks were presented against a background with a middle-sized annulus than when they were presented against a blank background. The illusory effect of the middle-size annulus presumably arises from the fact that the circles in the annulus were slightly larger than the equal disks. Thus, whereas the presence of a middle-sized annulus contributes to increasing manual estimation, it contributes to decreasing grip scaling. This dissociation shows that the presence of an annulus may have conflicting effects on perceptual estimate and on grip aperture.

Finally, Haffenden et al. (2001) went one step further. They presented subjects with three distinct Titchener circle displays one at a time, two of which are the traditional Titchener central disk surrounded by an annulus of circles either smaller than it is or larger than it is. In the former case, the *gap* between the edge of the disk and the annulus is 3 mm. In the latter case, the gap between the edge of the disk and the annulus is 11 mm. In the third display, the annulus is made of small circles (of the same size as in the first display), but the *gap* between the edge of the disk and the annulus is 11 mm (like the gap in the second display with an annulus of larger circles) (see Figure 7.9). What Haffenden et al. found was the following dissociation: In the perceptual task, subjects estimated the third display very much like the first display and unlike the second display. In the visuomotor task, subjects' grasping in the third condition was much more similar to grasping in the second than in the first

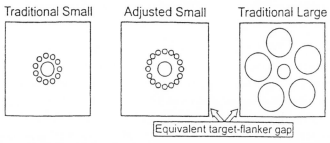

FIGURE 7.9. Leftmost square shows small distance between central disk and annulus of smaller circles in standard Titchener circles illusion. Rightmost square shows large distance between central disk and annulus of larger circles in standard Titchener circles illusion. Central square shows large distance between central disk and annulus of smaller circles in nonstandard Titchener circles illusion. *Source*: Reprinted from Angela M. Haffenden, Karen C. Schiff, and Melvyn A. Goodale (2001), The dissociation between perception and action in the Ebbinghaus illusion: Non-illusory effects of pictorial cues on grasp, *Current Biology*, 11 (February), 177–181.

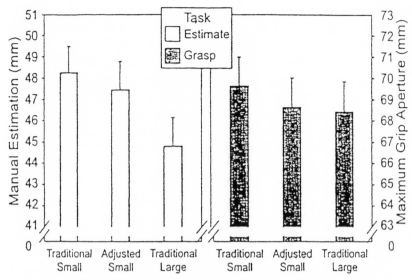

FIGURE 7.10. White bars show manual estimation for each of the three conditions displayed separately: traditional small, traditional large and adjusted small
(see Figure 7.9). Black bars show maximal grip aperture in grasping task in each
of three conditions. The experiment shows dissociation between manual estimation and grasping task: manual estimation of adjusted small is very much like
manual estimation of traditional small. But grasping of traditional small is
very much like grasping of traditional large. *Source*: Reprinted from Angela M.
Haffenden, Karen C. Schiff, and Melvyn A. Goodale (2001), The dissociation
between perception and action in the Ebbinghaus illusion: Non-illusory effects
of pictorial cues on grasp, *Current Biology*, 11 (February), 177–181.

condition (see Figure 7.10). Thus, perceptual estimate was far more sensitive to the size of the circles in the annulus than to the distance between
target and annulus. Conversely, grasping was far more sensitive to the
distance between target and annulus than to the size of the circles in the
annulus. The idea here is that the annulus is processed by the visuomotor
processing as a potential *obstacle* for the position of the fingers on the target disk. If so, then the visuomotor processing of objects can be fooled:
it can misrepresent (and thus represent) a 2D annulus as a 3D obstacle.

In Haffenden et al.'s (2001) experiment, two parameters are relevant:
the size of the diameter of the circles in the annulus and the distance
between the disk and the annulus. The suggested conclusion of their experiment is that what matters to the illusory visual percept is the contrast
between the size of the disk and the size of the circles in the annulus.
What matters to the visuomotor representation of the same display is the

distance between the target and the annulus. But Haffenden et al. consider only three conditions, not four. The question arises: does the two visual systems model lead to the prediction that the distance (between target and annulus) should produce the same dissociation between perceptual estimate and MGA when the disk is surrounded by larger circles? Given the difference in the size of the whole display, it is clear, I think, that the two visual systems model is *not* committed to this prediction. When the circles in the annulus and the distance between the disk and the annulus are smallest, then the size of the whole display is 56 mm. When the circles in the annulus and the distance between the disk and the annulus are largest, then the size of the whole display is 160 mm. If the circles in the annulus were large and the distance between the disk and the annulus were small, the size of the whole display would be 146 mm. What is true of a visuomotor response in the presence of a 56 mm display is not automatically true of the visuomotor response to a 146 mm display. Suppose that if the gap between the target and the potential obstacle is small, then the visuomotor system automatically programs the grasping of a 56 mm display. It does not entail that one produce the same visuomotor response to a 146 mm display.

Taken seriously, Haffenden et al.'s (2001) evidence shows that it is wrong to assume, as, for example, Milner and Goodale (1995) and Haffenden and Goodale (1998) have, that unlike the perceptual processing, the visuomotor processing of a visual display cannot 'afford the luxury' of being *fooled* by a visual display. Rather, the features of a display that can fool a visuomotor representation are not the same as those that can fool a perceptual representation. Automatic computation of the contrast between the sizes of elements of the display gives rise to a perceptual size-contrast illusion. Automatic computation of the distance between a 3D target and 2D surrounding elements can give rise to a visuomotor illusion (representing the 2D elements as if they were 3D).[12] If so, then presumably, visuomotor processing generates genuine visuomotor representations. Properly analyzed, the psychophysical data point, I think, away

---

[12] We normally grasp 3D targets against a 3D background. So it makes sense, as Kathleen Akins suggests, to assume that representing a 3D background is the default setting of the visuomotor representation of a target for action. Still, representing a 2D annulus as a 3D obstacle is misrepresenting it. What is important for my purpose is that an agent who automatically misrepresents a 2D annulus as a 3D obstacle as part of his or her visuomotor representation of a target may, of course, become perceptually aware that the annulus is a 2D display. But then the agent must switch from a visuomotor illusory representation to a correct perceptual representation of the annulus.

from what Franz et al. (2001) call the 'common representation' model
of human vision, and towards the 'separate representations' model, according to which selective tasks can prompt the elaboration of distinct
visual representations of one and the same visual stimulus.

6 Spatial Processing and Visual Awareness

One can think about, but one cannot see, entities that are not in space
(e.g., numbers). Arguably, the visual system cannot represent an object
perceptually or otherwise unless it locates the object in space. As preceding sections have made clear, the computational requirements made
respectively by the perceptual system and by the visuomotor system upon
the visual encoding of spatial information about objects are clearly different. There are at least two broad ways the location of an object can be
visually coded: It can be coded in egocentric and in allocentric coordinates. I will presently review evidence in support of the third of my five
assumptions: Representing the *spatial relations* among objects in a visual
scene is necessary for perceptual awareness of the other visual attributes
of objects. By contrast, in a visuomotor representation, the position of the
target (of a hand action) is coded in an egocentric frame of reference,
and it does not lead to perceptual awareness of the object's other visual
attributes.

6.1 Perceptual Proto-Objects

In an experiment based on a habituation/dishabituation paradigm designed by K. Wynn (1992a, 1992b), 4.5-month-old human infants see a
human hand bring one Mickey Mouse onto a stage. Then, the puppet
is hidden behind a removable screen. The babies see the empty hand
leave the stage. Then, the babies see the hand bring a second puppet
onto the stage with the screen up so that the babies do not see what is
behind. Then, the babies see the empty hand leave the stage one more
time. When the screen is removed, the babies can see either two puppets or one. By measuring the babies' looking time, Wynn found that
they look longer at one than at two puppets. The experiment shows that
4.5-month-old human babies can distinguish between two objects on the
basis of spatial and locational information, that is, from twice seeing a
hand that holds an object and then leaves empty.

In a set of experiments using the same habituation/dishabituation
paradigm, S. Carey (1995) and F. Xu and Carey (1996) showed 10-month-
old human infants alternatively a red metal car and a brown teddy bear.

The babies saw the red metal car appear from, and disappear behind, the left side of a screen. They saw the brown teddy bear appear from, and disappear behind, the right side of the screen. They never saw the two objects next to each other. When the screen was removed, the infants saw either two or only one object. Measuring the infants' looking time, Carey (1995) and Xu and Carey (1996) found that 10-month-old human babies were not more suprised to see only one object rather than two. They conclude that in the absence of visual spatial locational information, 10-month-old human infants fail to infer that they are being presented with two distinct objects on the basis of visual information about the colors, shapes, size and texture of objects. Featural information about colors, shapes, size and texture of pairs of objects presented separately was not sufficient to warrant in 10-month-old human infants the judgement that they are presented with two distinct objects.[13]

A. M. Leslie et al. (1998) argue further that objects can be visually 'indexed' by two distinct indexing mechanisms: object-individuation and object-identification. They claim that the dorsal pathway of the human visual system is the anatomical basis of the object-individuation mechanism that works by locational information. They claim that the ventral pathway is the anatomical basis of the object-identification mechanism that uses featural information (such as color, shape or texture). Finally, they argue that 10-month-old human infants can only make use of the object-individuation mechanism, not of the object-identification mechanism. If the experiments are to be trusted, then 10-month-old human infants use locational information, not featural information, in perceptual object-discrimination tasks.[14]

Z. Pylyshyn (2000a, 2000b) reports relevant experiments about so-called multiple object-tracking (MOT) in normal human adults. Subjects are shown eight identical circles at rest on a screen, four of which flicker briefly. Then, subjects see the eight circles move randomly on the screen for about 10 seconds. They are asked to keep track of the four circles that initially flickered. Normal human adults can keep track of four to five such distinct objects (or proto-objects). Importantly, Pylyshyn reports

---

[13] Experiments by L. Bonatti et al. (2002) suggest that 10-month-old human infants respond differently if presented with human faces or human-like stimuli.

[14] There is, however, conflicting evidence that 4.5-month-old infants' perceptual segregation of objects from the background can benefit from prior exposure to the object (see, e.g., Needham, 2001). So my claim about the role of locational information in infants' perception of proto-objects is conditional.

that subjects failed to notice changes in the colors and shapes of the proto-objects that they tracked by their relative locations.

Presumably, in MOT experiments, human adults use locational information, not featural information, to individuate and keep track of four to five randomly moving circles. At least, when a conflict arises between information about an object's location and featural information, locational information trumps information about the object's other visual properties. Now, keeping track of three to four proto-objects by locational information requires that the relative spatial positions of each proto-object be coded in an allocentric frame of reference centred on one of them. The very fact that adults engaged in MOT tasks fail to notice changes in colors, textures, shapes and other visual features of objects suggests that the object-individuation mechanism (in Leslie et al.'s 1998 sense) – or the system for representing spatial relations anmong proto-objects – must be a modular (informationally encapsulated) system designed for the visual perception of spatial relations among proto-objects. This modular system, which represents the spatial positions of proto-objects in allocentric coordinates, is impervious to the perception of colors and other featural visual attributes of objects.

6.2 *Unilateral Spatial Neglect*
Lesions located in the right inferior parietal lobe typically produce unilateral spatial neglect. Unlike lesions in the superior parietal lobe, which produce optic ataxia and which can be bilateral, lesions responsible for unilateral spatial neglect are located in the right hemisphere. Patients with unilateral spatial neglect are not perceptually aware of objects visually presented in their contralesional (i.e., left) hemispace. For example, when asked to mark line segments at different orientations, neglect patients will systematically fail to mark the segments lying in their contralesional hemifield. If asked to bisect a horizontal line, they will exhibit a strong ipsilesional bias revealing neglect of the part of the line falling within the neglected hemispace (see Milner et al., 1993).

Unlike blindsight patients whose primary visual cortex has been damaged, and to a lesser extent unlike visual form agnosic patients whose ventral stream has been impaired, neglect patients lack perceptual awareness on their affected side in spite of the fact that the visual pathway for processing the neglected visual information remains intact. Indeed, there is considerable evidence for covert processing of the neglected stimuli. For example, J. Marshall and P. W. Halligan (1994) showed a neglect patient drawings of two houses located on top of each other, one of which displayed brightly colored flames on its left side. When asked to make an explicit comparison between the two houses, the patient could report no difference. When asked, however, which of the two houses was preferable to live in, the patient pointed to the house without flames. This shows that the neglected stimuli are covertly processed in neglect patients even though this processing is not accompanied by perceptual awareness.

One important source of insight into neglect comes from studies of a phenomenon called 'extinction', reported by J. Driver and P. Vuilleumier (2001). Unlike blindsight patients, who typically will fail to notice the presence of a light presented in isolation to their left, patients with unilateral neglect in their contralesional left hemispace might easily detect an isolated stimulus on their left. If, however, they are presented with two *competing* stimuli, one farther to the left than the other, then they will typically fail to perceive the one farther to the left. Thus, extinction is the failure to perceive a stimulus presented farther within the contralesional hemispace, when the patient does perceive a competing stimulus presented more towards the ispsilesional hemispace. The stimulus located more towards the ipsilesional side of the lesion 'extinguishes' its competitor located more towards the contralesional side. As Driver and Vuilleumier (2001) emphasize, extinction reveals that neglect patients have a deep impairment in allocating attentional resources to competing stimuli according to their respective positions in the patient's hemispace.

Driver and Vuilleumier (2001: 52–54) report the following extinction study with a neglect patient. In some conditions, the stimulus was a Kanizsa white square whose subjective contours resulted from the removal of the relevant quarter-segments from four black circles appropriately located. In other conditions, the stimulus consisted of the four black circles in the same spatial position from which the same relevant quarter-segments had been removed, except for the persistence of narrow black arcs at the outer boundary of the removed quarter-segments that prevented the formation of the subjective contours of the Kanizsa white square (see Figure 7.11). The patient extinguished most left-sided presentations of the stimulus in bilateral trials when the narrow black arcs prevented the formation of the subjective contours of the Kanizsa white square. But extinction decreased significantly upon the perception of the Kanizsa square. Thus, neglect patients find it easier to allocate their perceptual attention to *one* object (the white Kanizsa square) than to *four* competing distinct objects (the four black circles each with

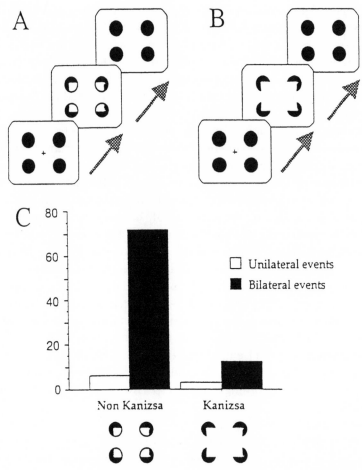

FIGURE 7.11. (A) shows high level of extinction by a neglect patient for presentations of a stimulus that fails to elicit a Kanizsa square illusion. (B) shows much lower extinction by the same neglect patient for presentations of a stimulus that does elicit a Kanizsa square illusion. (C) As the patient's fixation point was in the middle of the card ('bilateral events') in either A or B, she was visually aware of the Kanizsa square (low rate of extinction: black rectangle on the right); but she was not visually aware of the four black circles with white quarter segments (high rate of extinction: large black rectangle on the left). *Source*: Reprinted from J. Driver and P. Vuilleumier (2001), Perceptual awareness and its loss to unilateral neglect and extinction, *Cognition*, 79, 39–88, copyright 2001, with permission from Elsevier.

a partially removed quarter-segment) with four distinct relative spatial positions.[15]

Driver and Vuilleumier (2001: 54) further report a remarkable modulation of extinction according to the task. When asked to report the location of a stimulus (e.g., a small set of distinctive shapes) or when asked whether they perceive a small set of distinctive shapes at a given location, neglect patients extinguish left-sided stimuli in bilateral displays. However, changing the task decreases the extinction. If the task is to enumerate the members of the very same small set of shapes in bilateral displays, neglect patients succeed. For small numerosities (i.e., sets of three or perhaps even four objects), humans can provide precise numerical estimation by subitizing. The ability to subitize small numerosities is the ability to produce direct perceptual estimates of small numerosities. It seems to rely on the ability to build a visual representation of the distinct relative spatial positions of objects within a set (see Dehaene, 1997, ch. 3). Arguably, one condition for exploiting a subitizing procedure is to form a perceptual representation of a single set of small cardinality. If so, then the contrast between the two tasks is the contrast between attending to, for example, three competing distinct entities and attending to a single object involving three distinct elements, parts or components. In the first task, the stimuli compete for the patients' perceptual attention in the neglected hemispace and competition produces extinction. In the second task, the very same stimuli are transformed into elements of a single object with no competitor.

6.3 Awareness Respectively of Space and of Other Visual Properties

Joint work of Wynn (1992a, 1992b), Carey (1995), Xu and Carey (1996) and Leslie et al. (1998) suggests very strongly that 4.5-month-old human infants automatically open different object files on the basis of their perception of the relative locations of objects. If 10-month-old human infants are not provided perceptual information about the distinct locations of two distinct objects, then featural information alone (e.g., color, shape, orientation and texture) is not sufficient to warrant the opening of two distinct object-files. I call 'proto-objects' such object-files opened on the basis of relative locations. I think infants' proto-objects are quite similar to what human adults manage to track in Pylyshyn's (2000b) MOT tasks.

---

[15] There is evidence that activity in area V2 is necessary for perceiving the subjective contours of a Kanizsa figure (see Zeki, 1993: 316–317). If so, then area V2 is active in neglect patients.

Subjects of such experiments manage to attend to the random motions of four to five proto-objects identified by their relative positions on a screen, not by their colors, shapes or texture.

Lesions in the right inferior parietal lobe produce impairment in the ability to perceive spatial relations among objects in the patients' left contralesional hemispace. Driver and Vuilleumier (2001) adduce much evidence that in neglect patients, the neglected stimuli are actively processed in the brain areas (particularly, in the ventral pathway) that, in healthy human subjects, give rise to conscious visual experiences of objects' shapes, sizes, colors, texture and so on. As N. Block (2001: 198) rightly emphasizes, the condition of neglect patients is paradoxical: the brain areas (in the relevant portions of the ventral pathway) that should make neglect patients aware of the visual properties of extinguished stimuli in their contralesional side *are* active, but the patient remains unaware of these stimuli.

Driver and Vuilleumier (2001), however, suggest a solution to the paradox: the solution is the 'neuropsychological asymmetry' between awareness of the relative locations and awareness of other properties of visual stimuli. By losing visual awareness of the relative spatial locations of visual stimuli in their neglected side, neglect patients also lose visual awareness of their colors, sizes, shapes, orientations and textures. Although the properties of neglected stimuli are unconsciously processed in the relevant brain areas of neglect patients, nonetheless these patients remain visually unaware of them. In Driver and Vuilleumier's (2001: 74) terms, 'it thus appears that when the appropriate representation of a stimulus location is lost or degraded, as in neglect after parietal damage, then awareness of other stimulus properties (presumably coded elsewhere in the brain) is also lost'. As they note (2001: 75), the converse it not true: Loss of awareness of such visual properties of stimuli as their colors, shapes, sizes or orientations does not lead to unawareness of the relative locations of stimuli.[16]

Concluding Remarks on the Two Visual Systems Hypothesis

Human optic ataxic patients have a possibly bilateral lesion in the superior parietal lobe. As a result, their visuomotor transformation is deeply

---

[16] D. Dennett's (2001) emphasis on the role of competitive mechanisms in producing perceptual awareness is, it seems to me, on the right track. But he fails to notice the asymmetry between awareness of spatial relationships among objects and awareness of other visual attributes of objects.

altered: They cannot properly represent the size, shape and orientation of a target in a visuomotor representation in which its location is coded in egocentric coordinates centred on the patient's body. They can, however, form appropriate visual percepts of objects by means of which they can recognize the shapes, sizes and orientations of objects. Apperceptive visual form agnosic patients have a lesion in the inferotemporal areas (in the ventral pathway). As a result, they lose visual awareness of the shapes, sizes and orientations of objects. They can, however, accurately compute the size, shape and orientation of a target for the purpose of a visuomotor representation in which the location of the target is coded in egocentric coordinates. Neglect patients have a lesion in the right inferior parietal lobe but the rest of their visual cortex is intact. As a result of their inability to represent the respective spatial locations of objects in their neglected hemispace, they fail to reach awareness of the other visual attributes of these objects. As Driver and Vuilleumier (2001: 70) write: 'The pathology in [neglect] patients becomes more evident when several stimuli are presented simultaneously. . . . Interestingly, extinction in the patients can be determined less by the absolute location of a stimulus within the left or right hemifield than by its position relative to other competing stimuli'. Thus, such patients should not have serious trouble in reaching for one stimulus in their neglected hemispace, since they might be able to form a visuomotor representation of a target of prehension whose 'absolute' location is coded in egocentric coordinates. Indeed, there is evidence that neglect patients have such visuomotor capacities as reaching and grasping objects located in their left hemispace.

The neuropsychological picture suggests that from an anatomical point of view, the parietal lobe (which is part of the dorsal pathway of the human visual system) makes a twofold contribution to human vision: The superior parietal lobe makes a contribution to the visuomotor transformation. It allows the formation of a visuomotor representation of a target whose location is coded in egocentric coordinates centred on the agent's body. This system is still functioning in apperceptive visual form agnosic patient DF. The right inferior parietal lobe makes a decisive contribution to conscious visual perception: It allows the representation of the location of an object within allocentric coordinates centred on some item in the visual scene. Neglect patients show that unless the location of an object can be coded in allocentric coordinates, the object's other visual attributes are not available for visual awareness.

On the one hand, as apperceptive visual form agnosic patient DF shows, unless the relevant areas in the ventral pathway are active, the

TABLE 7.1.

	World-to-Mind Direction of Fit	Mind-to-World Direction of Fit	Localization of Target in Egocentric Coordinates	Localization of Object in Allocentric Coordinates	Awareness of Visual Attributes
Visuomotor representation	+	+	+	−	−
Visual percept	−	+	−	+	+

processing of shape, size and orientation of an object will not yield visual awareness of shape, size and orientation of an object. In particular, if such attributes are only processed for building a visuomotor representation of an object whose location is coded in egocentric coordinates, then conscious awareness fails to arise. On the other hand, as neglect patients show, active processing in the relevant areas in the ventral pathway for the shape, size and orientation of an object is not sufficient for visual awareness of the object's shape, size and orientation either. Representing the relative spatial locations of an object in allocentric coordinates is a necessary condition for awareness of the object's shape, size and orientation. Milner and Goodale (1995: 200) are right to claim that 'visual phenomenology . . . can only arise from processing in the ventral stream, processing that we have linked with recognition and perception'. But what they say may wrongly suggest that processing in the ventral stream is sufficient for visual phenomenology – which it is not. Block (2001: 199) asks the question: '[W]hat is the missing ingredient, X, which, added to ventral activation (of sufficient strength), constitutes conscious [visual] experience?'. To Block's question, I venture to answer that, for many of an object's visual attributes, X is: the coding of the object's spatial location within an allocentric frame of reference by the right inferior parietal lobe.[17]

In Jacob and Jeannerod (2003), we claim that for awareness of such visual attributes of an object as its shape, size and orientation to arise, the visual representation must satisfy what we call the constraint of contrastive identification. The visual processing of an object's shape, size

---

[17] For a slightly different view of the contribution of the parietal lobe to visual awareness, see N. Kanwischer (2001: 108) who writes: '[N]eural activity in specific regions within the ventral pathway is apparently correlated with the content of perceptual awareness, whereas neural activity in the dorsal pathway may be correlated instead with the occurrence of perceptual awareness in a completely content-independent fashion'. If awareness of spatial relationships among objects counts as 'completely content-independent perceptual awareness', then I think that I agree with Kanwischer.

and orientation must be available for comparison with the shapes, sizes and orientations of other neighbouring constituents of the visual array. It is of the essence of conscious visual perception that it makes automatic comparisons, which are the source of, for example, size-contrast perceptual illusions. Only if the spatial location of an object is coded in allocentric coordinates will the processing of the object's other visual attributes be available for visual awareness. This is exactly what is missing in a purely visuomotor representation of a target of prehension whose ('absolute') location is coded in egocentric coordinates. Table 7.1 summarizes the leading ideas.

Acknowledgments

First, I am grateful to Kathleen Akins, Steven Davis and especially Andrew Brook, who organized the wonderful Conference on Philosophy and the Neurosciences at Carleton University, October 17–20, 2002. Secondly, I am also grateful to Andrew Brook and especially to Kathleen Akins for penetrating comments on this chapter. Finally, I wish to register my deep gratitude to Marc Jeannerod, who introduced me to the cognitive neurosciences of vision and action.

References

Agliotti, S., Goodale, M., and Desouza, J. (1995) Size contrast illusions deceive the eye but not the hand. *Current Biology*, 5, 679–685.

Anscombe, G. E. (1957) *Intention*. Oxford: Blackwell.

Austin, J. (1962) *Sense and Sensibilia*. Oxford: Oxford University Press.

Bermudez, J. (1998) *The Paradox of Self-Consciousness*. Cambridge, Mass.: MIT Press.

Block, N. (2001) Paradox and cross purposes in recent work on consciousness. In Dehaene, S., and Naccache, L. (eds.), *The Cognitive Neuroscience of Consciousness*. Cambridge, Mass.: MIT Press.

Bonatti, L., Frot, E., Zangl, R., and Mehler, J. (2002) The human first hypothesis: Identification of conspecifics and individuation of objects in the young infant. *Cognitive Psychology*, 44, 388–426.

Boussaoud, D., Ungerleider, L., and Desimone, R. (1990) Pathways for motion analysis: Cortical connections of the medial superior temporal sulcus and fundus of the superior temporal visual areas in the macaque monkey. *Journal of Comparative Neurology*, 296, 462–495.

Bridgeman, B. (2002) Attention and visually guided behavior in distinct systems. In Prinz, W., and Hommel, B. (eds), *Common Mechanisms in Perception and Action*. Oxford: Oxford University Press.

Bridgeman, B., Lewis, S., Heit, G., and Nagle, M. (1979) Relation between cognitive and motor-oriented systems of visual position perception. *Journal of Experimental Psychology: Human Perception and Performance*, 5, 692–700.

Carey, D. P., Harvey, M., and Milner, A. D. (1996) Visuomotor sensitivity for shape and orientation in a patient with visual form agnosia. *Neuropsychologia*, 34, 329–337.

Carey, S. (1995) Continuity and discontinuity in cognitive development. In Osherson, D. (ed.), *An Invitation to Cognitive Science, Thinking*, vol. 3. Cambridge, Mass.: MIT Press.

Clark, A. (2001) Visual experience and motor action: Are the bonds too tight? *Philosophical Review*, 110, 495–519.

Dehaene, S. (1997) *The Number Sense: How the Mind Creates Mathematics*. New York: Oxford University Press.

Dennett, D. (2001) Are we explaining consciousness yet? In Dehaene, S., and Naccache, L. (eds.), *The Cognitive Neuroscience of Consciousness*. Cambridge, Mass.: MIT Press.

Dretske, F. (1969) *Seeing and Knowing*. Chicago: Chicago University Press.

Dretske, F. (1978) Simple seeing. In Dretske, F. (2000), *Perception, Knowledge and Belief*. Cambridge: Cambridge University Press.

Dretske, F. (1981) *Knowledge and the Flow of Information*. Cambridge, Mass.: MIT Press.

Driver, J., and Vuilleumier, P. (2001) Perceptual awareness and its loss to unilateral neglect and extinction. *Cognition*, 79, 39–88.

Evans, G. (1982) *The Varieties of Reference*. Oxford: Oxford University Press.

Farah, M. J. (1990) *Visual agnosia: Disorders of Object Recognition and What They Tell Us About Normal Vision*. Cambridge, Mass.: MIT Press.

Franz, V. H., Fahle, M., Bülthoff, H. H., and Gegenfurtner, K. R. (2001) Effects of visual illusion on grasping. *Journal of Experimental Psychology: Human Perception and Performance*, 27, 1124–1144.

Franz, V. H., Gegenfurtner, K. R., Bülthoff, H. H., and Fahle, M. (2000) Grasping visual illusions: No evidence for a dissociation between perception and action. *Psychological Science*, 11, 20–25.

Gallese, V., Murata, A., Kaseda, M., Niki, N., and Sakata, H. (1994) Deficit of hand preshaping after muscimol injection in monkey parietal cortex. *Neuroreport*, 5, 1525–1529.

Gentilucci, M., Chieffi, S., Deprati, E., Saetti, M. C., and Toni, I. (1996) Visual illusion and action. *Neuropsychologia*, 34, 5, 369–376.

Gibson, J. J. (1979) *The Ecological Approach to Visual Perception*. Boston: Houghton Mifflin.

Goodale, M. A. (1995) The cortical organization of visual perception and visuomotor control. In Osherson, D. (ed.), *An Invitation to Cognitive Science: Visual Cognition*, vol. 2. Cambridge, Mass.: MIT Press.

Goodale, M., and Humphrey, G. K. (1998) The objects of action and perception. *Cognition*, 67, 191–207.

Goodale, M. A., and Milner, A. D. (1992) Separate visual pathways for perception and action. *Trends in Neuroscience*, 15, 20–25.

Goodale, M. A., Jakobson, L. S., Milner, A. D., Perrett, D. I., Benson, P. J., and Hietanen, J. K. (1994) The nature and limits of orientation and pattern processing supporting visuomotor control in a visual form agnosic. *Journal of Cognitive Neuroscience*, 6, 46–56.

Goodale, M. A., Milner, A. D., Jakobson I. S., and Carey, D. P. (1991) A neurological dissociation between perceiving objects and grasping them. *Nature*, 349, 154–156.

Goodale, M. A., Pélisson, D., and Prablanc, C. (1986) Large adjustments in visually guided reaching do not depend on vision of the hand or perception of target displacement. *Nature*, 320, 748–750.

Haffenden, A. M., and Goodale, M. (1998) The effect of pictorial illusion on prehension and perception. *Journal of Cognitive Neuroscience*, 10, 1, 122–136.

Haffenden, A. M., Schiff, K. C., and Goodale, M. A. (2001) The dissociation between perception and action in the Ebbinghaus illusion: Non-illusory effects of pictorial cues on grasp. *Current Biology*, 11, 177–181.

Ingle, D. J. (1973) Two visual systems in the frog. *Science*, 181, 1053–1055.

Jacob, P., and Jeannerod, M. (2003) *Ways of seeing*. Oxford: Oxford University Press.

Jeannerod, M. (1986) The formation of finger grip during prehension: A cortically mediated visuomotor pattern. *Behavioural Brain Research*, 19, 99–116.

Jeannerod, M. (1994) The representing brain: Neural correlates of motor intention and imagery. *Behavioral and Brain Sciences*, 17, 187–245.

Jeannerod, M. (1997) *The Cognitive Neuroscience of Action*. Oxford: Blackwell.

Jeannerod, M., Decety, J., and Michel, F. (1994) Impairment of grasping movements following bilateral posterior parietal lesion. *Neuropsychologia*, 32, 369–380.

Kanwischer, N. (2001) Neural events and perceptual awareness. In Dehaene, S., and Naccache, L. (eds.), *The Cognitive Neuroscience of Consciousness*. Cambridge, Mass.: MIT Press.

Leslie, A. M., Xu, F., Tremoulet, P. D., and Scholl, B. J. (1998) Indexing and the object concept: Developing 'What' and 'Where' systems. *Trends in Cognitive Science*, 2, 1, 10–18.

Marshall, J., and Halligan, P. W. (1994) The yin and yang of visuospatial neglect: A case study. *Neuropsychologia*, 32, 1037–1057.

McDowell, J. (1994) *Mind and the World*. Cambridge, Mass.: Harvard University Press.

Millikan, R. G. (1996) Pushmi-pullyu representations. In Tomberlin, J. (ed.), *Philosophical Perspectives*, vol. 9. Atascadero, Calif.: Ridgeview Publishing.

Milner, A. D., and Goodale, M. A. (1995) *The Visual Brain in Action*. Oxford: Oxford University Press.

Milner, A. D., Harvey, M., Roberts, R. C., and Forster, S. V. (1993) Line bisection errors in visual neglect: Misguided action or size distorsion. *Neuropsychologia*, 31, 39–49.

Milner, A. D., Paulignan, Y., Dijkerman, H. C., Michel, F., and Jeannerod, M. (1999) A paradoxical improvement of misreaching in optic ataxia: New evidence for two separate neural systems for visual localization. *Proceedings of the Royal Society*, 266, 2225–2229.

Milner, A. D., Perrett, D. I., Johnston, R. S., Benson, P. J., Jordan, T. R., Heeley, D. W., Bettucci, D., Mortara, F., Mutani, R., Terazzi, E., and Davidson, D. L. W. (1991) Perception and action in 'visual form agnosia'. *Brain*, 114, 405–428.

Mountcastle, V. B., Lynch, J. C., Georgopoulos, A., Sakata, H., and Acuna, C. (1975) Posterior parietal association cortex of the monkey: Command functions for operations within extra-personal space. *Journal of Neurophysiology*, 38, 871–908.

Murata, A., Gallese, V., Luppino, G., Kaseda, M., and Sakata, H. (2000) Selectivity for the shape, size, and orientation of objects for grasping in monkey parietal area AIP. *Journal of Neurophysiology*, 79, 2580–2601.

Needham, A. (2001) Object recognition and object segregation in 4.5-month-old infants. *Journal of Experimental Child Psychology*, 78, 3–24.

Noë, A. (2002) Is the visual world a grand illusion? *Journal of Consciousness Studies*, 9, 5/6, 1–12.

O'Regan, J. K., and Noe, A. (2001) A sensorimotor account of vision and visual consciousness. *Behavioral and Brain Sciences*, 24, 939–1031.

O'Shaughnessy, B. (1980) *The Will*. Cambridge: Cambridge University Press.

O'Shaughnessy, B. (1992) The diversity and unity of action and perception. In Crane, T. (ed.), *The Contents of Experience*. Cambridge: Cambridge University Press, 216–266.

Pavani, F., Boscagli, I., Benvenuti, F., Rabufetti, M., and Farne, A. (1999) Are perception and action affected differently by the Titchener circles illusion? *Experimental Brain Research*, 127, 1, 95–101.

Peacocke, C. (1992) *A Study of Concepts*. Cambridge, Mass.: MIT Press.

Peacocke, C. (1998) Nonconceptual content defended. *Philosophy and Phenomenological Research*, 58, 381–388.

Peacocke, C. (2001) Does perception have nonconceptual content? *Journal of Philosophy*, 98, 239–264.

Piattelli-Palmarini, M. (ed.) (1980) *Language and Learning: The Debate Between Jean Piaget and Noam Chomsky*. Cambridge, Mass.: Harvard University Press.

Pisella, L., Gréa, H., Tilikete, C., Vighetto, A., Desmurget, M., Rode, G., Boisson, D., and Rossetti, Y. (2000) An 'automatic pilot' for the hand in human posterior parietal cortex: Toward a reinterpretion of optic ataxia. *Nature Neuroscience*, 3, 7, 729–736.

Pylyshyn, Z. (2000a) Situating vision in the world. *Trends in Cognitive Science*, 4, 5, 197–207.

Pylyshyn, Z. (2000b) Visual indexes, preconceptual objects and situated vision. *Cognition*, 80, 127–158.

Rossetti, Y., and Pisella, L. (2002) Several 'vision for action' systems: A guide to dissociating and integrating dorsal and ventral functions (tutorial). In Prinz, W., and Hommel, B. (eds.), *Common Mechanisms in Perception and Action*. Oxford: Oxford University Press.

Sakata, H., Taira, M., Murata, A., and Mine, S. (1995) Neural mechanisms of visual guidance of hand action in the parietal cortex of the monkey. *Cerebral Cortex*, 5, 429–438.

Schneider, G. E. (1967) Contrasting visuomotor functions of tectum and cortex in the golden hamster. *Psychologische Forschung*, 31, 52–62.

Searle, J. R. (1983) *Intentionality: An essay in the philosophy of mind*. Cambridge: Cambridge University Press.

Tanaka, K. (1993) Neuronal mechanisms of object recognition. *Science,* 262, 685–688.

Tanaka, K., Saito, H., Fukada, Y., and Moriya, M. (1991) Coding visual images of objects in the inferotemporal cortex of the macaque monkey. *Journal of Neurophysiology,* 66, 170–189.

Teuber, H. L. (1960) Perception. In Field, J., Magoun, H. W., and Hall, V. E. (eds.), *Handbook of Physiology: Section I, Neurophysiology.* Washington, D.C.: American Physiological Society, 89–121.

Tye, M. (1995) *Ten Problems of Consciousness.* Cambridge, Mass.: MIT Press.

Ungerleider, L., and Mishkin, M. (1982) Two cortical visual systems. In Ingle, D. J., Goodale, M. A., and Mansfield, R. J. W. (eds.), *Analysis of visual behavior.* Cambridge, Mass.: MIT Press, 549–586.

Weiskrantz, L. (1997) *Consciousness Lost and Found: A Neuropsychological Exploration.* Oxford and New York: Oxford University Press.

Wynn, K. (1992a) Addition and subtraction by human infants. *Nature,* 358, 749–750.

Wynn, K. (1992b) Evidence against empiricist accounts of the origins of numerical knowledge. *Mind and Language,* 7, 4, 315–332.

Xu, F., and Carey, S. (1996) Infants' metaphysics: The case of numerical identity. *Cognitive Psychology,* 30, 111–153.

Zeki, S. (1993) *A Vision of the Brain.* Oxford: Blackwell.

8

Action-Oriented Representation

Pete Mandik

Introduction

Often, sensory input underdetermines perception. One such example is the perception of illusory contours. In illusory contour perception, the content of the percept includes the presence of a contour that is absent from the informational content of the sensation. (By "sensation" I mean merely information-bearing events at the transducer level. I intend no further commitment, such as the identification of sensations with qualia.) I call instances of perception underdetermined by sensation "underdetermined perception."

The perception of illusory contours is just one kind of underdetermined perception (see Figure 8.1). The focus of this chapter is another kind of underdetermined perception: what I shall call "active perception." Active perception occurs in cases in which the percept, while underdetermined by sensation, is determined by a combination of sensation and action. The phenomenon of active perception has been used by several to argue against the positing of representations in explanations of sensory experience, either by arguing that no representations need be posited or that far fewer than previously thought need be posited. Such views include, but are not limited to, those of J. Gibson (1966, 1986), P. S. Churchland et al. (1994), T. Jarvilehto (1998), and J. O'Regan and A. Noë (2001). In this chapter, I argue for the contrary position that active perception is actually best accounted for by a representational theory of perception. Along the way, this will require a relatively novel conception of what to count as representations. In particular, I flesh out a novel account of *action-oriented*

FIGURE 8.1. Illusory contours. *Source*: Figure drawn by Pete Mandik.

representations: representations that include in their contents commands for certain behaviors.[1]

Examples of Active Perception

A somewhat famous and highly fascinating example of active perception is shown in the experiences of subjects trained in the use of P. Bach-y-Rita's (1972) Tactile Visual Sensory Substitution System (TVSS). The system consists of a head-mounted video camera that sends information to an array of tactile stimulators worn pressed against the subject's abdomen or back. The subjects can aim the camera at various objects by turning their heads and can adjust the zoom and focus of the camera with a handheld controller. Blindfolded and congenitally blind subjects can utilize the device to recognize faces and objects. Especially interesting are the ways in which the TVSS approximates natural vision. Subjects reported losing awareness of the tingles on their skin and instead saw through the tactile array, much in the same way that one loses awareness of the pixels on a television screen and instead sees through it to see actors and scenery. Bach-y-Rita reports an incident in which someone other than the subject wearing the device increased the camera's zoom. The subject ducked, since the zoom effect made objects seem as if they were heading toward the subject. Bach-y-Rita notes that these sorts of reports only occurred for subjects whose training with the TVSS involved the active control of the camera's direction, focus, and zoom. In conditions in which the subjects had no control over these features and instead only passively received the video-driven tactile information, the subjects never reported the phenomenon of seeing through the tingles on their skin to locate the perceived object in the external environment. For these reasons, then, experiences with the TVSS count as instances of active perception. Information provided at the skin by the tactile stimulators is insufficient to determine the perception of distal objects. The determination of the percept occurs only when certain contributions from action are combined with the tactile input.

---

[1] I did not coin the term "action-oriented representation," although I am unsure of what its first appearance in the literature was. See Clark (1997) and Colby (1999) for discussions of action-oriented representation.

An even simpler, "chemically pure," example of this sort of TVSS-based active perception is reported by C. Lenay et al. (1997) and S. Hanneton et al. (1999). Subjects use a tactile-based device to identify simple two-dimensional forms, such as broken lines and curves. The subjects wear a single tactile stimulator on a fingertip. The stimulator is driven by a magnetic pen used in conjunction with a graphic tablet. A virtual image in black-and-white pixels is displayed on a screen that only the experimenter is allowed to see. The subject scans the pen across the tablet and thus controls a cursor that moves across the virtual image. A stimulus is delivered to the fingertip only when the cursor is on pixels that make up the figure and not on background pixels. Subjects with control over the pen are able to identify the images. Subjects who merely passively receive the tactile information cannot.

One caveat should be stated concerning the proposal that both of these cases count as instances of underdetermined perception. If the sensory inputs are only the tactile inputs, then these are relatively clear cases of underdetermined perception. However, the contribution of action may be sensational if the contribution is exhausted by sensory feedback from the muscles. If this latter possibility obtains, then we have cases of relative, not absolute, underdetermined perception, since the percept would be underdetermined only relative to the tactile input. However, if the contribution of action is, say, an efference copy instead of sensory feedback from the muscles, then the cases are absolute cases of underdetermined perception. I postpone for now further discussion of the distinction between relative and absolute active perception.

A Challenge Posed to Representational Theories

What I am calling active perception has been alleged by others to undermine, either partially or totally, the representational theory of sensory perception. But how, exactly, is this undermining supposed to take place? Before answering this question we must first answer another: what is the representational theory of perception?

Many and various things have been written about the representational theory of perception – enough, perhaps, to render suspicious any claims that there is such a thing as *the* representational theory of perception.[2] However, the theory I sketch here will have sufficient detail to both serve

---

[2] The theory is also known in the philosophical literature as the causal theory of perception. See, for example, Grice (1961), Oakes (1978), and Hyman (1992).

the purposes of the current chapter and do justice to the main features common to typical explications of perception in representational terms. The representational theory of perception may be crudely characterized as the view that one has a perceptual experience of an *F* if and only if one mentally represents that an *F* is present and the current token mental representation of an *F* is causally triggered by the presence of an *F*.[3] There are thus two crucial components of this analysis of perception: the representational component and the causal component. The purpose of the representational component is to account for the similarity between perception, on the one hand, and imagery and illusion, on the other. As is oft noted at least since Descartes, from the first-person point of view accurate perceptions can be indistinguishable from dreams and illusions. This similarity is classically accounted for by the hypothesis that veridical mental states such as perceptions are representational. They are thus hypothesized to differ from their nonveridical counterparts (dreams and hallucinations) not in whether they are representations but in whether they are *accurate* representations.[4] The causality component in the account of perceptual experience is a further articulation of the idea that in spite of similarities, there are crucial differences between perceptions and other representational mental phenomena. It is thus part of the normal functioning of perceptions that they are caused by the things that they represent. Simply having a mental representation of, say, a bear is insufficient for perceiving the bear. The relevant mental representation must be currently caused by a bear to count as a percept of a bear.[5] Further, the causal component will have much to do with the specification of sensory modality. So, for example, if the causal processes intervening between the percept and the bear have largely to do with sound waves, then the perceptual event counts as hearing the bear, and if the causal

---

[3] More can be added to this analysis, of course. For example, if someone sneaks up behind me and hits me on the head with a hammer and this causes me to have a visual hallucination of a hammer, this wouldn't count as a visual perception of the hammer in spite of being a hammer-caused mental representation of a hammer. Additional criteria for perception would include, for example, specifications of the normal channels of causation, which were bypassed in the hammer example. However, attending to this level of detail in the analysis of perception is unnecessary for my present purposes.

[4] See J. Austin 1964 for a classic discussion and, of course, criticism of this line of thought oft referred to as the *argument from illusion*. I will not here review all of the various objections to this argument. My focus is instead to defend the representational theory of perception from attacks predicated on active perception.

[5] See Grice (1961) for an expanded discussion of the topic, including these sorts of causal conditions in the analysis of perception.

processes instead largely involve reflected light, then the perceptual event counts as seeing the bear.[6]

Typically, the notion of representation employed in the representation component is explicated in terms of the kinds of causal processes specified in the causal component. Certain causal relations that obtain between the percept and the thing that it is a percept *of* are thus brought in to explicate what it is for the percept to count as a representation. This is not to say that anyone believes in "The Crude Causal Theory" (Fodor 1987) that says a state represents *F*s if and only if it is caused by *F*s. It is instead to say that being caused by *F*s is going to be an important part of the story of what it is to represent *F*s. The typical kind of story of which the causal relation is a part is a kind of teleological story in which what it means to represent *F*s is to be in a state that is supposed to be caused by *F*s or has the function of being caused by *F*s or has been naturally selected to be caused by *F*s or is caused by *F*s in biologically optimal circumstances. (See, for example, Dretske 1995.)

This view of representation is perhaps the most widespread notion of representation used in the neurosciences. It underlies talk of detectors and instances in which something "codes for" a perceptible environmental feature. For example, there are claimed to be edge detectors in visual cortex (Hubel and Wiesel 1962) and face detectors in inferotemporal cortex (Perrett et al. 1989). Magnocellular activity codes for motion and parvocellular activity codes for color (Livingstone and Hubel 1988). Thus, from the neural point of view, being a representation of *F*s is being a bit of brain "lit up" as a causal consequence of the presence of such *F*s. The teleological element is brought on board to explain how *F*s can be represented even in situations in which no *F*s are present. The lighting up of the relevant brain bit represents *F*s because in certain normal or basic cases, *F*s would cause the lighting up of that bit of brain. This sort of view shows up in neuroscience in the popular account of imagery as being the off-line utilization of resources utilized on line during sensory perception. Thus, for example, the brain areas utilized in forming the mental image of an *F* overlap with the brain areas utilized in the perception of an *F*. S. Kosslyn et al. (2001) report that early visual cortex (area 17) is active in perception as well as imagery and that parahippocampal place area is active in both the perception and imagery of places. K. O'Craven and N. Kaniwisher (2000) report fusiform face area activation for both the

---

[6] For an extended discussion of the individuation of sensory modalities, see B. Keeley (2002).

imagery and perception of faces. In these sorts of cases, the main difference between imagery and perception of *F*s is that in imagery, unlike perception, no *F* need be present.

Another way in which this teleofunctional view of representation emerges in neuroscience is in explanations of underdetermined perception, such as the perception of illusory contours. Neuroimaging studies in humans show that illusory contours activate areas in striate and extrastriate visual cortex similar to areas also activated by real contours (Larsson et al. 1999). Additionally, orientation-selective neurons in monkey V2 also respond to illusory contours with the same orientation (von der Heydt et al. 1984, Peterhans and von der Heydt 1991).

The teleofunctional explanations of both imagery and illusory contours amount to what I shall term the nervous system's employment of a "recruitment strategy": Processes whose primary and original functions serve the perception of real contours get "recruited" to serve other functions. (S. Gould [1991] calls such recruitment "exaptation.") Viewing the nervous system as employing the recruitment strategy thus involves viewing it as conforming to the classical empiricist doctrine that nothing is in the mind that is not first in the senses. Further, it supplies an outline in neural terms of how that which is first in the senses can come to serve other cognitive processes.

The representational explanation of the perception of illusory contours helps to show that the representational theory has the resources to explain at least some cases of underdetermined perception. But the question arises of whether it has the resources to explain all cases of underdetermined perception, especially cases of active perception. Some theorists, such as Gibson (1966, 1986) and O'Regan and Noë (2001), have urged that it does not. O'Regan and Noë (2001) reject representational theories of vision: "Instead of assuming that vision consists in the creation of an internal representation of the outside world whose activation somehow generates visual experience, we propose to treat vision as an *exploratory activity*" (p. 940). According to O'Regan and Noë's alternative – their "sensorimotor contingency theory" – all visual perception is characterized as active perception, or, in their own words, "*vision is a mode of exploration of the world that is mediated by knowledge of what we call sensorimotor contingencies*" (p. 940, emphasis in original). I presume that O'Regan and Noë intend the *knowledge* of sensorimotor contingencies to not involve the *representation* of sensorimotor contingencies.

I will not here rehearse Gibson's or O'Regan and Noë's case against the representational theory, but instead sketch some general reasons why viewing perception as active might be thought to pose a threat. In brief, the problem is that active perception highlights the importance of output while the representational story is told in terms of inputs. Recall that the notion of representation is explicated in terms of states that have the function of being caused by environmental events. Thus, the basic case of a representation is neural activation that occurs as a response to some sensory input. Active perception, however, is a kind of underdetermined perception – that is, perception underdetermined by sensory inputs. Further, what does determine the percept in active perception is a combination of the inputs with certain kinds of outputs. Since output seems to be bearing so much of the load, there seems to be little hope for a story told exclusively in terms of inputs. A further problem arises when we consider that it is not clear that the recruitment strategy is as readily available for active perception as it is for other kinds of underdetermined perception.

Illusory contour perception is subjectively similar to the perception of real contours. The reactivation of brain areas responsible for the perception of real contours gives rise to a subjective appearance similar to what is experienced when real contours are present. This is part of what it means to call illusory contour perception "illusory." If something similar were occurring in active perception, then we would expect an analogous tactile illusion. However, in the pen and tablet version of TVSS, the percept does not involve tactile illusion; that is, the subject doesn't feel the portions of the contours that are not currently being scanned. Given these sorts of considerations, the threat of active perception to the representational theory of perception seems to be two-pronged: The first prong criticizes the representational theory for being overly reliant on the contributions of input, and the second prong criticizes the representational theory for being overly reliant on the recruitment strategy.

Meeting the Challenge

Active perception poses an apparently serious threat to the representational theory of perception. However, this apparent seriousness should not be confused with hopelessness. On the contrary, a rather minor revision of the representational theory will suffice to ward off the threat. The revision concerns the conditions on being a representation and will include a role for output as well as input for determining representational contents.

To get the clearest possible grasp on this account of the representational basis of perception, it will be useful to consider the simplest possible examples of a creature undergoing a fully determined visual perception. Imagine a creature that moves about a planar surface and utilizes a pair of light sensors – mounted on the creature's left and right, respectively – to orient toward sources of illumination. Sunlight is beneficial to various creatures in various ways, and thus positive phototaxis is a common example of an adaptive response to an environmental stimulus. In the two-sensor creature that we are imagining, activity in each sensor is a linear function of the local light intensity, and given a constant light source, degree of activation in the sensor represents proximity to the light source. Thus, the difference in the activity between the two sensors encodes the location of the light source in a two-dimensional egocentric space. Information encoded by the sensors can be relayed to and decoded by motor systems responsible for steering the creature. For example, left and right opposing muscles might have their activity directly modulated by contralateral sensors so that the greater contraction corresponds to the side with the greatest sensor activity, thus steering the creature toward the light. More complex uses of the sensory inputs would involve having them feed into a central processor that gives rise to a perceptual judgment that, say, the light is to the right. The example sketched so far constitutes an example of determined perception on the following grounds. If the perception is a state of the organism specifying the location in two-dimensional egocentric space of the light source, then this is a percept fully determined by the information encoded at the sensory transducers.

To see a simple example of underdetermined perception, in particular, an example of active perception, let us contrast the aforementioned case with a creature forced to make due with only a single light sensor. The single sensor only encodes information regarding proximity to the light source, and thus encodes information about only one dimension of egocentric location of the source. However, this does not prevent the creature from coming to know or coming to form a percept of the two-dimensional egocentric location of the distal stimulus. One way in which the creature might overcome the limitations of a single sensor is by scanning the sensor from left to right while keeping track of the direction in which it has moved the sensor. By comparing the reading of the sensor when moved to the right to the reading of the sensor when moved to the left, the creature thereby has access to information similar to the creature with two sensors. Here, two-dimensional location is encoded not in the difference between two sensors but, instead, in the difference between the activity occurring at two different times within the same sensor. In order to make use of this information, however, the scanning creature needs some way of knowing when the sensor is in the left position and when the sensor is in the right position.

There are two general conditions in which the creature can accomplish this. In the first condition – the feedback condition – the creature receives sensory feedback regarding the states of its muscles. Thus, in the feedback condition, while the percept may be underdetermined by the input from the light sensor, it is not underdetermined by sensation altogether, since sensory input from the muscles, combined with the light sensor input, determines the percept. Thus, the feedback condition is only a case of relative, not absolute, underdetermined perception. In the second condition – the efference copy condition – the creature knows the position of the scanning organ by keeping track of what commands were sent to the scanning organ. Thus, in the efference copy condition, the percept is genuinely underdetermined by sensation, since what augments the sensory input from the light sensor is not some additional sensory input from the muscles, but instead a record of what the outputs were – that is, a copy of the efferent signal. If it is a requirement on active perception that it be underdetermined by sensation altogether, and not just underdetermined relative to some subset of the sensory inputs, then only the efference copy condition constitutes a genuine case of active perception. Thus, if so-called active perception is only relatively underdetermined, then it doesn't pose the kind of threat to the representational theory outlined earlier. There ultimately is adequate input information for the determination of the percept. However, as I will argue, even genuine (efference copy–based) active perception can be explained in terms of the representational theory of perception.

The representational theory of perception, although not defeated by active perception, will nonetheless require an adjustment. The adjustment required is to acknowledge that there are occasions in which outputs instead of inputs figure into the specification of the content of a representational state. I propose to model these output-oriented – that is, action-oriented – specifications along the lines utilized in the case of inputs. When focusing on input conditions, the schematic theory of representational content is the following: A state of an organism represents *F*s if that state has the teleological function of being caused by *F*s. I propose to add an additional set of conditions in which a state can come to represent *F*s by allowing that a reversed direction of causation can suffice. A state of an organism represents *F*s if that state has the teleological function of causing *F*s. Thus, in the single-sensor creatures described earlier, the motor command to scan the sensor to the left is as much an adequate representation that something is happening to the left as is a sensory input caused by something happening to the left. Efference copies inherit their representational contents from the motor commands that they are copies of. Efference copies thus constitute an action-oriented version of the recruitment strategy. We are now in a position to define "action-oriented representation" as any representation whose content is determined, in whole or in part, by involving states whose teleofunction is to be the causal antecedents of actions. Another way to state the definition is that action-oriented representations are any representations that have, in whole or in part, imperative content.[7] Active perception thus does not threaten the representational theory of perception. Instead, it forces us to acknowledge that action-oriented representations can contribute to the representational content of perception, and further, that percepts themselves may sometimes be action-oriented representations.

I should note a point of contrast between the account of spatial content I articulate here and elsewhere (Mandik 1999, 2001, 2002, and 2003) and other action-involving accounts, such as G. Evans (1985) and especially R. Grush (2001, this volume). On Grush's "skill theory" of spatial content, certain behavioral dispositions are necessary for a mental state such as a percept to have spatial representational content. According to this view, it would thus be impossible for an organism to perceive a stimulus as being to the left without at the same time being able to orient toward that stimulus. On such a view, states at the input side of the cognitive system cannot by themselves carry spatial content; only states appropriately engaged with motor outputs count as genuinely representing spatial properties and relations. In contrast, though I grant that certain output-involving processes (such as motor commands and efference copies) are sufficient for spatial content, I reject the claim that they are thereby necessary. There are many varieties of spatial representation, only some of which significantly engage motor processes. (See Mandik 2003 for a longer discussion of these varieties of representation.)

Now that the representational account of active perception has been sketched, I devote the rest of the chapter to the following three questions. First, is the solution sketched feasible? That is, is it possible to employ it as an engineering solution to the problem of utilizing action to

---

7 This contrasts with the way A. Clark (1997) defines action-oriented representations. For Clark, action-oriented representations always have both imperative and indicative content and are thus the same as what R. Millikan (1996) calls "Pushmi-Pullyu Representations." On my definition, even representations with only imperative content (e.g., motor commands) are action-oriented representations.

compensate for impoverished inputs? Second, is the solution sketched evolvable? Given the reliance on evolution in typical versions of the tele-ofunctional portion of the story sketched earlier, it remains a serious question whether the sort of incremental adaptations posited in most evolutionary scenarios could possibly give rise to such a solution. Third, even if feasible and evolvable, are such solutions actually instantiated in human nervous systems?

Is the Action-Oriented Solution Feasible? A Reply from Robotics

If the action-oriented representation solution described so far is indeed feasible, then it ought to be possible to construct a robotic model that employs such principles to exhibit perceptually guided adaptive behaviors. In Mandik (1999), I discuss a thought experiment about an imaginary robot named Tanky that traverses a planar surface by means of tank treads. Various patches of the surface are considered either nutritious or noxious to Tanky, and while poised over one of these patches, Tanky's chemoreceptors can indicate as much. However, the chemoreceptors are alone insufficient to give Tanky much information about the spatial arrangement of the various proximal and distal chemical patches in its environment. Tanky's perceptual contact with the spatial features of its environment is mediated through the tank treads, thus implementing a form of odometry. There are two general ways in which this odometry might be accomplished to give Tanky knowledge of, for instance, the distance between the chemical patch it currently perceives and the last patch it visited. The first way is akin to the feedback solution described earlier, whereas the second way is akin to the efference copy solution. On the feedback solution, distance estimates measured in numbers of tank tread revolutions are updated in virtue of information from a sensor that counts actual tank tread revolutions. In contrast, the efference copy solution forgoes sensor information and instead involves the counting of the number of commands sent to revolve the tank treads.

In Mandik (1999), I hypothesized that both solutions would be equally adequate to provide Tanky with a perception of the spatial arrangement of chemical patches in its environment. Since 1999, however, I have had many occasions to experiment with real robots and discovered, among other things, that odometry and tank tread locomotion don't mix very well due to the high degree of slippage where "rubber meets the road" necessary to effect steering in a treaded vehicle. Nonetheless,

FIGURE 8.2. The robot "Tanky Jr."

real robots offer ample opportunities to demonstrate the viabilities of the feedback and efference copy solutions to spatial perceptual underdetermination. I constructed Tanky Jr. (depicted in Figure 8.2) using the LEGO® MINDSTORMS™ robot kit and programmed the robot using David Baum's (2002) third-party programming language, NQC (Not Quite C).[8] Tanky Jr. is an experimental platform for implementing strategies of positive phototaxis utilizing a single light sensor combined with the kinds of scanning strategies described earlier. Tanky Jr. has three motors: Two drive the left and right wheels, respectively, and the third is utilized to scan Tanky Jr.'s single light sensor left and right.

To implement a feedback strategy to monitor the position of the scanning light sensor, Tanky Jr. has as additional inputs two touch sensors mounted to the left and right of the light sensor. When the robot is first turned on, its wheels remain stationary while it performs a scanning procedure. The first part of the scanning procedure is to scan the light sensor to the right until the touch sensor dedicated to that side is activated. The program then updates a variable that serves as a record of the light sensor activity at that position. Next, the sensor is scanned in the opposite direction until the other touch sensor is activated. The reading of the light sensor in this position is then compared to the previous reading. If the difference in the light readings from the left and right positions are relatively negligible, the robot then moves straight ahead a short distance; otherwise, the robot will turn a bit in the direction of the greatest light reading before making its forward motion. The robot then stops and begins another run of the scanning procedure. The alternating repetition

---

[8] For a nice overview of the philosophical uses of robots as tools for both research and pedagogy with special focus on the LEGO® MINDSTORMS™ system, see J. Sullins (2002).

of these steps is quite effective in getting the robot to move toward a light stimulus, such as a spot of light shone on the floor from a flashlight.

Equally successful is a strategy that forgoes sensory feedback in favor of efference copies. In this latter condition, Tanky Jr.'s touch sensors are removed and the program is altered so that the commands involved in the scanning procedure do not specify that the scanning motion be ceased when the touch sensors are activated, but instead be ceased after a fraction of a second. The left and right light sensor variables are updated not as a response to touch sensor feedback but instead as a response to a record of what commands have been sent. Thus is this latter strategy describable as implementing a system that utilizes efference copies. The equivalence in performance of the efference copy and feedback solutions shows that the efference copy solution is no less representational than the feedback solution.

Is the Action-Oriented Solution Evolvable?
A Reply from Artificial Life

Tanky Jr. shows the feasibility of the solution, although the neural feasibility has not yet been addressed. Also unaddressed until now is the question of whether the efference copy solution is evolvable. In Mandik (2002, 2003), I discuss several artificial life experiments I have conducted to evolve various kinds of neural network controllers for artificial organisms solving simple yet representationally demanding perceptual tasks. Typical experiments involved the modeling of legged land creatures traversing a planar surface. Survival and other estimations of fitness depend on the capacities of the creatures to utilize sensor information to find food distributed through the environment. In Mandik (2003, pp. 118–122), I describe experiments designed to coax the evolution of action-oriented representations in these neural controllers. The artificial creature "Radar" utilized in these latter experiments had the general body structure depicted in Figure 8.3 and the general neural network topology depicted in Figure 8.4. Body structure and neural topology were specified by hand. An evolutionary algorithm was employed to evolve specifications of the neural weights. Radar's forward locomotion is effected by four limbs, and steering is effected by a single bending joint in the middle of its body. Food is detected utilizing a single sensor mounted on a scanning organ that moves left and right in a manner similar to the scanner used by Tanky Jr.

The top layer in figure 8.4 depicts the portion of Radar's nervous system serving as a central pattern generator that sends a sinusoidal signal to

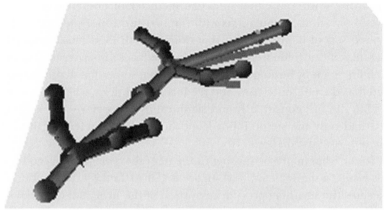

FIGURE 8.3. The artificial life creature "Radar."

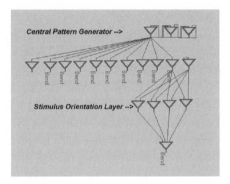

FIGURE 8.4. Radar's nervous system.

the muscles responsible for the forward walking motions as well as the left and right scanning of the sensor organ. Stimulus orientation is effected by a three-layer feed-forward network consisting of a two-neuron input layer, a four-neuron hidden layer, and a single-neuron output layer. The two inputs are the single food sensor and sensory feedback concerning the state of the scanning muscle. The four-unit hidden layer then feeds into the single orientation muscle. A second version of Radar's neural topology replaces the muscular feedback with an efference copy. Instead of receiving feedback from the scanning muscle, the hidden layer of the orientation network receives as input a copy of the command that the central pattern generator sends to the scanning muscle. Neural weights were evolved for three kinds of controller topologies: The first had orientation layer inputs from the sensor and the muscular feedback, the second had inputs from the sensor and an efference copy, and the third had only sensor input without either muscular feedback or efference copies. On several occasions, populations with the first two topologies successfully evolved sets of neural weights that utilized both food sensor input and muscular input (either efferent copy or feedback) in order to maximize their life spans by finding food. However, I was somewhat disappointed to find that the efference copy and feedback conditions, while equally successful, did not consistently and significantly outperform the creatures that had only the single input from the food sensor feeding into the stimulus orientation layer.

To see what might be missing in the neural topologies to account for this result, it is instructive to compare Radar to Tanky Jr. When Tanky Jr. executes the scanning-procedure portion of the program, a crucial step involves using a single sensor to take two different readings – the left and right readings, respectively – of the local light levels. After the second of the two readings is taken, it is compared to a *memory record* of the first reading. This employment of a memory is, I suggest, the crucial difference between Tanky Jr. and Radar. The stimulus orientation network in Radar's nervous system is a three-layer *feed forward* network that lacks recurrent connections or any other means of instantiating a memory. In other words, it lacks the means of being sensitive to information spread out over time. But the task of comparing left and right readings gathered with a single scanning sensor is, crucially, a process that occurs over time. Therefore, future versions of Radar must incorporate some means (such as recurrent connections) of storing information about a previous sensor reading long enough for it to be compared to a current sensor reading.

While I have not yet experimented with versions of Radar that incorporate memory into the scanning procedure, in Mandik (2003, pp. 111–118), I discuss creatures that I have evolved to utilize memory in a similar task, namely, the comparison of a past and current stimulus. In these simulations, creatures with a single sensor did not scan it left and right but, however, did utilize it in a comparison between past and current stimuli by routing the sensor signal through two channels in the stimulus orientation network. One of the two channels passed its signal through more neurons, thus constituting a memory delay. The portion of the network that had to effect a comparison thus compares the current signal to a delayed signal. This can be part of an adaptive strategy for food-finding insofar as it, in combination with the tacit assumption that the creature is moving forward, allows the creature to draw something like the following inference: If the current value is higher than the remembered value, then the creature must be heading toward the stimulus and should thus continue doing so, but if the current value is lower than the remembered value, then the creature must be heading away from the stimulus and must thus turn around. Such a use of memory has been shown to be used by *E. coli* bacteria to navigate up nutrient gradients (Koshland 1977, 1980).[9] This initial success with these artificial life simulations helps bolster the claim of the evolvability of the kinds of action-oriented representation solutions implemented in the robot Tanky Jr. Much remains open, however – in particular, the question to which I now turn: *Do human nervous systems utilize any action-oriented representations?*

Is the Action-Oriented Solution Instantiated in Human Nervous Systems? A Reply from Neuroscience

One especially promising line of evidence concerning whether efference copy–based action-oriented representations are employed in human nervous systems comes from research on visual stability during saccadic eye movements. The phenomenon to be explained here is how it is that we don't perceive the world to be jumping around, even though our eyes are constantly moving in the short jerky movements known as saccades. Hermann Helmholtz (1867) hypothesized that efference copies are used in the following manner. When the eye moves, there is a shift in the array of information transduced at the retina. When the eye movement is caused in the normal way – that is, by self-generated movements due to commands sent to ocular muscles – an efference copy is used to compute the amount to compensate for the anticipated shift in the retinal image. The amount of movement estimated on the basis of the character of the efference copy is thus used to offset the actual shift in retinal information, giving rise, ultimately, to a percept that contains no such shift.

This hypothesis implies that there should be a perception of a shift in cases in which the eye is moved in the absence of efference copies, as well as in cases in which efference copies are generated but no eye movement is produced. The first sort of case may be generated by eye movements produced by tapping or pushing on the eye. A quick way to verify this is to take your own (clean!) finger and gently push the side of your eye. Your eye is now moving with respect to the visual scene in a manner actually less extreme than in many saccadic motions. However, the instability of

---

9 For further discussion of artificial life simulations involving neural representation of information about the past, see Mandik 2002, pp. 14–15, and Mandik 2003, pp. 111–118.

the visual scene – it jumps dramatically as you gently nudge your eye with your finger – far outstrips any visual instabilities that rapid saccades might occasion (Helmholtz 1867). The second sort of case arises when subjects have their ocular muscles paralyzed by a paralytic such as curare. When the subjects attempt to move their eyes, they perceive a shift in the visual scene even though no physical movement has actually occurred. (Mach 1885 and Stevens et al. 1976).

C. Colby (1999) hypothesizes that the lateral intraparietal area (LIP) constitutes the neural locus for the efference copy–based updating of the visual percept. LIP neural activity constitutes a retinocentric spatial representation. However, this activity reflects not just current retinal stimuli but also a memory record of previous stimulation. Additionally, the memory representation can be shifted in response to efference information and independently of current retinal stimulation. Colby (1999, pp. 114–116) reports experiments on monkeys in which LIP neural responses to a stimulus flashed for only 50 msec get remapped in response to a saccade. (The remapping is the shift of receptive fields from one set of neurons to another.) The duration of the stimulus was insufficiently short to account for the remapping; thus, the remapping must be due to the efference copy.

The previously discussed evidence concerning the role of efference copies in perceptual stability during saccades points to some crucial similarities between, on the one hand, complicated natural organisms such as humans and monkeys and, on the other hand, extremely simple artificial organisms such as Tanky Jr. and Radar. Both the natural and the artificial creatures actively scan their environments, and the content of the percept is, while underdetermined by sensory input, determined by the combined contribution of sensory input and efference copy information concerning motor output.

I turn now to consider a possible objection to my account. The account I'm offering here sees action-oriented representations as determining the character of many instances of perceptual experience. J. Prinz (2000) and A. Clark (2002) raise a concern about accounts such as this that postulate relatively tight connections between the determinants of action and the content and character of perceptual experience.[10] The worry stems from consideration of D. Milner and M. Goodale's (1995) hypothesis that visually guided action is localized primarily in the dorsal stream (cortical

---

[10] Such accounts include Grush 1998, Cotterill 1998, Hurley 1998, and O'Regan and Noë 2001.

areas leading from V1 to the posterior parietal area), whereas conscious perception is localized primarily in the ventral stream (cortical areas leading from V1 to inferotemporal cortex). The worry that Prinz and Clark raise is that action cannot be too closely coupled to perception, since the work of Milner and Goodale serves to show a dissociation between the processes that are most intimately involved in action and the processes that are most intimately involved with perceptual consciousness.

I have two responses to this worry. The first is that, unlike, say, R. Cotterill (1998) and O'Regan and Noë (2001), I am *not* saying that the sorts of contributions that action sometimes makes to perception will be either necessary or sufficient for a perceptual state to count as a conscious mental state. I am arguing merely that action-oriented processes sometimes contribute to the representational contents of perceptual consciousness. What contributes to the content of a conscious state need not be one and the same as what makes that state a conscious mental state. Indeed, there are plenty of accounts of consciousness that dissociate the conditions that make a state have a particular content and the conditions that make that state conscious. Two prominent examples are M. Tye's (1995) Poised Abstract Non-conceptual Intentional Content (PANIC) theory and D. Rosenthal's (1997) Higher-Order Thought (HOT) theory. Further, the theories of consciousness that Clark (2000a, 2000b) and Prinz (2000, 2001, this volume) advocate are consistent with this general sort of dissociation.

My second, and not unrelated, response to the stated worry is that there is evidence that activity in the dorsal stream does influence conscious perception. Such evidence includes the evidence previously described concerning parietal processing of efference copies for visual stability during saccades. Additionally, see V. Gallese et al. (1999) for a brief review of various imagery studies implicating parietal areas in conscious motor imagery. M. Jeannerod (1999) similarly questions whether dorsal stream activity should be regarded as irrelevant for conscious perception. He describes PET studies by I. Faillenot et al. (1997) that implicate parietal areas in both an action task involving grasping objects of various sizes and a perception task involving matching the objects with each other.

Conclusion

Perception oft involves processes whereby the perceiver is not a passive receptacle of sensory information but actively engages and explores the perceptible environment. Acknowledging the contributions that action makes to perception involves a certain rethinking of perception. However, we are not thereby forced to abandon the view that perception is a representational process. Indeed, the impact of action on the mind is mediated through representations of action. In cases in which transducer input is insufficient to provide the requisite representations of action, efference copies of motor commands may be substituted, since they themselves are representations of action. Efference copies are examples of action-oriented representations, and insofar as they contribute to the makeup of perceptual contents, our perceptual states themselves become action-oriented representations.

Acknowledgments

This work was supported in part by grants from the McDonnell Foundation (administered through the McDonnell Project in Philosophy and the Neurosciences) and the National Endowment for the Humanities. For helpful feedback I thank the audiences of oral presentations of this work at the Carleton/ McDonnell Conference on Philosophy and Neuroscience and the William Paterson University Psychology Department Colloquium Series. I am especially grateful for comments from Patricia Churchland, Chris Eliasmith, Pierre Jacob, Alva Noë, and Ruth Millikan.

References

Austin, J. (1964). *Sense and Sensibilia.* New York: Oxford University Press.
Bach-y-Rita, P. (1972). *Brain Mechanisms in Sensory Substitution.* New York and London: Academic Press.
Baum, D. (2002). *The Definitive Guide to LEGO MINDSTORMS, Second Edition.* Berkeley, CA: Apress.
Churchland, P. S., Ramachandran, V. S., and Sejnowski, T. J. (1994). "A critique of pure vision." In C. Koch and J. L. Davis (eds.), *Large-Scale Neuronal Theories of the Brain.* Cambridge, MA: MIT Press, pp. 23–60.
Clark, A. (1997). *Being There.* Cambridge, MA: MIT Press.
Clark, A. (2000a). "A case where access implies qualia?" *Analysis* 60: 1: 30–37.
Clark, A. (2000b). "Phenomenal immediacy and the doors of sensation." *Journal of Consciousness Studies* 7 (4): 21–24.
Clark, A. (2002). "Visual experience and motor action: Are the bonds too tight?" *Philosophical Review* 110: 495–520.
Colby, C. (1999). "Parietal cortex constructs action-oriented spatial representations." In N. Burgess, K. J. Jeffery, and J. O'Keefe (eds.), *The Hippocampal and Parietal Foundations of Spatial Cognition.* New York: Oxford University Press, pp. 104–126.
Cotterill, R. (1998). *Enchanted Looms: Conscious Networks in Brains and Computers.* Cambridge: Cambridge University Press.

Dretske, F. (1995). *Naturalizing the Mind.* Cambridge, MA: MIT Press.

Evans, G. (1985). "Molyneux's question." In Gareth Evans, *The Collected Papers of Gareth Evans.* London: Oxford University Press.

Faillenot, I., Toni, I., Decety, J., Gregoire, M. C., and Jeannerod, M. (1997). "Visual pathways for object-oriented action and object identification: Functional anatomy with PET." *Cerebral Cortex* 7: 77–85.

Fodor, J. (1987). *Psychosemantics.* Cambridge, MA: MIT Press.

Gallese, V., Craighero, Laila, Fadiga, Luciano, and Fogassi, Leonardo (1999). "Perception through action." *Psyche* 5 (21): http://psyche.cs.monash.edu.au/v5/psyche-5-21-gallese.html.

Gibson, J. (1966). *The Senses Considered as Perceptual Systems.* Boston: Houghton Mifflin.

Gibson, J. (1986). *The Ecological Approach to Visual Perception.* Hillsdale, NJ: Lawrence Erlbaum.

Gould, S. (1991). "Exaptation: A crucial tool for evolutionary psychology. *Journal of Social Issues* 47: 43–65.

Grice, H. (1961). "The causal theory of perception." *Proceedings of the Aristotelian Society,* sup. vol. 35: 121–152.

Grush, R. (1998). "Skill and spatial content." *Electronic Journal of Analytic Philosophy* 6: http://ejap.louisiana.edu/EJAP/1998/grusharticle98.html.

Grush, Rick (2001). "Self, world and space: On the meaning and mechanisms of egocentric and allocentric spatial representation." *Brain and Mind* 1 (1): 59–92.

Hanneton S., Gapenne, O., Genouel, C., Lenay, C., and Marque C. (1999). "Dynamics of shape recognition through a minimal visuo-tactile sensory substitution interface." *Third International Conference on Cognitive and Neural Systems,* pp. 26–29.

Helmholtz, H. (1867). *Handbuch der Physiologischen Optik.* In G. Karsten (ed.), *Allgemeine Encyklopädie der Physik,* vol. 9. Leipzig: Voss.

Hubel, D. H., and Wiesel, T. N. (1962). "Receptive fields, binocular interaction, and functional architecture in the cat's visual cortex." *Journal of Physiology* 195: 215–243.

Hurley, S. (1998). *Consciousness in Action.* Cambridge, MA: Harvard University Press.

Hyman, J. (1992). "The causal theory of perception." *Philosophical Quarterly* 42 (168): 277–296.

Jarvilehto, T. (1998). "Efferent influences on receptors in knowledge formation." *Psycoloquy* 9 (41): http://www.cogsci.soton.ac.uk/cgi/psyc/newpsy?9.41.

Jeannerod, M. (1999). "A dichotomous visual brain?" *Psyche* 5 (25): http://psyche.cs.monash.edu.au/v5/psyche-5-25-jeannerod.html.

Keeley, B. (2002). "Making sense of the senses: Individuating modalities in humans and other animals." *Journal of Philosophy* 99: 5–28.

Koshland, D. (1977). "A response regulator model in a simple sensory system. *Science* 196: 1055–1063.

Koshland, D. (1980). "Bacterial chemotaxis in relation to neurobiology." In W. C. Cowan et al. (eds.), *Annual Review of Neurosciences 3.* Palo Alto: Annual Reviews, pp. 43–75.

Kosslyn, S., Ganis, G., and Thompson, W. (2001). "Neural foundations of imagery." *Nature Reviews Neuroscience* 2: 635–642.

Larsson J., Amunts, K., Gulyas, B., Malikovic, A., Zilles, K., and Roland, P. (1999). "Neuronal correlates of real and illusory contour perception: Functional anatomy with PET." *Eur J Neurosci.* 11 (11): 4024–4036.

Lenay, C., Cannu, S., and Villon, P. (1997). "Technology and perception: The contribution of sensory substitution systems." In *Second International Conference on Cognitive Technology.* Aizu, Japan, and Los Alamitos, CA: IEEE, pp. 44–53.

Livingstone, M., and Hubel, D. (1988). "Segregation of form, color, movement and depth: Anatomy, physiology and perception." *Science* 240 (4853): 740–749.

Mach, E. (1885). *Die Analyse der Empfindungen.* Jena: Fischer.

Mandik, P. (1999). "Qualia, space, and control." *Philosophical Psychology* 12 (1): 47–60.

Mandik, P. (2001). "Mental representation and the subjectivity of consciousness." *Philosophical Psychology* 14 (2): 179–202.

Mandik, P. (2002). "Synthetic neuroethology." In T. W. Bynum and J. H. Moor (eds.), *CyberPhilosophy: The Intersection of Philosophy and Computing.* New York: Blackwell, 2003, pp. 8–25.

Mandik, P. (2003). "Varieties of representation in evolved and embodied neural networks." *Biology and Philosophy* 18 (1): 95–130.

Millikan, R. (1996). "Pushmi-pullyu representations." In L. May, M. Friedman, and A. Clark (eds.), *Minds and Morals.* Cambridge, MA: MIT Press, pp. 145–161.

Milner, D., and Goodale, M. (1995). *The Visual Brain in Action.* Oxford: Oxford University Press.

Oakes, R. (1978). "How to rescue the traditional causal theory of perception." *Philosophy and Phenomenological Research* 38 (3): 370–383.

O'Craven, K., and Kanwisher, N. (2000). "Mental imagery of faces and places activates corresponding stimulus-specific brain regions." *Journal of Cognitive Neuroscience* 12: 1013–1023.

O'Regan, J., and Noë, A. (2001). "A sensorimotor account of vision and visual consciousness." *Behavioral and Brain Sciences* 24 (5): 939–1011.

Perrett, D., Mistlin, A., and Chitty, A. (1989). "Visual neurons responsive to faces." *Trends in Neurosciences* 10: 358–364.

Peterhans, E., and von der Heydt, R. (1991). "Subjective contours – bridging the gap between psychophysics and physiology." *Trends in Neurosciences* 14: 112–119.

Prinz, J. (2000). "The ins and outs of consciousness." *Brain and Mind* 1 (2): 245–256.

Prinz, J. (2001). "Functionalism, dualism, and the neural correlates of consciousness." In W. Bechtel, P. Mandik, J. Mundale, and R. Stufflebeam (eds.), *Philosophy and the Neurosciences: A Reader.* Oxford: Blackwell.

Rosenthal, D. (1997). "A theory of consciousness." In Ned Block, O. Flanagan, and G. Guzeldere (eds.), *The Nature of Consciousness.* Cambridge, MA: MIT Press.

Stevens, J., Emerson, R., Gerstein, G., Kallos, T., Neufield, G., Nichols, C., and Rosenquist, A. (1976). "Paralysis of the awake human: Visual perceptions." *Vision Research* 16: 93–98.

Sullins, J. (2002). "Building simple mechanical minds: Using LEGO® robots for research and teaching in philosophy." In T. W. Bynum and J. H. Moor (eds.), *CyberPhilosophy: The Intersection of Philosophy and Computing.* New York: Blackwell, 2003, pp. 104–116.

Tye, M. (1995). *Ten Problems of Consciousness: A Representational Theory of the Phenomenal Mind.* Cambridge, MA: MIT Press.

von der Heydt, R., Peterhans, E., and Baumgartner, G. (1984). "Illusory contours and cortical neuron responses." *Science* 224: 1260–1262.

COLOR VISION

9

Chimerical Colors: Some Novel Predictions from Cognitive Neuroscience

Paul M. Churchland

1 Introduction

The qualitative character of subjective experience is often claimed to be beyond the predictive or explanatory powers of any physical theory.[1] Almost equally often, conclusions are then drawn concerning the physical irreducibility and the metaphysical distinctness of the subjective qualia at issue. Resistance to such Dualist themes has typically focused on the dubious legitimacy of the inference just displayed.[2] The present chapter, by contrast, focuses on the premise from which the inference is drawn. My burden here is to show that this premise is false.

I will illustrate its falsity by drawing a number of novel, counterintuitive, and, in some cases, patently paradoxical predictions concerning the qualitative character of certain highly unusual visual sensations, sensations produced under some highly unusual visual circumstances, sensations you have probably never had before. These predictions will be drawn, in a standard and unproblematic way, from the assumptions of what deserves to be called the Standard Model of how color is processed and represented within the human brain.[3] I am thus posing only as a

[1] Huxley, T. H., *Elementary Lessons in Physiology* (London: Macmillan, 1866); Nagel, T., "What Is It Like to Be a Bat?" *Philosophical Review*, 83, no. 4, 1974: 435–450; Jackson, F., "Epiphenomenal Qualia," *Philosophical Quarterly*, 32, no. 127, 1982: 127–136; Chalmers, D., *The Conscious Mind* (Oxford: Oxford University Press, 1996).

[2] Churchland, P. M., "Reduction, Qualia, and the Direct Introspection of Brain States," *Journal of Philosophy*, 82, no. 1, 1985: 8–28; Churchland, P. M., "The Rediscovery of Light," *Journal of Philosophy*, 93, no. 5, 1996: 211–228; Bickle, J., *Psychoneural Reduction: The New Wave* (Cambridge, Mass.: MIT Press, 1998).

[3] Hurvich, L. M., *Color Vision* (Sunderland, Mass.: Sinauer, 1981); Hardin, C. L., *Color*

consumer of existing cognitive neuroscience, not as an advocate of new theory. But standard or not, this familiar "color-opponency" theory of chromatic information processing has some unexpected and unappreciated consequences concerning the full range of neuronal activity possible, in an extreme, for the human visual system. From there, one needs only the tentative additional assumption of a systematic identity between *neuronal coding vectors*, on the one hand, and *subjective color qualia*, on the other – a highly specific material assumption in the spirit of the classical Identity Theory, and in the spirit of intertheoretic reductions generally – to formally derive the unexpected but qualia-specific predictions at issue.

Accordingly, these predictions provide no less than an empirical test of the Identity Theory itself, in one of its many possible (physically specific) guises. We may therefore approach with interest the question of whether the weird predictions promised here actually accord with the data of subjective experience, in addition to the question of whether and how those predictions arise in the first place. The several color figures provided with this chapter, plus some experimental procedures to be described as we proceed, will allow you to test the relevant predictions for yourself. The aim is to produce in you color sensations that you have (almost certainly) never experienced before, sensations whose highly specific descriptions in commonsense terms are, by prior semantic lights, flatly self-contradictory. Nonetheless, those nonstandard sensations are real, their paradoxical descriptions are accurate, and the Standard Model predicts them all, right out of the box.

2 The Standard Model: The Color Spindle and the Hurvich Net

The many colors perceivable by humans bear a complex set of similarity and dissimilarity relations that collectively position each color uniquely within a continuous manifold. The global structure of that manifold has been known since Albert Munsell first pieced it together over a century ago. (See Plate 3 for a slightly oversimplified rendition. A more accurate rendition would have both cones bulging outwards somewhat.)[4] The agreeably simple Hurvich net is a recent attempt to *explain* that

---

for *Philosophers: Unweaving the Rainbow* (Indianapolis: Hackett, 1988); Clark, A., *Sensory Qualities* (Oxford: Oxford University Press, 1993).

[4] Strictly speaking, Munsell intended his solid to represent the relations between the various external or *objective* colors. But it serves equally well as a representation of the similarity and difference relations between our internal color *sensations* as well. That is the use to which it is here being put. That is, the spindle-shaped solid represents our *phenomenological color space*.

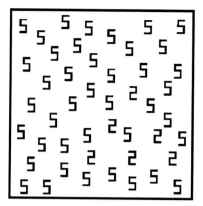

PLATE 1. *Source*: Redrawn from V. S. Ramachandran and E. M. Hubbard (2001), Synaesthesia: A window into perception, thought and language. *Journal of Consciousness Studies* 8(12): 3–34.

PLATE 2. *Source*: Redrawn from V. S. Ramachandran and E. M. Hubbard (2001), Synaesthesia: A window into perception, thought and language. *Journal of Consciousness Studies* 8(12): 3–34.

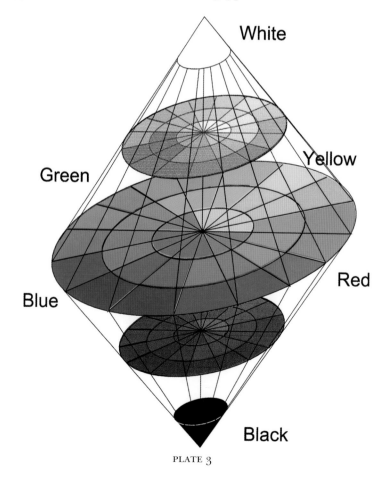

PLATE 3

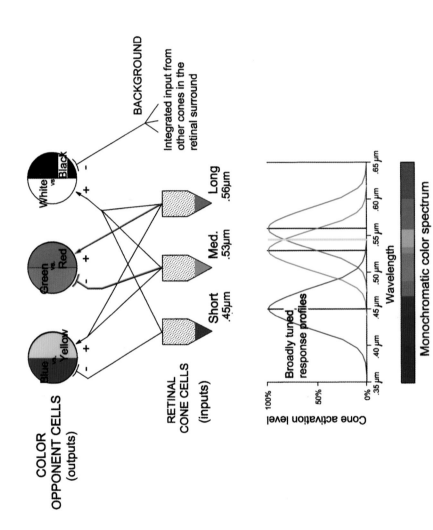

PLATE 4

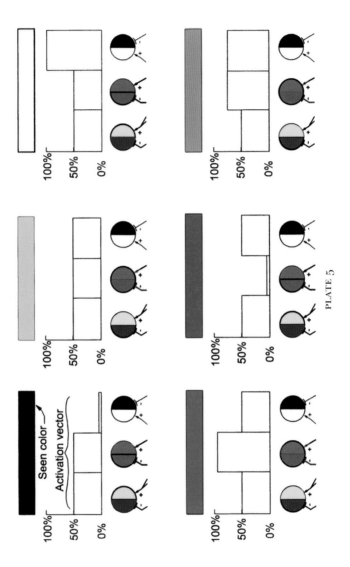

PLATE 5

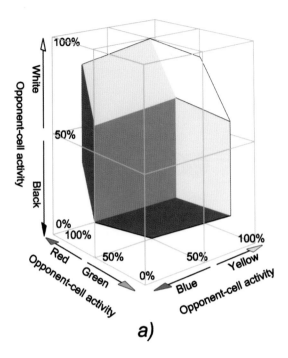

a)

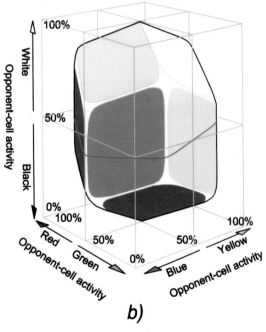

b)

PLATE 6

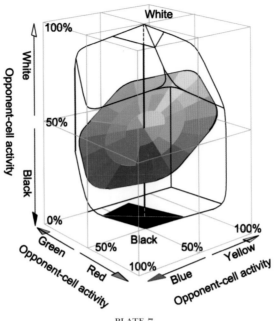

PLATE 7

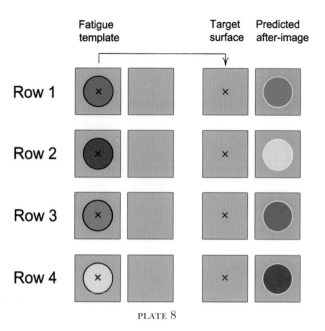

Fatigue template | Target surface | Predicted after-image

Row 1

Row 2

Row 3

Row 4

PLATE 8

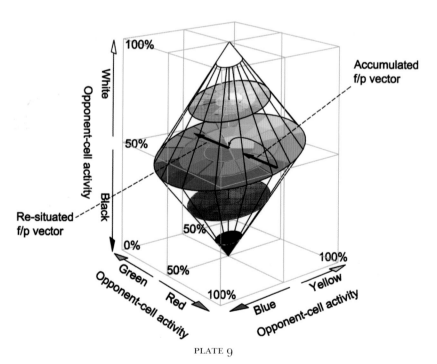

100%

White

Opponent-cell activity

50%

Black

0%

Accumulated f/p vector

Re-situated f/p vector

50%

Green

Opponent-cell activity

Red

50%

100%

Blue

100%

Yellow

Opponent-cell activity

PLATE 9

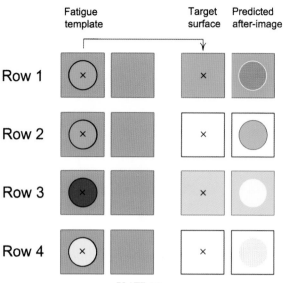

PLATE 10

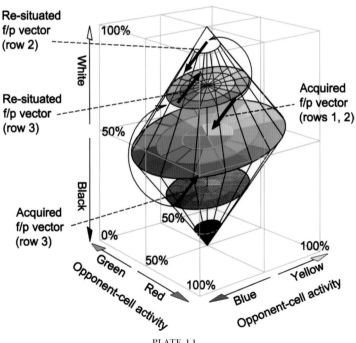

PLATE 11

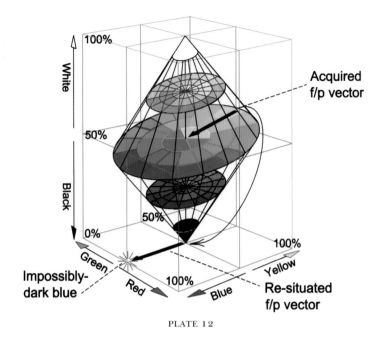

PLATE 12

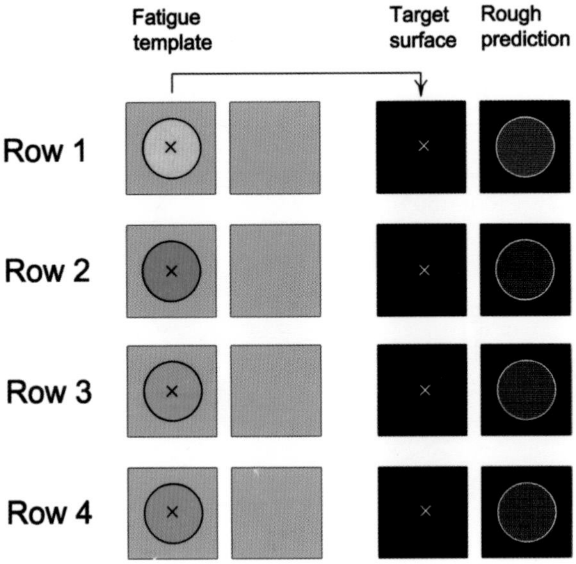

Row 1

Row 2

Row 3

Row 4

Fatigue
template

Target
surface

Rough
prediction

PLATE 13

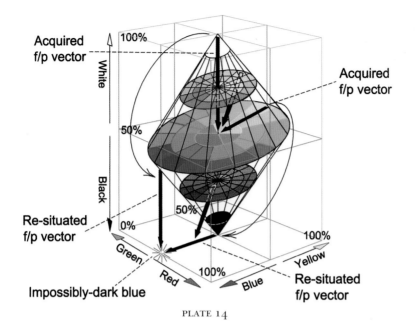

PLATE 14

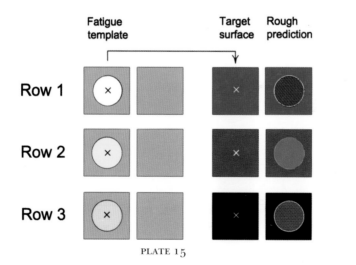

PLATE 15

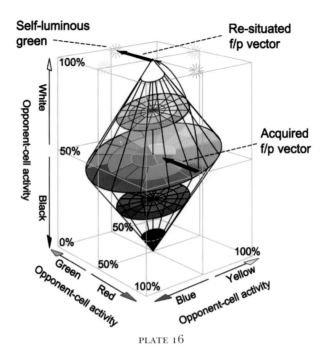

PLATE 16

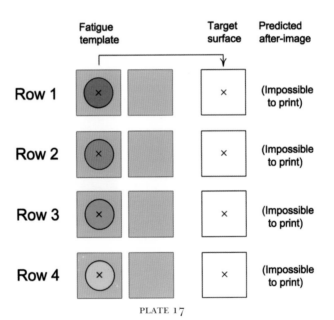

PLATE 17

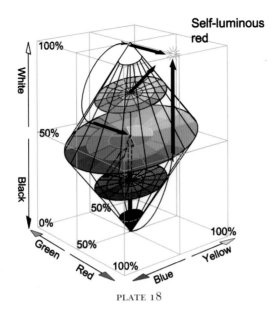

Self-luminous
red

White

Black

100%

50%

50%

0%

50%

50%

Green

Red

100%

100%

Blue

Yellow

PLATE 18

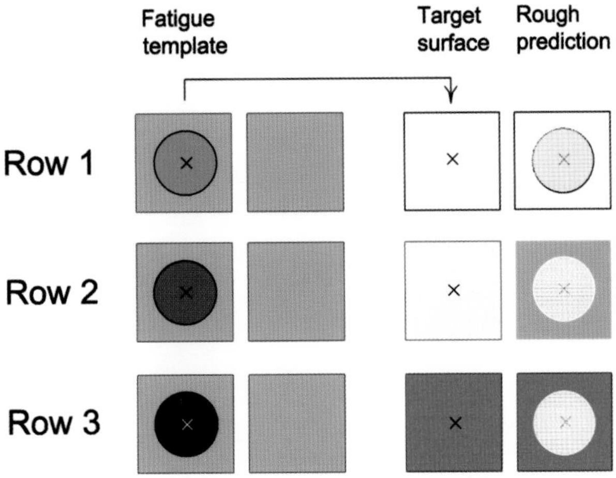

Fatigue
template

Target
surface

Rough
prediction

Row 1   ×

× ×

Row 2   ×

× ×

Row 3   ×

× ×

PLATE 19

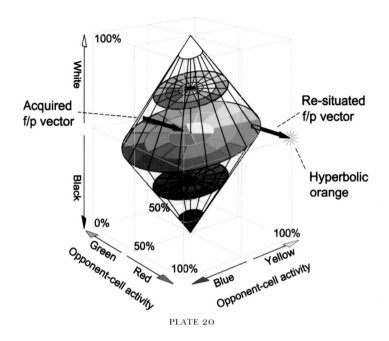

PLATE 20

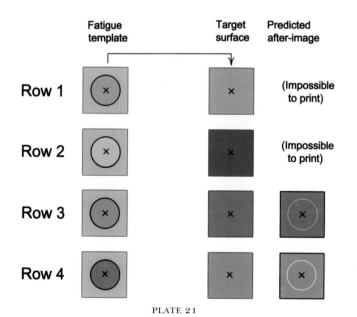

PLATE 21

global structure in terms of the known elements of the human visual system. It begins with the three types of cone-cells distributed across the retina, cells broadly tuned to three distinct regions of the visible spectrum, conventionally dubbed the short, medium, and long wavelength (S, M, and L) cones, respectively. And it ends with three kinds of color-coding cells at its output layer (corresponding, perhaps, to neurons in the lateral geniculate nucleus of the primate visual system), cells whose activity levels code for the simultaneous position of any visual stimulus along a blue-to-yellow axis, a green-to-red axis, and a white-to-black axis (see Plate 4).

On this model, the resting levels of electrical activity in the three input cones are postulated to be zero, with a maximum level of 100%.[5] By contrast, the resting levels of electrical activity in the three second-rung output cells are postulated to be 50% of their maximum possible activation levels (the full range is zero spikes/sec to roughly 100 spikes/sec). Excursions above and below that midway default level are induced by whatever excitatory or inhibitory inputs happen to arrive from the various cone-cells below. Each output cell is thus the site of a tug-of-war between various input cones, or coalitions of input cones, working with and against each other to excite or to inhibit, above or below the spontaneous resting level of 50%, the particular output cell to which they severally project.

The so-called Green/Red cell is the simplest case, since its activation level registers the relative preponderance of the long-wavelength light over/under the medium-wavelength light arriving to the cones at the tiny area of the retina that contains them. A local preponderance of long over medium excites the L-cones more than the M-cones, which yields a net *stimulation* at the Green/Red cell (note the "+" and "−" signs next to the relevant synaptic connections, indicating excitatory and inhibitory connections, respectively). This net stimulation sends its activation level *above* 50% by an amount that reflects the degree of the mismatch between the excitatory and the inhibitory signals arriving from the L- and M-cones. That Green/Red opponency cell will then be coding for something in the direction of an increasingly saturated red. Alternatively, if the local preponderance of incoming light favors the medium wavelengths over the

---

[5] As an aside, human cone-cells respond to light with smoothly varying *graded* potentials (voltage coding), rather than with the varying frequencies of spiking activity (frequency coding) so common in the rest of the nervous system. This wrinkle is functionally irrelevant to the first-order model, which is why cellular activation levels are expressed *neutrally*, in what follows, as a simple percentage of maximum possible activation levels.

long, then the net effect at the Green/Red cell will be *inhibitory*. Its activation level will be pushed below the default level of 50%, and it will then be coding for something in the direction of an increasingly saturated green.

The story for the Blue/Yellow opponency cell is almost identical, except that the rightmost two cone-cells (the M- and the L-cones), the ones that are jointly tuned to the *longer* half of the visible spectrum, here join forces to test their joint mettle against the inputs of the S-cone that is tuned to the *shorter* half of the spectrum. A predominance of the latter over the former pushes the Blue/Yellow opponency cell *below* its 50% default level, which codes for an increasingly saturated blue; and a predominance in the other direction pushes it *above* 50%, which (if the inputs from the M- and L-cells are roughly equal) codes for an increasingly saturated yellow.

Finally, the White/Black opponency cell registers the relative preponderance of light at any and all wavelengths arriving at the tiny area of the retina containing the three cone-cells at issue, over/under the same kind of undifferentiated light arriving to the larger retinal area that surrounds the three cones at issue. Accordingly, the White/Black cell performs an assay of the hue-independent *brightness* of the overall light arriving at the tiny retinal area at issue, relative to the brightness of the light arriving at the larger retinal area surrounding it. If that tiny area is much brighter than its surround, the White/Black cell will be pushed above its 50% default level, which codes in the direction of an increasingly bright white. Alternatively, if the tiny area is much *darker* than its comparatively bright surround, then the cone-cells within that larger surround will inhibit the White/Black cell below its 50% default level, which codes for an increasingly dark black.

Collectively, the elements of the Hurvich net are performing a systematic assay of the *power distribution* of the various wavelengths of light incident at its tiny proprietary area on the retina, and an auxiliary assay of the total power of that light relative to the total power of the light incident on its background surround. Distinct objective colors (which implies distinct power distributions within the incident light), when presented as inputs to the net, yield distinct assays at the net's output layer. Six such assays – represented as six histograms – are presented in Plate 5 as are the six landmark colors that produce them. To each color, there corresponds a unique assay, and to each assay there corresponds a unique color.

As you can see, the Hurvich net converts a four-tuple of inputs – S, M, L, and B (for the level of Background illumination) – into a *three*-tuple of outputs: $A_{B/Y}$, $A_{G/R}$, and $A_{W/B}$. Given the several functional relations described in the preceding paragraphs, the activation values of these three output cells can be expressed as the following arithmetic functions of the four input values:

$$A_{G/R} = 50 + (L - M)/2$$
$$A_{B/Y} = 50 + ((L + M)/4) - (S/2)$$
$$A_{W/B} = 50 + ((L + M + S)/6) - (B/2)$$

These three equations are uniquely determined by the requirement 1) that each of the three second-rung cells has a resting or default activation level of 50%, 2) that the activation levels of every cell in the network range between 0% and 100%, 3) that the different polarities of the several synaptic connections are as indicated in top part of Plate 4, and finally 4) that each of the three tugs-of-war there portrayed is an even contest.

Very well, but what is the point of such an arrangement? Why convert, in the manner just described, positions in a four-dimensional retinal input space into positions in a three-dimensional opponent-cell output space? The answers start to emerge when we consider the full *range* of possible activation points in the original 4-D retinal input space, and the range of their many transformed *daughter* activation points within the 3-D opponent-cell output space. To begin, you may observe that those daughter points are all confined within a trapezoid-faced *sub*space of the overall cube-shaped opponent-cell activation space (Plate 6*a*). Points outside that space cannot be activated by *any* combination of retinal inputs, so long as the network is functioning normally, and so long as all inputs consist of reflected ambient light. As written, the three equations preclude any activation triplets outside that oddly shaped subspace.

The cut-gem character of that subspace reflects the fact that the three equations that jointly define it are simple linear equations. A somewhat more realistic expression of how $A_{B/Y}$, $A_{G/R}$, and $A_{W/B}$ vary as a function of L, M, S, and B would multiply the entire right-hand side of each of the three equations by a nonlinear, sigmoid-shaped squashing function, to reflect the fact that each of the three output cells is easily nudged above or below its default level of 50%, in the regions close to that level, but is increasingly *resistant* to excursions away from 50% as each approaches the two extremes of zero percent and 100% possible activation levels. This wrinkle (for simplicity's sake, I'll suppress the algebra) has the effect of rounding off the sharper corners of the trapezoidal solid, yielding something closer to the spindle-shaped solid with tilted "equator" portrayed in Plate 6*b*.

Famously, this peculiar configuration of possible *coding vectors* is structurally almost identical to the peculiar configuration, originally and independently reconstructed by Munsell, of possible *color experiences* in normal humans. If one maps the white/black axis of the Munsell solid onto the ‹50, 50, 100›/‹50, 50, 0› vertical axis of the Hurvich spindle we have just constructed, and the green/red Munsell axis onto the ‹50, 0, 50›/‹50, 100, 50› horizontal axis of the Hurvich spindle, and the blue/yellow Munsell axis onto the ‹0, 50, 35›/‹100, 50, 65› tilted axis of the Hurvich spindle, then the family of distance relations between all of the color experiences internal to the Munsell space is roughly identical with the family of distance relations between all of the coding triplets internal to the Hurvich spindle.

Note the deliberately color-coded interior of the Hurvich spindle, which also fades smoothly to white at the top and to black at the bottom, as portrayed in Plate 7, and compare it to the interior of Plate 3. From precisely such global isomorphisms are speculative thoughts of intertheoretic identities likely to be born. The systematic parallels here described – though highly improbable on purely a priori grounds – become entirely nonmysterious if human color experiences (at a given point in one's visual field) simply *are* the output coding vectors (at a suitable place within some topographical brain map of the retina) produced by some neuronal instantiation of the Hurvich net. Such coding vectors presumably reside in the human LGN, or perhaps in cortical area V4, two brain areas where color-opponent cells have been experimentally detected,[6] areas whose lesion or destruction is known to produce severe deficits in color discrimination.

This isomorphism of internal relations is joined by an isomorphism in external relations as well. For example, the visual experience of white and the opponent-cell coding vector ‹50, 50, 100› are both caused by sunlight reflected from such things as snow, chalk, and writing paper. The experience of yellow and the coding vector ‹50, 100, 65› are both caused by sunlight reflected from such things as ripe bananas, buttercups, and canaries. And so on for the respective responses to *all* of the objective colors of external objects. The a priori probability of these assembled external

---

[6] Zeki, S., "The Representation of Colours in the Cerebral Cortex," *Nature*, 284, 1980: 412–418.

coincidences is as low as that of the internal coincidences just noted, and their joint (a priori) probability approximates an infinitesimal. These facts most certainly do not entail that the two spaces, and their respective elements, be numerically identical. Other explanations are possible. But we can be forgiven for exploring this most salient possibility.

Two further virtues will complete this brief summary of the Hurvich net's claim to capture the basics of human color vision. The first additional virtue is the network's capacity for accurately representing the same objective color across a wide range of different levels of ambient illumination. From bright sunlight, to gray overcast, to rain-forest gloom, to a candlelit room, a humble gray-green object will look plainly and persistently gray-green to a normal human, despite the wide differences in energy levels (across those four conditions of illumination) reaching the three cone-cells on the retina. The Hurvich net displays this same indifference to variations in ambient brightness levels. Thanks to the tug-of-war arrangement described earlier, the network cares less about the absolute *levels* of cone-cell illumination than it does about the positive and negative *differences* between them.

For example, a cone input pattern of $\langle L, M, S \rangle = \langle 5, 40, 50 \rangle$ will have exactly the same effect at the opponent-cell output layer as a cone input pattern of $\langle 15, 50, 60 \rangle$, or $\langle 25, 60, 70 \rangle$, or $\langle 43, 78, 88 \rangle$, namely, a stable output pattern, at the second layer, of $\langle A_{B/Y} = 36.25, A_{G/R} = 32.5, A_{W/B} = 50 \rangle$, for all four inputs. At least, it will do so if (and only if) the increasing activation levels just cited are the result of a corresponding increase in the level of *general* background illumination. For in that case, the absolute value of B (which codes for background brightness) will also climb, in concert, by 10, 20, and then 38 percentage points as well. This yields incremental increases in inhibition that exactly cancel the incremental increases in stimulation, from the three focal cones, on the White/Black opponent cell. A color representation that might otherwise have climbed steadily whiteward, into the region of a pastel chartreuse toward the upper apex of the spindle, thus remains accurately fixed at the true color of the external object – a dull middle green. And this same stability or light-level independence will be displayed for any of the other colors that the network might be called upon to represent.

To cite a second virtue, the Hurvich net is also roughly stable, in its color representations, across wide variations in the *wavelength profile* of the ambient illumination. At least, it will be thus stable if its constituting cells are assigned the same tendencies to fatigue and to potentiation shown by neurons in general. The human visual system shows the same tendencies and the same stability. This second form of stability is a little slower to show itself, but it is real. Consider, for example, a nightclub whose ceiling lights emit the lion's share of their illumination at the *long* wavelengths in the visible spectrum. This will provide a false-color roseate tilt to every object in the club, no matter what that object's original and objective color. Upon first entering the club, a normal human will be struck by the nonstandard appearance of every (non-red) object in sight. But after several minutes of adjustment to this nonstandard illumination, the objective colors of objects begin to reassert themselves, the roseate overlay retreats somewhat, and something close to our normal color recognition and discrimination returns, the ever-reddish ceiling lights notwithstanding.

The principal reason for this recovery is that our Green/Red opponent cells, and only those opponent cells, all become differentially *fatigued*, since the nightclub's abnormal ambient illumination forces them all to assume, and to try to maintain, a chronic level of excitation well *above* their normal resting level of 50% – a level of 70%, for example. (That is why nothing looks exactly as it should: every activation triplet produced in this condition, by normal objective colors, will have an abnormal 20% surplus in its $A_{G/R}$ component.) But that 70% level cannot be chronically maintained, because the cells' energy resources are adequate only for relatively brief excursions away from their normal resting level. With their internal resources thus dwindling under protracted demand, the increasingly exhausted Green/Red opponent cells gradually slide *back* toward an activation level of 50%, despite the abnormal ambient light. The artificial 20% surplus in the $A_{G/R}$ component of every coding triplet thus shrinks back toward zero, and the visual system once again begins to represent objective colors with the same coding triplets produced when normal light meets a normal (i.e., unfatigued) visual system. The fatigue accumulated in the system thus compensates, to a large degree, for the nonuniform power distribution across the wavelengths in the ambient illumination.

A symmetric compensation will occur when the ambient light – this time dominated by green, let us suppose – forces the opponent cell to an activation level *below* its normal 50% – to a steady 30%, for example. In that condition, the cell's internal energy resources are not consumed at the normal rate, and they begin to accumulate, within the cell, to abnormal levels. The cell is thus increasingly *potentiated*, rather than fatigued, and its activation levels begin to creep back up toward its default level of 50%, the chronic inhibition of the unusual background light notwithstanding. The same general story holds for the Blue/Yellow opponent cells as well. Taken collectively, these automatic negative and positive compensations allow the Hurvich net, and us humans, to adapt successfully to all but the most extreme sorts of pathologies in the power spectrum of the ambient light. Such compensations are not perfect, for reasons we shall here pass over.[7] But they are nontrivial. And one need not visit garish nightclubs to have a need for it. In the course of a day, the natural background light of unadorned nature varies substantially in its power spectrum as well as in its absolute power levels. For example, the ambient light under a green forest canopy with a mossy green floor strongly favors the medium wavelengths over the short and the long. But we adjust to it automatically, as described here. And the ambient light from the setting sun favors the longer wavelengths only slightly less than did our roseate nightclub (because the shorter wavelengths are increasingly scattered away as the sinking sun's light is forced to take a progressively longer path through the intervening atmosphere). But we adjust to this late-afternoon power-spectrum tilt as well.

Colors look roughly normal to us under these two conditions, but they would look arrestingly abnormal without the compensating grace of the adjustments here described. For example, color *photographs* taken in these two conditions will subsequently provide you with an *un*compensated portrayal of the relevant scene (photographic film does not fatigue or potentiate in the manner at issue). Such photos will thus be chromatically more hyperbolic – and thus visually more striking – than were your original (fatigue-compensated) visual experiences under the two conditions cited. And if you return to the roseate nightclub for 10 minutes or so, there is a further way to appreciate directly just how much your Green/Red opponent cells' current default levels have been pushed away from their normal level of 50%. Simply step outside the nightclub and behold the world in normal daylight. For 10 or 20 seconds, the entire world will have an eerie *greenish* cast to it. This is because every opponent-cell coding triplet, generated in you by light from every observed surface, will have a 20% deficit in its $A_{G/R}$ component, as compared to what an unfatigued system would produce. And such deficits code for green.

3 Opponent-Cell Fatigue and Colored Afterimages

This last point will serve to introduce the topic of *colored afterimages*. If a specific area within your visual field (a small circular area, for example)

---

[7] Among other things, the input *cones* also become differentially fatigued, but these input cells display a different pattern of compensation. Since their resting activation level is zero, they can display no potentiation, but *only* fatigue.

is made deliberately subject to chromatic fatigue or potentiation (by your fixating on a saturated red circle, for example, on a gray background under normal light, for 20 seconds or so), then those differentially fatigued/potentiated opponent cells will yield an appropriately circular afterimage when one's gaze is relocated to a uniformly middle-gray background surface. But the apparent color of that afterimage will be the color complement of the red of the original circular area. That is to say, its apparent color will be at or toward the *antipodes* of the color-spindle position of the red of the original circular stimulus: it will be decidedly green. To illustrate this, look at the first row of Plate 8. Fixate for 20 seconds on the small × within the red circle of the leftmost gray square, and then quickly refixate on the 3 in the middle-gray square two jumps to its right. You will see there a circular green afterimage hovering against that gray background, roughly as portrayed in the final square to the right.[8]

This happens because when the (now fatigued) opponent cells representing the circular red stimulus are suddenly asked to fall back to representing a less-demanding middle-gray stimulus (as in the third square), they overshoot the required ‹50, 50, 50› coding vector by an amount equal to whatever fatigue or potentiation has been acquired in each of the three coding dimensions during the protracted exposure to the original red stimulus. That original red stimulus produced an initial coding vector of ‹50, 95, 50›, but during protracted fixation, that initial vector slowly inches back to something like a vector of ‹50, 55, 50›, thanks to the accumulated minus-40% fatigue in its middle or $A_{G/R}$ element.

Accordingly, when the opponent cells in the fatigued area are suddenly asked to represent an objectively middle-*gray* stimulus, they can only manage to produce a vector of ‹50, *10*, 50› – the coding triplet for an obvious middle green – instead of the ‹50, *50*, 50› they would normally produce.

[8] The immediate point of placing the colored circle against a middle-gray background square is to ensure that only the visual area comprehending the circle itself is subjected to opponent-cell fatigue or potentiation. The immediate point of placing a second, uniformly gray square immediately to the right of the first is to ensure that this square visual area also suffers no opponent-cell potentiation or fatigue. The ultimate point of thus avoiding any fatigue or potentiation in those areas is that, when one's gaze is subsequently refixated on the × within the third square, *everything* within the third and fourth squares will be seen normally, *except* the circular area, within the third square, where the induced afterimage is situated. The point of the final or rightmost square with its colored circle is to provide a *prediction* of the shape and expected color of the induced afterimage, next door to it in the third square, so that you may compare directly and simultaneously the reality with the prediction.

For the $A_{G/R}$ cells in the affected circular area are, temporarily, too tired to respond normally. They produce a coding vector with a much-reduced middle component, an abnormal vector that represents green, not gray. (See the fatigue arrows in Plate 9.) Thus, if you fatigue a small part of your visual system by prolonged fixation on a small red circle, you will subsequently see, when you relocate your gaze on a middle-gray surface, a small green circle as an afterimage.

The behavior displayed in this red/green example can be generalized to any color whatever, except middle gray itself.[9] Given any position on or toward the outer surface of the Munsell/Hurvich spindle, a protracted activation triplet starting at that position will slowly creep back toward the middle-gray position at the central core of the spindle. Put another way, for any extremal activation triplet whatever, across the second-rung opponent cells of the Hurvich network, a *fatigue/potentiation vector* gradually accumulates, a directed line whose arrowhead always points toward the ‹50, 50, 50› center of the color spindle, and whose tail is always located at the original, extremal activation triplet. When the network is then suddenly given a middle-gray stimulus, the activational result across the opponent cells is always equal to ‹50, 50, 50› *plus* whatever fatigue/potentiation vector ‹$f_{B/Y}$, $f_{G/R}$, $f_{W/B}$› has accumulated during the protracted fixation on the original extremal color.[10] The abnormal coding triplet (= color spindle position) that finally results will thus always be directly opposite the original protracted coding triplet, at a distance from the spindle's (gray) center that is equal to the length of the accumulated fatigue/potentiation (f/p) vector. Return to the final three rows of Plate 8,

---

[9] Note well that an activation level of 50% of maximum produces neither fatigue nor potentiation in the relevant opponent cell. For under normal conditions, 50% just is the spontaneous resting level of any such cell. Absent any net stimulation or inhibition from the retinal cones, the opponent cells will always return, either immediately or eventually, to a coding vector of ‹50, 50, 50›, that is, to a middle gray.

[10] Note that each element of this f/p vector – ‹$f_{B/Y}$, $f_{G/R}$, $f_{W/B}$› – can have either a negative value (indicating fatigue for that cell) or a positive value (indicating potentiation for that cell). Note also that the length of that f/p vector will be determined by i) how *far away* from the spindle's middle-gray center was the original fixation color, and by ii) *how long* the opponent cells were forced to (try to) represent it. A brief fixation on any color close to middle gray will produce an f/p vector of negligible length. By contrast, a protracted fixation on any color far from middle gray will produce an f/p vector that reaches almost halfway across the color spindle. Strictly speaking then, the three equations for $A_{B/Y}$, $A_{G/R}$, and $A_{W/B}$ cited earlier should be amended by adding the appropriate f/p element to the right-hand side of each. This might seem to threaten extremal values below zero or above 100, but the (suppressed) squashing function mentioned earlier will prevent any such excursions. It asymptotes at zero and 100, just as required.

and repeat the experiment for each of blue, green, and yellow circles as the initial fatigue inducer.

A simple rule will convey the point visually (see again, Plate 9). For any protracted color stimulus, pick up its accumulated f/p vector as if it were a small arrow pointing rigidly in a constant direction in absolute space, and then place the tail of that arrow at the center of the color spindle, which is the proper coding point for middle gray. The tip of the arrow will then be pointing precisely at the color of the afterimage that will be seen when one's gaze is redirected to a middle-gray surface. Repeat the exercise for a protracted fatigue/potentiation on blue. That f/p arrow will point from the blue periphery of the spindle toward middle gray. Now pick it up and place its tail on middle gray. That arrow's head will come to rest on yellow, which will be the color of the afterimage that results from fixation on blue.

Quite evidently, if a middle-gray surface is the default background against which any colored afterimage is evaluated, then the apparent color of the afterimage must always be located toward a point on the color spindle that is exactly antipodal to the original color stimulus. In any case, we have here one further family of predictions, well known to visual scientists, where the predictive power of the Hurvich net comes up roses.

4 Afterimages Located on Non-Gray Backgrounds

We have seen how the f/p vector produced by protracted fixation on any non-gray stimulus will produce the full range of antipodal afterimages when our gaze is subsequently redirected to a neutral or middle-gray background surface. But there is no reason to limit ourselves to locating and evaluating our afterimages against that background alone. We can locate an acquired f/p vector against a background of any color we like, and we will get a differently colored afterimage for each such differently colored background. The rule established for middle-gray backgrounds holds for any background whatever. Simply add each element of the relevant f/p vector, $\langle f_{B/Y}, f_{G/R}, f_{W/B}\rangle$, to the corresponding element of the opponent-cell activation vector, $\langle A_{B/Y}, A_{G/R}, A_{W/B}\rangle$, that would *normally* be produced (in an unfatigued visual system) by the now-colored background surface at issue. That sum will characterize the apparent color of the afterimage as projected against that particular colored background.

For the same reasons, the visual trick cited earlier can also be trusted to give the appropriate predictions here. For any given case, simply pick up the rigidly pointing f/p "arrow" located within the color spindle, and relocate its tail at the color of the chosen background. The arrow's head will then lie exactly at the apparent color of the afterimage that will appear against that chosen background.

Several illustrations of this background sensitivity in the apparent color of an afterimage are displayed in Plate 10. The case of row 1 may surprise you slightly. Fixate at length on the × in the pink circle, and then suddenly refixate on the × in the identically pink square two jumps to the right. You will there find a distinctly *gray* circle hovering over the pink square, much as portrayed in the fourth (predictive) square to its right. This illustrates directly the general principle that acquired f/p vectors always point toward middle gray (see Plate 11). As you fixate at length on the original pink circle, you don't realize that your chromatic representation for that area is slowly fading. But you can see instantly that it has indeed faded when it is suddenly relocated against a larger unfatigued background of exactly the same original color. The color *contrast* is obvious.

In row 2, the same f/p vector (again acquired during fixation on the pink circle) is subsequently overlaid on a white background square. The relevant prediction is given by picking up that vector and relocating its tail at the uppermost tip of the color spindle (i.e., at white), as in Plate 11. The head of the arrow will then rest at a very pale green. And that will be the apparent color of the circular afterimage.

In row 3 of Plate 10, a new f/p vector, generated this time by fixation on a dark blue, is to be relocated on a square background of light blue. Placing the tail of the relevant f/p vector at the light-blue position leaves its arrowhead pointing close to the top of the spindle (i.e., to white), but not quite getting there, as in Plate 11. Thus, the off-white afterimage that results.

In row 4, a final f/p vector, generated by a bright yellow, is to be relocated on a white background. Moving that vector's tail to the top of the spindle leaves its arrowhead resting at pale blue. And pale blue is the color of the afterimage. (Row 4's vectors – both acquired and resituated – have been left out of Plate 11 to avoid clutter.)

Evidently, there are a great many more predictions implicit in the model network at issue. If one considers only a very coarse partitioning of the color spindle (5 gray-scale positions on the vertical axis, 12 hue stations around the maximally saturated equator, 12 stations of dullish hue just inside the equator, 12 stations of pastel hue just above it, and 12 stations of darkish hue just below it), one is looking at a total of 53 distinct colors on which to fixate at length, and 53 possible colors on which to locate the resulting afterimage. The possible combinations total 53^2, or

2,809 distinct experiments, each one a test of the Hurvich hypothesis about human color coding. I have personally tested 280 (or 10%) of them. The Hurvich net's predictive performance is both systematic and strikingly accurate. But there is more to come.[11]

5 Breaking out of the Color Spindle: Chimerical Colors

It was remarked in section 2 that the equations governing the Hurvich net guarantee that any activation triplet within the opponent-cell activation space would be strictly confined to the subspace that constitutes the classical color spindle, no matter what the combination of cone-cell activities that produced that triplet. As those equations are written, that observation is correct, and it serves to explain the gross shape of Munsell's spindle, including its tilted equator that makes saturated blue a much darker color than saturated yellow.[12] But, you may still want to ask, what about all that unused space in the several upper and lower corners of the opponent-cell activation cube? What would be the significance of a possible activation triplet *outside* the classical color spindle, a triplet somewhere in that fairly considerable volume of unused opponent-cell activation space?

[11] The wary reader may have noticed that I am assuming that it is the *opponent-cell* fatigue/potentiation, as opposed to *retinal cone-cell* fatigue, that is primarily responsible for the chromatic appearance of our afterimages. Why? For three reasons. First, we still get strongly colored afterimages even at modest light levels, when the cone-cells are not put under stress, but the delta-sensitive opponent cells regularly are. Second, as we noted earlier, the opponent cells code by variations in high-frequency spiking, which is much more consumptive of energy than is the graded-voltage coding scheme used in the retina. Accordingly, the opponent cells are simply more *subject* to fatigue/potentiation. Finally, the Hurvich theory entails one pattern of afterimage coloration if cone-cell fatigue is the primary determinant, and a very different pattern of afterimage coloration if opponent-cell fatigue/potentiation is the primary determinant. The observed pattern of coloration agrees much more closely with the latter assumption. Colored afterimages, it would seem, are primarily an opponent-cell phenomenon.

[12] The explanation is obvious. Recall that to produce an opponent-cell triplet for maximum white requires that *all three* of the S, M, and L cones have activation levels of 100. To produce a triplet for maximum black requires those same cells all to be at zero. To produce a triplet for maximum yellow requires that S be at zero, while M and L are *both* at 100. Accordingly, the retinal input for yellow already takes you two-thirds of the way up the opponent-cell cube toward white (recall that white requires S, M, and L all to be at 100). Similarly, the input for maximum blue requires that S be at 100, while both M and L are at zero. Accordingly, the retinal input for blue already takes you two-thirds of the way toward black (which requires that S, M, and L are all at zero). Hence, blue is darker than yellow, and the equator of maximum hue saturation must be tilted so as to include them both. (Note that red and green display no such brightness asymmetry.)

Well you might ask. In particular, you might ask after the *phenomenolog-ical* significance of such an extra-spindle activation vector. Would it still be a color appearance of some sort, but chromatically distinct from any-thing within the spindle? Would it still follow or respect the basic rules of color similarities and differences that hold within the spindle? What would it "be like" to have such an activation vector realized in one's own opponent cells? If the Hurvich account of things is even roughly correct, you are about to find out.

Inserting stimulating/inhibiting electrodes directly into some substan-tial population of opponent-cell neurons in the human LGN would af-ford us the opportunity to produce directly, and independently of the peculiar connectivity of the Hurvich net, any activation vector that we choose. In principle, it could be done. But capturing a sufficiently large *number* of cells simultaneously (after all, the anomalous chromatic area within one's subjective visual field has to be large enough for one to dis-criminate) is currently beyond our experimental technologies, and the cranial invasion would needlessly threaten the health of our subject, in any case.

Fortunately, there is a much easier way to produce the desired result. In the last several figures, you have already been introduced to the re-quired technology, namely, selective fatigue/potentiation by prolonged fixation on some suitable color stimulus. Recall that the opponent-cell activation vector that would *normally* result from a given retinal stimulus is subject to substantial *modification* by the addition of whatever f/p vector has accumulated in the system immediately prior to that external reti-nal stimulus. Adding those two vectors together can indeed yield a vector that reaches well outside the classical spindle. For example, let the system fatigue on yellow, as indicated in Plate 12. Then present the system with a maximally *black* stimulus. The resulting vector will reach out from the bottom tip of the spindle, along the floor of the opponent-cell activation space, to a place directly but distantly underneath the standard coding triplet for a maximally saturated blue, as also indicated in Plate 12.

Extrapolating from what we already know about the coding signifi-cance of the three major dimensions of the color spindle and of the Hurvich opponent-cell activation space, that anomalous activation triplet must code for a color appearance that is

1) fully as dark as the darkest possible black (it is, after all, on the maximally dark *floor* of the opponent-cell activation space), but nevertheless is of

2) an obvious and distinctive hue (it is, after all, on a radius quite far from the hueless central axis of the opponent-cell activation space), a hue that must be

3) more similar to blue than to any other hue around the spindle's equator (it is, after all, closer to blue than to the position of any other such color).

On the face of it, the joint satisfaction of these descriptive conditions might seem to be impossible, for no *objective* hue can be as dark as the darkest possible black and yet fail to *be* black. As the original Munsell color spindle attests, to get to anything that has an objective hue, one must leave the hueless central axis in some horizontal direction or other. But to do that, one must come up that brightness axis, at least some distance, if one is to escape the bottommost singularity of maximal black.

However, we are not talking about the objective colors of real objects at this point. We are talking about an anomalous *color representation* within the opponent-cell activation space. And these anomalous representations are robustly possible, as you can discover for yourself in row 1 of Plate 13. A 22nd fixation on the × at the center of the yellow circle will produce in you precisely the f/p vector at issue. And your subsequent fixation on the × at the center of the maximally black square to its right will produce in you precisely the anomalous, extra-spindle coding vector here under discussion. (It fades away after a few seconds, of course, as the relevant cells progressively recover from their induced fatigue/potentiation.)

The final black square to the far right of row 1 contains, as before, a (very rough) prediction of what your circular afterimage will look like. But here my prediction image is doomed to be inaccurate, for the very dark blue circle there inscribed is still objectively and detectably brighter than its black surround. (It could not be otherwise without losing its blue hue entirely.) The anomalous afterimage, by contrast, presents a circular patch that is every bit as dark as its black surround, and yet appears decidedly blue-ish in some unfamiliar way. It is also visibly *darker* than the dark-blue "predictive" circle to its immediate right. That afterimage meets, while the (roughly) predictive objective image does not meet, all of conditions 1 to 3 listed here. This provides you with an experience of what might be called a "chimerical color" – a color that you will absolutely never encounter as an objective feature of a real physical object, but whose qualitative character you can nonetheless savor in an unusually produced illusory experience.[13]

Rows 2 to 4 of Plate 13 provide the resources for three more "chimerically colored" afterimages. The second yields an impossibly dark but still vivid green. The third yields an impossibly dark but still somehow vivid red. And the fourth yields, with maximum implausibility, a yellow that is as dark as the darkest possible black, and yet is still not black. You may well judge it to be some kind of unfamiliar *brown*, rather than yellow. (In general, the visual system sees things as brown exactly when it is confronted with an external stimulus that returns the same *wavelength profile* as yellow, or orange, but which is also judged, by contrast effects, to have a very low overall intrinsic reflectance. This fits the case at hand.)

The theory predicts, of course, the existence of impossibly dark versions of all of the hues around the classical spindle's equator, not just the canonical four we have examined here. Fixate at length on any saturated hue of your choosing; locate the resulting afterimage over a maximally black background, and the "seen color" will be an impossibly dark version of the color complement of the original circular color stimulus.

The theory also predicts that there exists more than one way to produce a given chimerical color sensation. Plate 14 illustrates the point. One can produce a chimerical dark blue by fatiguing on a saturated yellow stimulus and then locating the afterimage over a black background (as we have already seen in Plate 13, row 1). But one can produce the same result by fatiguing on a bright *white* stimulus and then locating the afterimage over a saturated *blue* background. You get to the same place, but by a very different route. Indeed, one can produce (almost) the same result by fatiguing on a pastel yellow stimulus and then locating the afterimage on a *dark gray* background. You may test all three of these predictions simultaneously by fixating at length on the pastel yellow stimulus in row 2 of Plate 15. This will produce in you three afterimages at once, arranged in

---

[13] You may have to resist an initial temptation to judge the anomalous afterimage to be at least slightly brighter than its black surround *on the grounds that* anything with a detectable hue *must* be somewhat brighter than black. But the principle that would normally license that inference is valid only for objective colors, not for internal color representations. Repeated examination of the circular afterimage will reveal that it is indeed as dark as its maximally dark surround, despite its vivid saturation, and that it is indeed darker than the (inadequate) dark-hue prediction to its immediate right. In sum, these weird afterimages are definitely outside the classical spindle. They are not just pressing at its periphery.

a vertical line. When you then locate the middle row's afterimage over the dark-gray background to the right, the first and third rows' afterimages will be located over the blue and the black backgrounds, respectively. You can then compare the qualitative character of all three afterimages at once, at least for the few seconds before they begin to fade.

The theory predicts that the afterimages in rows 1 and 3 will be the darkest and the most strikingly blue. The afterimage in the middle should be less dramatic, both in its darkness and in its blue hue, for the pastel yellow stimulus that originally produced it was inadequate to produce the maximal fatigue achieved by the other two stimuli. The theory also entails a different pattern of *fading* for each of the three afterimages. The afterimage in row 1 will slowly fade in its degree of darkness, but not in its degree of blueness. (Its fatigue lay in the black/white dimension.) By contrast, the afterimage in row 3 will progressively fade in its blue hue, but not in its degree of darkness. (Its fatigue lay in the blue/yellow dimension.) Finally, the afterimage in row 2 will progressively fade in both its hue *and* its darkness. (Its fatigue lay in both of those dimensions.) You may repeat this exercise, of course, with any other hue around the spindle's equator.

Such fine-grained predictive prowess, concerning both these un-usual qualitative characters and the various dynamical changes displayed therein, is noteworthy. But there is still more to come.

6 Out of the Spindle Again: Self-Luminous Colors

Let us not forget the *upper* regions of the opponent-cell activation space (see Plate 16). Prolonged fixation on a red circle will produce an f/p vector which, when its tail is relocated to the upper tip of the spindle, will reach out horizontally, across the ceiling of the space, to a point that represents a color that is as bright as the brightest white (after all, it is on the ceiling). But it cannot be white (after all, it is some distance away from the hueless central axis). Instead, it must be some implausibly luminous cousin of green. See for yourself in row 1 of Plate 17. As before, fixate on the central × for at least 20 seconds, and then refixate on the central × of the white target square to its immediate right. Here you will notice that the bright-green(ish) afterimage seems positively *self-luminous*, as if it were a colored lightbulb or a colored LED (light emitting diode). The impression of faint self-luminosity here is entirely understandable, for no physical object with a detectable objective hue could be possibly be as bright as a maximally white surface unless it were in fact *self*-luminous, *emitting* even more light energy than a white surface could possibly reflect under the ambient lighting.

Row 2 displays the same phenomenon, but with a blue fatigue template. In this case, the afterimage is an apparently self-luminous yellow. Other instances of apparently self-luminous afterimages can be produced by prolonged fixation on any hue whatever, so long as the hue's A_{BW}-component is at or below 50%. (Otherwise, the acquired f/p vector will have a nonzero downward component that will force the resulting afterimage down and away from the maximally bright ceiling of the opponent-cell activation space. The illusion of self-luminosity will progressively fade.) Rows 3 and 4 will complete the chromatic quartet of apparently self-luminous afterimages.

Once again, we are contemplating color qualia whose location in qualia space (= opponent-cell activation space) lies well outside the classical color spindle. Their existence is not quite the curiosity that their impossibly dark basement-floor cousins were, but that is because we have all encountered them in common experience. Munsell's original concerns were confined to the colors of *non*-self-luminous Lambertian (light-scattering) surfaces. Chromatic phenomena may begin there, but they do not end there. Self-luminous colors occur when an object emits (rather than merely reflects) a nonuniform wavelength profile at an energy level that is too high (i.e., too bright) to be accounted for by the maximum reflection levels possible given the ambient or background illumination (i.e., by a white surface). Such unusual stimuli must therefore be coded at the absolute ceiling of the opponent-cell activation space, but (because of the nonuniform wavelength distribution) it must be coded at some appropriate distance away from the hue-neutral center of that ceiling.

A normal Lambertian surface, meeting a normal visual system, will never produce such a coding triplet, no matter what the ambient illumination. But a self-luminous colored object will certainly do so, even in a normal visual system, for the stimulation levels it produces in the cones exceeds anything that the ambient or background light levels could possibly produce. Such an extra-spindle coding triplet is thus a signature sign of self-luminance in the environment. And that is why, when we produce such ceiling-dwelling coding triplets artificially, as in the fatigue/potentiation experiments of this section, the immediate impression is of a self-luminous colored object.

Note also that one can produce a sensation of the same anomalously bright color in more than one way, as illustrated in Plate 18. Fixate on a green circle, for example, and then look at a white surface, as in Row 1 of

Plate 19. Alternatively, fixate on a black circle, and then look at a red surface, as in row 3. Both procedures produce the same afterimage. Evidently, the family of predictions explored near the floor of the opponent-cell activation space is reflected in a similar family of predictions concerning the behavior of sensations at its ceiling. These, too, test out nicely.

7 Out of the Spindle One Last Time: Hyperbolic Colors

Return your attention to the central plane of the color spindle. Note that the coding triplets for the maximally saturated versions of the four primary hues – green, red, blue, and yellow – are all hard-pressed against the outer walls of the all-inclusive opponent-cell activation cube. By contrast, the four intermediate hues – yellow-green, orange, purple, and blue-green – all look out on some normally unused space in the four corners of the central horizontal plane. Using the techniques already explored, we can contrive to activate a coding triplet in any one of those four extremal corners. If one fixates at length on a pale blue-green stimulus, as in Plate 20, and then refixates on an *already* maximally saturated orange surface, then the resituated f/p vector will yield an activation triplet within the farthest corner of the cube, beyond the limits of the classical spindle. The Hurvich theory of our internal color representations entails that one should there find a circular afterimage of a *hyperbolic* orange, an orange that is more "ostentatiously orange" than any (non-self-luminous) orange you have ever seen, or ever will see, as the objective color of a physical object. Row 1 of Plate 21 will allow you, once more, to test such a prediction for yourself.

Row 2 provides access to a similarly hyperbolic version of purple. Rows 3 and 4 jointly provide a *non*hyperbolic contrast to the first two rows. Here we are set up to try to produce a hyperbolic red and a hyperbolic green, respectively. But here the theory says that there is little or no room inside the opponent-cell activation cube, beyond saturated red and saturated green, where any such hyperbolic activation triplet might locate itself. The coding triplets for saturated red and saturated green are *already* at, or close to, the relevantly extremal positions. And so here the theory predicts that our attempts to find a hyperbolic red and green must fail – or, at least, find a much feebler success than we found in the cube's more capacious corners. See what you think about the relative vividness of the afterimages in the last two rows, relative to those achieved in the first two rows.

8 The Consequences for Current Debates

The reader will note that despite the nontrivial (but wholly defeasible) case laid out here, in support of the strict identity of human visual color qualia, on the one hand, and human opponent-cell coding triplets, on the other, at no point did we establish, or even try to establish, that there is any sort of *necessary connection* between the two. I did not argue, nor claim, that the former are "logically supervenient" upon the latter.[14] I did not argue, nor do I believe, that the identity at issue is blessed by any form of "metaphysical necessity."[15] Nor did I suggest that there is any form of "lawlike" or "nomological" connection between the two.[16] As I have argued elsewhere, all of these quasi-modal relations are philosophical extravagances or confusions imposed, post facto, on successful cases of historical intertheoretic reductions, all of which were achieved without the help of such modal relations, and *none of which displays any one of them*.[17] Here, as in those other cases from our scientific history, the principal intellectual motive for embracing the systematic color-qualia/coding-vector identities proposed is simply the extent and quality of the predictive and explanatory unity that the relevant reduction provides.

But that basic motive was already in place, independently of the experimental predictions of the present chapter. If those predictions are correct, they provide an *additional* motive for embracing the proposed reduction of color qualia to coding vectors. For it was no part of the motives – for Hurvich's original reductive proposal – that these particular experimental predictions be a part of the explanatory target. They were unanticipated, and they are faintly paradoxical on their face. They thus provide some "excess empirical content" beyond the original explanatory target, namely, our familiar experiences of the mundane colors of external objects.

Such excess empirical contents are familiar from the history of science. In the latter part of the 19th century, the assumption that light was identical with electromagnetic waves entailed that there should be such a thing as *invisible* light (an apparent contradiction, note well). Specifically, there should be light with a wavelength longer than the red end

---

[14] Cf. Chalmers, *The Conscious Mind*.

[15] Cf. Kripke, S., "Naming and Necessity," in Davidson, D., and Harman, G., eds., *Semantics of Natural Language* (Dordrecht, Holland: D. Reidel, 1972): 253–355.

[16] Cf. Davidson, D., "Mental Events," in Foster, L., and Swanson, J., eds., *Experience and Theory* (Amherst: University of Massachusetts Press, 1970).

[17] Cf. Churchland, P. M., *Scientific Realism and the Plasticity of Mind* (Cambridge: Cambridge University Press, 1979): 80–88. See also Churchland, 1985, 1996, op. cit.

(namely, infrared light), and there should be light with a wavelength shorter than the violet end (namely, ultraviolet light), of the visible spectrum (.40μm to .65μm). Despite this clear violation of then-normal semantic expectations, the existence of light – not faint light, but very bright light – *outside* the visible spectrum was subsequently confirmed by John Herschel, Heinrich Hertz, and Wilhelm Roentgen.[18] The elusive and apparently singular nature of light[19] was thus brought under the broad umbrella of electromagnetic phenomena in general, not to its detriment but to its welcome illumination.

The parallel assumption, that human color representations or color qualia are identical with opponent-cell coding triplets in a neuronal instantiation of the Hurvich network, yields a similarly implausible prediction. There should be color qualia outside the qualitative range of the classical color spindle, qualia whose perfectly accurate descriptions violate our normal semantic expectations. And the Hurvich theory further suggests how to produce such chimerical qualia – through opponent-cell fatigue/potentiation – so that we may test those unexpected predictions against our own experience.

Let me now point out that just as in the case of light, the new theory provides a wealth of *explanatory* power commensurate with its extensive predictive power. Why, in the case of Plate 13, row 1, is one's circular afterimage of something similar to *blue*? Because the coding vectors for the opponent cells in that part of your visual field all have an $A_{G/R}$ component that is neutrally balanced at 50%, and an $A_{B/Y}$ component that is well below 50%. Such vectors are precisely those that code for the various blues.

Why is that $A_{B/Y}$ component so unusually low? Because the relevant cells were antecedently fatigued on a maximally yellow stimulus, which forced them to try to maintain an $A_{B/Y}$ component close to 100%. Their subsequent response in that dimension, to any external stimulus, will thus be much lower than normal, at least for a short time.

---

[18] Herschel placed the bulb of a mercury thermometer just outside the redmost edge of the spectral "rainbow" image produced by directing sunlight through a prism. The mercury level shot up. Hertz confirmed the existence of light at much longer wavelengths with his seminal radio transmitter and radio receiver. Roentgen, you will recall, stumbled across X rays while playing with a cathode-ray tube, and correctly characterized them, after a week or two of sleuthing, as light of much-shorter-than-visible wavelengths.

[19] John Milton, writing in the same century, characterized light thus: "Ethereal, quintessential, pure, the first of things" (*Paradise Lost*). Milton's antiphysicalist presumptions here reach back a long way.

Why is the seen blue so curiously and implausibly *dark?* Because in this case, the $A_{W/B}$ component of the relevant activation vector is close to zero percent, which makes it similar in its darkness to a sensation of maximal *black*.

Why does that impossibly dark blue afterimage fade over 5 seconds or so, as I gaze at the black square target? Because the B/Y opponent cells (no longer under input pressure to maintain an extreme value) slowly recover from their fatigue, and slowly return to representing the black background for what it really is: a maximal black.

Why does the initial saturation level of the "impossibly" blue afterimage depend on *how long* I stared at the yellow circle? Because the degree of fatigue induced in the B/Y cells is a function of how long they were forced (by a maximally yellow stimulus) to try to maintain an unusually high level of activation. The greater the fatigue, the more abnormally *low* will be their subsequent response to any stimulus. And the farther from the 50% neutral point they fall, the greater is the saturation of the stygian blue therein represented.

Evidently I could go on illustrating the Hurvich net's explanatory virtues – concerning the qualitative characters and qualitative behaviors of colored afterimages – but you can now see how to deploy the explanatory virtues of that model for yourself. The point of the preceding paragraph is to underscore the point that the theory here deployed has just as much explanatory power as it has predictive power.

A caution: as colorful as they might be, one's individual reactions to the several tests set out in this chapter (Plates 8, 10, 13, 15, 17, 19, and 21) should not be regarded as adequate grounds for believing the theory here deployed. The Hurvich theory has an independent authority, derived from many prior tests over the past two decades, as the first half of this chapter attempts to summarize. And the issue of chimerically colored afterimages, in particular, wants attention from competent visual psychologists who can bring the appropriate experimental procedures to bear on a sufficiently large population of naive subjects. But you may appreciate, from the simple tests here provided, why such experimental work might be worth doing. And you may also appreciate my own expectations in this matter. Indeed, by now you may share them.

Withal, how those experiments actually turn out is strictly beside the philosophical issue that opened this chapter. We began by confronting the philosophical claim that no physical theory could ever yield specific predictions concerning the qualitative nature of our subjective experience. And yet here we have before us a theroretical initiative that yields

precisely the sorts of predictions – in qualitative detail – that were supposed to be impossible. Moreover, the predictions at issue concern genuinely novel phenomena, phenomena beyond our normal qualitative
experience. And at first blush, it seems that those predictions might even
be true.

That would indeed be interesting. But it is not strictly the point. The
point is that a sufficiently fertile theory of chromatic information processing in the human visual pathway (i.e., the Hurvich model), coupled with
a sufficiently systematic grasp of the structure of the explanatory target
domain (i.e., the Munsellian structure of our phenomenological quality
space), can, when fully explored, yield predictions that could never have
been anticipated beforehand, predictions that would have been summar-
ily dismissed as "semantically odd" or outright impossible, even if they had
been.

This lesson is as true, and as salutary, in the present and rather more
modest case of subjective color qualia as it was in the 19th-century case
of light. Apparent "explanatory gaps" are with us always and everywhere.
Since we are not omniscient, we should positively expect them. And here,
as elsewhere, apparent (repeat: *apparent*) qualitative "simples" present an
especially obvious challenge to our feeble imaginations.[20] But whether
an apparent gap represents a mere gap in our current understanding and
imaginative powers, or an objective gap in the ontological structure of reality, is always and ever an empirical question – to be decided by unfolding
science, and not by preemptive and dubious arguments a priori. In light of
the Hurvich network's unexpectedly splendid predictive and explanatory
performance across (indeed, *beyond*) the entire range of possible colors,
the default presumption of some special, nonphysical ontological status
for our subjective color experiences has just evaporated. Our subjective
color experiences – the chimerical ones, included – are just one more
subtle dimension of the labyrinthine material world. They are activation
vectors across three kinds of opponency-driven neurons. This should occasion neither horror nor despair. For while we now know these phenomenological roses by new and more illuminating names, they present
as sweetly as ever. Perhaps even more sweetly, for we now appreciate why
they behave as they do.

I conclude by addressing a final objection to the specific identities
here proposed: "We can see why you propose to identify the subjective

---

[20] On this point in particular, see Churchland, 1996, section V, "Some Diagnostic Remarks
on Qualia," 225–228.

qualia of saturated redness with an opponent-cell activation vector of ‹50, 100, 50› (as in Plate 5), and so on for all of the other elements of the proposed mapping. It is because this mapping has the virtue that all of the proximity (similarity) relations within qualia space are successfully mirrored in the assembled proximity (similarity) relations within the relevant activation vector space. But a problem remains. To begin, there is no guarantee that this particular mapping is the *only* mapping that would achieve that end. Perhaps there are others, as contemplated in the familiar class of 'inverted spectrum' thought-experiments (cf. Chalmers, 1996). More specifically, the account proposed in the preceding pages fails to give an adequate explanation, or indeed *any* explanation, of why an activation vector of ‹50, 100, 50› should have, or produce, or be associated with, a qualitative character of *this* particular nature (I here advert to what I have learned to call 'a sensation of *red*'), as opposed to any of the other available color qualia. In the absence of such an explanation, the account of the preceding pages has uncovered nothing more than a systematic but still-puzzling empirical *correlation* between qualia, on the one hand, and opponent-cell activation vectors, on the the other. What qualia might be, in themselves, remains a mystery."

This deflationary complaint is seductive, but it betrays a fundamental misunderstanding of what is going on in any proposed intertheoretic reduction, and of the requirements the reduction must meet in order to be successful. The demand for an *explanation,* as outlined in the preceding paragraph, is ill-conceived for precisely the case of the intertheoretic identities at issue. This is not hard to see. To ask for an explanation of why a given qualia is correlated with a given activation vector is to ask for some natural law or laws that somehow *connect* qualia of that kind with activation vectors of the relevant kind. But there can be such a natural law only if the quale and the vector are *distinct things,* things fit for enjoying nomic connections with one another. In the case at issue, however, the proposal is that the qualia and the vectors are not distinct things at all: they are one and the same thing; they are identical. An explanation of the kind demanded is thus impossible, and the demand that we provide it is misconceived from the outset. As well demand a substantive explanation for the curious and universal co-occurrence of the substance *snow* and the substance *neige.*

Granted, the two background conceptual frameworks that embed the notions of qualia, on the one hand, and activation vectors, on the other, are much more different from one another than are the respective English and French conceptions of snow. But that is precisely why the identifications proposed in the present chapter, and those proposed in intertheoretic reductions generally (recall "light = electromagnetic waves," and "temperature = mean molecular kinetic energy"), are so much more *informative* than are the identifications made in the humdrum case of closely synonymous translations. They bring new explanatory resources to bear on an old and familiar domain, and they provide novel empirical predictions unanticipated from within the old framework, as this essay has illustrated. To demand a substantive explanation of the "correlations" at issue is just to beg the question against the strict identities proposed. And to find any dark significance in the absence of such an explanation is to have missed the point of our explicitly reductive undertaking.

Nor need the specter of various possible qualia inversions across distinct individuals, or within a given individual over time, trouble the reductive account here proposed. For the Hurvich account of subjective color experiences not only allows for the possibility of such inversions; it also specifies exactly how to produce them. For example, if you wish to produce a global green/red inversion in your subjective qualitative responses to the external world (while holding the black/white and blue/yellow dimensions unchanged), simply change the polarity of all of the L-cone projections (to the green/red opponent cells) from excitatory to inhibitory, and change the polarity of all of the M-cone projections (to the green/red opponent cells) from inhibitory to excitatory, and change nothing else, especially in the rest of your visual system downstream from your now slightly rewired opponent cells (see again Plate 4). That will do it. Upon waking from this (strictly fanciful) microsurgery, everything that used to look red will now look green, and vice versa.

But there is no metaphysical significance in this empirical possibility, nor in the many other possible inversions and gerrymanderings that similar rewirings would produce. For we are here producing systematic *activation-vector inversions* relative to the behavior of activation vectors in a normal (i.e., unrewired) Hurvich network. Given the strict identities proposed, between specific qualia and specific activation vectors, it is no surprise that changes in the response profile of either one will be strictly "tracked" by changes in the other. Of course, it remains an a priori possibility that our color qualia might vary *independently* of the physical realities of the Hurvich network. But this is just another way of expressing the permanent a priori possibility that the Hurvich account of our color experiences might be factually mistaken, and this is something to which everyone must agree. But do not confuse this merely a priori issue with a closely related empirical issue. If the Hurvich account of our color experiences is correct, then it is empirically *im*possible to change the profile of our subjective color responses to the world without changing, in some way, the response profile of our opponent-cell activation vectors, as outlined, for example, in the preceding paragraph. From this reductive perspective, sundry "qualia inversions" are indeed possible, but not without the appropriate rewirings within the entirely physical Hurvich net that embodies and sustains all of our color experience. If we wish to resist this deliberately reductive account – as well we may – let us endeavor to find some real *empirical* fault with it. Imaginary faults simply don't matter.

References

Bickle, J. (1998). *Psychoneural Reduction: The New Wave*. Cambridge, Mass.: MIT Press.

Chalmers, D. (1996). *The Conscious Mind*. Oxford: Oxford University Press.

Churchland, P. M. (1979). *Scientific Realism and the Plasticity of Mind*. Cambridge: Cambridge University Press, 80–88.

Churchland, P. M. (1985). Reduction, Qualia, and the Direct Introspection of Brain States. *Journal of Philosophy*, 82, no. 1, 8–28.

Churchland, P. M. (1996). The Rediscovery of Light. *Journal of Philosophy*, 93, no. 5, 211–228.

Clark, A. (1993). *Sensory Qualities*. Oxford: Oxford University Press.

Davidson, D. (1970). Mental Events. In Foster, L., and Swanson, J., eds., *Experience and Theory*. Amherst: University of Massachusetts Press.

Hardin, C. L. (1988). *Color for Philosophers: Unweaving the Rainbow*. Indianapolis: Hackett.

Hurvich, L. M. (1981). *Color Vision*. Sunderland, Mass.: Sinauer.

Huxley, T. H. (1866). *Elementary Lessons in Physiology*. London: Macmillan.

Jackson, F. (1982). Epiphenomenal Qualia. *Philosophical Quarterly*, 32, no. 127, 127–136.

Kripke, S. (1972). Naming and Necessity. In Davidson, D., and Harman G., eds., *Semantics of Natural Language*. Dordrecht, Holland: D. Reidel, 253–355.

Nagel, T. (1974). What Is It Like to Be a Bat? *Philosophical Review*, 83, no. 4, 435–450.

Zeki, S. (1980). The Representation of Colours in the Cerebral Cortex. *Nature*, 284, 412–418.

10

Opponent Processing, Linear Models, and the Veridicality of Color Perception

Zoltán Jakab

1 Two Contrasting Intuitions About Color

In roughly the last 15 years, it has become increasingly clear that philosophy and color science have some conflicting views about the nature of color. A philosophical view of color emerged in color science more or less as a collateral effect of the practice in that discipline. In philosophy first arose the need to precisely formulate what many believed was *the* commonsense view of color. This did happen, and later on the theory became a crucial part of a rather popular theory of mind. Other philosophers, however, were more interested in rigorously framing the view that they saw emerging from color science.[1] Let us take a first look at the two approaches.

It seems to be a fairly widely held view among color scientists that color is at least as much "in the eye of the beholder" as on the surfaces of objects or in the light reaching our retinas. In other words, features of our color experience are crucially determined, and correspondingly explained, by the neuronal processing in our visual systems, as opposed to the environmental properties themselves that are the canonical causes of color experience. There exist, to be sure, color stimuli, or "physical colors," that reliably and predictably evoke experiences as of color. However, color science has been taken by certain philosophers and color scientists to suggest that these stimuli offer little help in understanding the nature of color experience. For instance, the fact that summer tree leaves reflect a lot of

---

[1] Including both applied color science and the psychology of color perception.

light between 500 and 600 nm, but much less at the two ends of the visible spectrum, seems to offer no explanation of why summer leaves evoke in us the kind of color experience that we associate with the color name "green." Here is another example. In daylight, an ordinary reflecting surface that is a long-pass cutoff filter at 600 nm looks (predominantly) red; a long-pass cutoff filter at 550 nm looks orange, whereas a long-pass cutoff filter at 500 nm looks yellow (see Meyer, 1988; MacAdam, 1997, pp. 36, 38–39 for illustration). But how does this observation help in understanding the apparent psychological fact that orange is a perceptual mixture of red and yellow (i.e., it is a binary hue), whereas red and yellow are not perceptual mixtures (they are unique hues)? It seems that the opponent processing theory of color vision (Jameson and Hurvich, 1955; DeValois et al., 1958; Werner and Wooten, 1979; Hunt, 1982; Hardin, 1988, pp. 34–35; DeValois and DeValois, 1997) offers us more insight into at least some of these problems.[2] Larry Hardin (1988) developed these observations into a powerful argument, starting from relevant empirical findings and reaching a philosophical theory of color and color experience. However, at roughly the same time, the philosophical view of color that is the diametric opposite of Hardin's was also announced (Hilbert, 1987). D. Hilbert was the defender of the allegedly commonsense-embracing theory.[3] The two authors started what has since become a very active and passionate debate about color and color experience, extending from empirical issues to theories of consciousness (Hilbert, 1987, 1992; Hardin, 1988; Thompson et al., 1992; Dretske, 1995, ch. 3; Thompson, 1995, 2000; Byrne and Hilbert, 1997, 2003; Stroud, 2000; Tye, 2000, ch. 7; Jackson, 2000; Hilbert and Kalderon, 2000; Akins and Hahn, 2000; McLaughlin, 2003).

In his book, Hardin argued that objects in our environment do not have the kind of properties that color perception presents them as having, and that are, at the same time, the canonical causes of color experience – therefore, object colors do not exist. Color perception is a pervasive visual illusion. Hardin's view is called *eliminativism* or *subjectivism* about color. Hilbert, however, contended that object colors are real, causally effective

---

[2] The opponent-processing theory offers a solid explanation of the second-mentioned phenomenon: the unique-binary division. However, the first (and more basic) question is not captured by this theory (not at least by its present version).

[3] For some other authors who agree with Hilbert about what the commonsense view of color is, see Campbell, 1997; Tye, 2000, pp. 147–148; Stroud, 2000.

physical properties of surfaces: types of surface reflectance.[4] In Hilbert's view, to every perceived color there corresponds such an environmental physical type, captured in terms of *triplets of integrated reflectances* (McCann et al., 1976, pp. 446–451; Hilbert, 1987, p. 111). Years later, Hilbert's approach was supplemented by the idea that what it is like to see the colors – the subjective aspect of our experience of color – is a straightforward consequence of the nature of colors themselves (Dretske, 1995; Tye, 1995, 2000; Byrne and Hilbert, 1997). Seeing red is like what it is because it is the seeing *of red* (mutatis mutandis for other colors). Colors crucially determine what it is like to see them. Our brains still play an important role in generating color experience: The visual system exhibits a specific, and quite selective, sensitivity to types of surface reflectance. The "crudeness" of trichromatic color vision (i.e., having only three different wavelength-selective photoreceptor types, as opposed to, say, the cochlea that has a large number of hair cells, each sensitive to different sound wave frequencies) sets a limit on what broad types of surface reflectance our vision can single out, and hence what color experiences it is capable of undergoing. The colors we cannot sense we cannot subjectively experience, says Hilbert.

This view of color and color experience naturally became part of some theories of mind, namely, those whose ambition was to account for phenomenal consciousness within a dominantly externalist theory of mental representation (Dretske, 1995; Tye, 1995, 2000; Byrne and Hilbert, 1997, 2003; see also Davies, 1997). These theories have the goal of uniting the two fundamental philosophical problems about the mind, intentionality and phenomenal consciousness, into one, namely intentionality, and thereby "dispelling the mystery of consciousness." On this approach, conscious experience is a matter of how the mind represents the world (this is representationalism), and representation is a matter of how our minds/brains (and bodies) are situated in the environment – how certain states of our brain are related to states of affairs in the external world (this is externalism). If this project succeeds, then conscious experience need not be viewed as some intrinsic property of brain states having strange attributes like inherent subjectivity and ineffability, and perhaps causal inefficacy (Flanagan, 1992; Block et al., 1997), or as a phenomenon with a special status that, according to some philosophers, should force materialists

---

[4] Or, in the case of emitting surfaces, some related property that Byrne and Hilbert also attempted to define (Hilbert, 1987, pp. 132–134; Byrne and Hilbert, 2003; see section 3 in the main text).

to reconsider their views of the physical world (Chalmers, 1996).[5] Along these lines, we can avoid the so-called *explanatory gap* (Levine, 1983), the problem of how to understand the link between neuronal processes in the brain and conscious experience (Tye, 2000, ch. 2).

Alas, this particular project about conscious experience is unlikely to succeed. Why? At least because the theory of color that it crucially assumes is wrong. Moreover, it is wrong for empirical reasons. In this chapter, I shall concentrate on theories of color, and, as far as possible, abandon further discussion of consciousness. All I want to note is that in my opinion, the problem is not with representationalism (or intentionalism) about consciousness but, rather, with the externalist version of representationalism (on this distinction, see McLaughlin, 2003). My goal here is to pinpoint certain empirical problems that face the theory of color proposed by A. Byrne, D. Hilbert, and M. Tye (Byrne and Hilbert, 1997, 2003; Tye, 2000, ch. 7; see also Matthen, 1988, pp. 24–25). In particular, I shall focus on perceived color similarity (or unity) and the unique-binary distinction. This problem has received a lot of philosophers' attention (Hardin, 1988; Johnston, 1992; Thompson, 1995, pp. 122–133; Byrne and Hilbert, 1997, 2003; Matthen, 1999; Hilbert and Kalderon, 2000; Tye, 2000, ch. 7; Bradley and Tye, 2001). In order to enrich future philosophical discussion of the topic, I shall look at some relevant empirical work and theories about color similarity and color space (Maloney, 1986, 1999; Maloney and Wandell, 1986; Wuerger et al., 1995).

The rest of this chapter will proceed as follows. Section 2 introduces the opponent processing theory of color perception. Section 3 is a more detailed presentation of the reflectance theory of color, and an overview of the major problems that face it. Section 4 is an introduction of L. Maloney and B. Wandell's theory of color constancy that is also an account of how surface color properties are represented in the human visual system. Section 5 will discuss unity and the unique-binary distinction, in light of relevant empirical work. For a philosophical moral, I shall suggest that the idea of perceived color similarity and the unique-binary distinction being veridical representations of certain stimulus features (Tye, 2000, pp. 162–165; Hilbert and Kalderon, 2000; Bradley and Tye, 2001; Byrne and Hilbert, 2003) is thoroughly misguided, and that even though

---

[5] The idea that phenomenal characters are determined internally by our brain states is currently held by a number of philosophers (see, for instance, Kirk, 1994; Block, 1997; McLaughlin, 2003). Moreover, this is an old idea that was proposed, among others, by Johannes Müller (Müller, 1826a, 1826b, 1833–1837, 1843; for a selection of the relevant passages in Müller's work, see Diamond, 1974).

colors are real physical properties of environmental surfaces, color perception in some respects systematically misrepresents them. Objects do have colors, but what it is like for us to see the colors seems to depend more on our minds than on the colors themselves (see McLaughlin, 2003, for a similar view). Finally, section 6 places the discussion of color in a broader context, arguing that systematic misrepresentation in certain respects is not a problem from a representationalist perspective, nor is it a problem from the point of view of evolution.

2 Opponent Processing

The opponent-processing theory of color vision is an abstract model that helps to explain the organization of color experience in light of some basic empirical facts about color vision.[6] Humans have three different, wavelength-selective retinal receptor types called cones. The psychological similarity space of perceived color (the so-called color space) has three dimensions (hue, saturation, and lightness). However, psychologically, there are four primary chromatic color categories, those of the unique hues (red, green, yellow, and blue). These primaries appear perceptually unmixed, as opposed to experiences of so-called binary hues, which appear to be mixtures of two primaries. There are two achromatic primitives as well (black and white), and achromatic grays can be understood as perceptual mixtures of these two primitives. Two dimensions of color space, saturation and lightness, are straight; that is, along these dimensions, perceptual qualities gradually change from one extreme to another, the two ends being most different. The hue dimension is circular because it does not have two extremes. Rather, the two ends of the visible spectrum (wavelength range of light) appear more similar in hue to one another than any of them to the middle of the spectrum. Psychologically there is no convenient end of the hue dimension; in one direction, reds are followed by reddish yellows (oranges); then come yellows, yellowish greens, greens, bluish greens, blues, reddish blues (violets, purples), and again reds. Somewhere between violets and slightly yellowish reds lie the physical ends of the visible spectrum (400 nm light looks violet; 700 nm light looks slightly yellowish red). Pure wavelengths of light never give rise to unique red color appearance, though unique blue, green, and yellow can be seen in spectral lights. Spectral lights also

---

[6] On the theory, see Jameson and Hurvich, 1955; DeValois et al., 1958; Werner and Wooten, 1979; Hunt, 1982; Hardin, 1988, pp. 34–35; Wandell, 1995, chs. 4 and 9; DeValois and DeValois, 1997.

cannot give rise to a broad range of bluish and reddish colors (purples and magentas).[7]

Certain pairs of chromatic hues are such that when they are mixed in appropriate proportions, they "cancel" one another, resulting in the perception of achromatic gray. Such pairs are called opponent colors. Red and green are opponent pairs; so are yellow and blue (at least approximately). In addition, for any orange there is some bluish green that is its opponent color; opponent colors of purples are yellowish greens.

To explain such facts about color perception, different schemas were proposed earlier in the history of color-vision research, for instance by Thomas Young (1802) and Hermann Helmholtz (1911). These accounts failed on a number of counts (Hurvich, 1981, pp. 128–129) like predicting psychological mixing and opponent colors, or a sufficient account of color deficiencies. The current version of the opponent-processing model was developed by Dorothea Jameson and Leo Hurvich (Hurvich and Jameson, 1955; Jameson and Hurvich, 1968; Hurvich, 1981). This model can account for all the aforementioned phenomena (chromatic and achromatic primaries; mixing and opponent relations; the dimensions of color space and the circularity of the hue dimension).

The opponent-processing theory describes the ways in which signals coming from the three retinal cone types combine to shape color appearance. A rather simplified version of the model is as follows. The three cone types have the following peak sensitivities: 440–450 nm ("short-wave – S – cones"), 535–545 nm ("middle-wave – M – cones"), and 560–570 nm ("long-wave – L – cones"). The visual system computes the difference between the L and the M cones (at different points of the retina) to obtain a red-green (RG) signal. If the difference $L-M$ is positive, that corresponds to the perception of some red (or reddish) hue; negative difference corresponds to green(ish) perception (positive and negative signs here are merely a matter of convention). The formula $L+M-S$ predicts the yellow-blue (YB) hue component: Positive values correspond to yellow(ish) perception, negative output to blue(ish) perception. In accordance with empirical observations, this model predicts that there is no perceived mixture of reds and greens, nor is any such mixture of yellows and blues. If the red-green channel outputs a positive value (red), whereas the yellow-blue channel a negative one (blue), the result is a binary hue perception: a perceptual mixture of red and blue like

---

[7] Color space can also be represented by three straight dimensions: lightness, reddishness-greenishness, and yellowishness-bluishness. See DeValois and DeValois, 1997; Sivik, 1997, p. 178.

violet, purple, or magenta (depending on the ratio of the outputs of the
two channels). The lightness (L) dimension arises as the sum of the L
and M cone outputs. Saturation is correlated with the absolute values of
the red-green and yellow-blue channel outputs. For instance, perception
of magenta – highly saturated, slightly bluish red – arises if the red-green
channel outputs a large positive value and the yellow-blue channel out-
puts a small negative one; thus, the proportion $|RG|/|YB|$ (the proportion
of the absolute values of the outputs) is substantially greater than 1. If this
proportion stays the same while both $|RG|$ and $|YB|$ decrease, the result
will be pink: unsaturated, slightly bluish red. This happens when L−M
(reddishness) decreases while L+M (lightness) stays the same and S also
decreases to an appropriate degree (thus L+M−S, though still negative,
will be closer to zero).

   This model offers a straightforward explanation of two phenomena of
color perception: *unity*, and the already mentioned *unique-binary distinction*. "Unity" refers to the perceived similarity relations of the colors, the
feature that *uniform color spaces* were constructed to capture. For example,
reds are perceptually more similar to magentas (slightly bluish reds) than
to purplish (i.e., slightly reddish) blues. Yellowish greens look more similar to oranges (both are yellowish) than to purples (not at all yellowish).
Uniform color spaces are those in which (1) the points correspond to
color percepts of the standard trichromat observer;[8] and (2) Euclidean
distances between the points approximately correspond to perceived differences between the corresponding colors. In the examples, points of
the yellow-green region are closer to those of the orange region than to
points corresponding to purples. Similarly, points representing reds are
closer to those representing magentas than to those representing purplish blues. The unique-binary distinction is the idea that most colors
look like mixtures of two other chromatic color components (e.g., orange is both reddish and yellowish), whereas some colors (the unique
hues, or the six Hering primaries: red, green, yellow, blue, black, and
white) are perceptually unmixed.

   Things are never even this simple, however. In order to support the
view I offer, first I briefly present two versions of the opponent-processing

---

[8] The CIE (Commission Internationale de l'Éclairage) standard observer is a theoretical
construct: The color-vision parameters of many actual observers are measured and av-
eraged (e.g., spectral sensitivities of the cones, transparency of ocular media, etc.), and
these averages constitute the standard observer. Due to individual variation, however, just
as virtually no one earns the average salary, no trichromat is exactly like the standard
observer (see Hardin, 1988; McLaughlin, 2003).

model that are less simplified than the introductory version presented so far; then I address a critique of the theory.

J. Werner and B. Wooten (1979) examined the relation between three levels of color vision: (i) estimated spectral sensitivities of the three photoreceptor types, (ii) chromatic response functions resulting from hue cancellation experiments,[9] and (iii) hue naming of spectral lights, and subjects' location of unique blue, green, and yellow location in the spectrum. The naming method estimated the percentages of the four chromatic hue components (reddishness, greenishness, yellowishness, and bluishness) in spectral lights (Sternheim and Boynton, 1966; Werner and Wooten, 1979, p. 423).

Werner and Wooten used three observers for data collection. Their results showed substantial individual differences in the chromatic response functions.[10] The authors' main questions were (1) whether the empirical chromatic response functions can be derived as linear combinations of the cone spectra (as proposed by Jameson and Hurvich, 1968; Werner and Wooten, 1979, p. 426), and (2) whether subjects' estimated percentages of the four hue components in spectral lights can be predicted from their chromatic response functions. Werner and Wooten were able to predict their hue-naming data from the chromatic response functions, thus demonstrating a linear relation between these two levels.[11] As concerns the relation between receptor activity and opponent mechanisms, Werner and Wooten confirmed that for the red-green opponent mechanism, chromatic response functions can be derived as linear combinations of the three cone absorption spectra (Werner and Wooten, 1979, pp. 424, 428–433). However, they argued that spectral sensitivities of the cones

---

[9] See Hurvich, 1997. In hue cancellation experiments, the stimuli are monochromatic lights of different wavelengths, and subjects are asked to mix another, fixed-wavelength light to the stimulus to cancel one of the hue components in the stimulus. For instance, the stimulus can be 595 nm (orange) light, and the subject, by operating a control, can add 468 nm (blue) light to it until the blue light exactly cancels the yellowish appearance from the orange, leaving a pure red perception. Repeating this for the whole visible spectrum (say, at 10 nm steps), with appropriate instructions for the different regions (canceling reddishness, greenishness, yellowishness, or bluishness), and measuring the energy of the canceling light at each point gives rise to the chromatic response functions.

[10] The shapes of both the red-green and yellow-blue response functions showed individual variation. The wavelengths of peak sensitivity for the red-green response function were in agreement among the three observers, whereas the blue-yellow response function's peak sensitivities proved quite variable (Werner and Wooten, 1979, p. 426).

[11] They did this by converting the chromatic response functions resulting from hue cancellation into hue percentages using Jameson and Hurvich's hue coefficient (Hurvich and Jameson, 1955; Werner and Wooten, 1979, p. 423).

are not, in general, linearly related to the blue-yellow channel, since the blue-yellow response function cannot arise as a linear combination of the cone spectra. They proposed that the nonlinearity applies to the original Jameson and Hurvich (1968) model (Werner and Wooten, 1979, p. 428). That model is close to the formulas of the simplified opponent model previously presented. In terms of the simplified model, Werner and Wooten propose that to express the activity in the blue-yellow channel, the formula $(k_1L + k_2M)^n - k^3S$ (instead of L+M−S) should be used. That is, we have to include, in addition to differential weighting of the cone inputs $(k_1\ k_2, k_3)$, a power transformation (n). In addition, the exponent n varies substantially: In some individuals it is between 3 and 4, whereas in others it is close to 1, which means that even the linear or nonlinear nature of the blue-yellow channel is a matter of individual variation.

R. Hunt's goal (Hunt, 1982, p. 95) was also to describe a mathematical relation between cone spectral sensitivities and color appearance data. As representations of color appearance data, he used the Munsell Color System and the Natural Color System (NCS) color space. Since the latter is based entirely on color-appearance judgments, whereas the former is based on a mixture of color-appearance considerations and an attempt to handle the distribution of color differences, in some respects[12] Hunt gave priority to NCS data (Hunt, 1982, pp. 95, 104–105). Without going into the details of this model, here is a quick comparison between it and Werner and Wooten's schema. Hunt's model is more complex; in particular, it assumes four different chromatic channels (instead of two), and one achromatic one – arguing (p. 111) that this is still a neurophysiologically feasible mechanism. It assumes a nonlinearity different from what Werner and Wooten propose. Namely, the cone signals are supposed to undergo a power transformation *before* summing (Werner and Wooten apply the power transformation after summing), and the same exponent is applied to all four chromatic channels.[13]

In addition to these improvements, some key ideas of the opponent-processing theory have been recently criticized (Jameson and D'Andrade, 1997) on empirical grounds. K. Jameson and R. D'Andrade note that in order to construct a model of color space, a great variety of methods for scaling similarity were employed, and the overall pattern

---

[12] Especially in fitting his model's predictions with appearance data in the red-blue range, with respect to which the Munsell system and the NCS space differ the most (Hunt, 1982, p. 105).

[13] The four chromatic channels in Hunt's model are (1) $R^{0.5} - G^{0.5}$; (2) $G^{0.5} - B^{0.5}$; (3) $B^{0.5} - R^{0.5}$; (4) $R^{0.5} + G^{0.5} + B^{0.5}$, where R, G, and B correspond to cone signals (long-, middle-, and short-wave ones, respectively).

emerging from these studies confirmed the structure of the Munsell color space (Indow and Aoki, 1983; Indow, 1988; Jameson and D'Andrade, 1997, p. 299). However, in the Munsell hue circle, red and green do not occupy opposing positions, nor do yellow and blue (Jameson and D'Andrade, 1997, pp. 298–299). Therefore, red-green and yellow-blue do not appear to be straight dimensions of color space. Another important observation is that the so-called *additive complements* (pairs of lights that, when mixed in appropriate proportion, yield an achromatic stimulus) do not include unique red–unique green pairs, nor unique yellow–unique blue ones (Jameson and D'Andrade, 1997, pp. 307–308). Also, the principle of linear additivity for light mixtures seems to be violated in certain cases (op. cit., pp. 306, 317, note 8). For example, if two red lights, L_1 and L_2, produce no output on the blue-yellow channel separately, L_1+L_2 may well produce a positive output (i.e., a yellowish appearance), due to the Bezold-Brücke effect.

The two authors make some proposals to solve these problems. One is that the relation between the CIE tristimulus space for color mixtures of spectral lights, on the one hand, and color space, on the other, becomes a complicated nonlinear one (p. 306, and earlier in this section). The other is that the unique hues arise from the irregular structure of color space, and not from opponent-processing neurophysiology (pp. 310–314). This way, perceptual irreducibility (i.e., of the unique hues) is detached from color opponency.

In response, we might note that at least some methods of color-similarity assessment seem to result in straight red-green and blue-yellow dimensions of color space (Sivik, 1997, pp. 178–179). In constructing the Natural Color System, L. Sivik and his colleagues estimated color similarity from subjects' judgments of the relative hue components (redness, greenness, yellowness, blueness, whiteness, and blackness), not combining these data with the psychophysics of color mixing (op. cit., pp. 168–169). The arising regular structure of color space appears consistent with Jameson and D'Andrade's idea that the relation between tristimulus space and color space may be rather complex (but, of course, not with their suggestion that red-green and blue-yellow are not straight dimensions of an irregular color space).[14] Finally, I would like to note that another argument for a genuine "high-level opponency" of the unique

---

[14] I mean, there has to exist some mapping that takes tristimulus space into NCS color space. The ultimate question is which structure of color space, together with the corresponding mapping, gains more support – the regular one assumed in NCS, or the irregular one proposed in Jameson and D'Andrade's paper. On the structure of color space, see section 5.

hue pairs (red-green and blue-yellow) is that they do not mix perceptually. This observation is quite independent from their not being additive complements, and it seems consistent with Jameson and D'Andrade's proposal that the perceptually irreducible unique hues arise as higher-level representational elements for color.

3 The Reflectance Theory of Color

3.1 From an Empirical Point of View

Inspired by Edwin Land's retinex theory of color constancy (McCann et al., 1976; Land, 1977); Hilbert (1987) developed an empirically based philosophical account of color, which identifies colors with surface reflectances. A key problem that such an account has to handle is metamerism: the fact that many, widely different surface reflectances look to us the same in color under normal circumstances; therefore, at first glance it is unclear which color should be identified with which surface reflectance.[15] Hilbert gets around this problem in the following way. In his view, different surface reflectances are all different colors. However, since trichromat human color vision is a pretty crude reflectance-measuring device (it has only three different types of sensors), we cannot nearly see all different colors *as different*. Still, all those surfaces that look to us the same in color have identifiable reflectance characteristics. These characteristics can be expressed in terms of *triplets of integrated reflectances* (TIRs): All those surfaces that look to us the same in color under a given illuminant have the same TIR. The TIR of a surface is obtained by integrating its reflectance separately over the three (overlapping) sensitivity ranges of the cone types.[16]

In his book, Hilbert does not provide a single example of a particular color, or color category, defined in terms of TIRs. Byrne and Hilbert (1997) do offer one such example: that of green. In their new version of

---

[15] There is already a simplification here: Metamerism is illumination-dependent. Two surfaces with different reflectance that look the same in color under one illuminant can look quite different under another illuminant.

[16] In fact, Land (1977) and McCann et al. (1976, Part II) used integrals of reflectance weighted by the sensitivity curves of the corresponding cone types. Hilbert (1987, p. 111) mentions this, but seems not to consider the philosophical problems which this feature raises for his theory. Briefly, if color sensations correlate well, not with integrated reflectance as such but with *reflectance times retinal sensitivity, integrated* (and further scaled by a psychophysical sensitivity factor: see McCann et al., 1976, p. 449–450), then no *local*, or "inherent" property of surfaces is specified that is a correlate of perceived color. This is a problem for Hilbert's theory. Tye (2000, pp. 160) and Bradley and Tye (2001) offer a solution to it that I'll discuss in the main text (see also Byrne and Hilbert, 2003, secs. 3.2.2. and 3.2.3.).

the reflectance theory, they combine the simplified opponent-processing model (outlined earlier) with Hilbert's original idea of TIRs.[17] Tye (2000, pp. 159–162) works out the details of this idea, offering a schema for characterizing, in terms of surface reflectance, the four unique hues (red, green, yellow, blue), the four broad binary hue categories (orange, purple, yellowish green, and bluish green), and achromatic grays including black and white. Since according to the opponent-processing model, perceived reddishness and greenishness derive from a comparison, in color vision, of the L and M cone outputs, Tye boldly proposes that for a reflecting surface to be reddish is for it to be disposed to reflect, under some "normal" illuminant, more light in the sensitivity range of the L cones than in that of the M cones (Tye, 2000, pp. 160–161).[18] A surface is greenish if it is disposed to reflect, under normal illuminants, more light in the M range than in the L range, and it is neither reddish nor greenish if it is disposed to reflect approximately the same amount of light in these two wavelength ranges.[19] Tye's proposal preserves Hilbert's idea that colors are to be identified with triplets of integrated reflectances.[20]

By the same coin, since the blue-yellow channel arises from the comparison of the S cone outputs with the sum of the L and M cone outputs, Tye proposes that an object is bluish just in case it is disposed to reflect, under normal illuminants, more light in the sensitivity range of the S cones[21] than in that of the L and M cones together. Yellowishness in objects is to be disposed to reflect more light in the L and M ranges together than in the S range. Achromatic grays are those surfaces that reflect approximately the same amount of light in the L and M ranges, and also in the S and (L + M) ranges.

---

[17] The definition of green thus proceeds as follows in their paper: A surface is (unique) green if it reflects more light in the middle waveband than in the long one, and approximately as much light in the short waveband as in the other two wavebands together.

[18] Here I am largely clarifying Tye's claim (see Tye, 2000, p. 161). By "normal" illuminant, Tye would probably mean some CIE illuminant, or an illuminant whose color rendering is sufficiently close to average daylight. Another ideal that defenders of type physicalism about color have in mind is an "equal energy illuminant": illuminant with perfectly even spectral power distribution (see Byrne and Hilbert, 2003, sec. 3.2.3.).

[19] Again, the details are missing from Tye's text, but the only plausible way to fill them in seems to be that the L range is 450–680 nm, and the M range is 435–640 nm, that is, roughly, the sensitivity ranges of the two corresponding cone types. Tye uses the symbols L^*, M^*, and S^* to express *the amount of light, reflected in the sensitivity ranges of the three cone types* (Tye, 2000, pp. 159–161).

[20] Byrne and Hilbert (2003) propose a somewhat improved, but still thoroughly problematic, version of this kind of reflectance schema.

[21] That is, roughly 400–525 nm.

There still remain problems, however. Taken as it is, this schema is simply wrong. For example, neither red nor green surfaces are such that in daylight (or incandescent light, or fluorescent tube light) they reflect approximately the same amount of light in the S range than in the L and M ranges together. The schema also misclassifies achromatic grays with even reflectance distribution as yellows, and purples and yellows as oranges. For this reason, Tye adds (2000, p. 160) that his schema has to be appropriately "qualified" by multiplying reflectance (or better, the color signal arising from surfaces) with the spectral sensitivities of the cones, and possibly by further parameters deriving from human color vision.[22] No matter how complex the schema becomes, there will always be some (admittedly quite complex) surface reflectance property for each and every type of color experience such that that reflectance property reliably correlates with the corresponding color experience type,[23] in broadly normal circumstances, of perception.[24]

If such a qualified version of the schema is correct, then it provides some account of metamerism. For instance, all metamers of a particular shade of green are such that they reflect more light in the M range than in the L range (under normal illuminants), despite the differences in the particular reflectance curves. Moreover, all green surfaces belong to this general reflectance type. Narrower subcategories like lime green or olive green can be further specified within this general schema (e.g., lime greens are lighter; that is, they have higher average reflectance than olive greens). Tye's schema also provides an objectivist account of the unique-binary distinction (Tye, 2000, pp. 162–165). If an object exhibits simultaneously the reflectance property of reddishness and yellowishness, then it is orange; thus, oranges are objectively mixtures of reddishness

---

[22] In a later paper (Bradley and Tye, 2001, p. 482), the authors cite Hunt (1982) and Werner and Wooten (1979) in making the same point.

[23] Types of color experience can be characterized by correspondence to points in a color space that makes explicit the opponent organization of color vision (like the CIELAB space), or by the experimental method introduced by C. E. Sternheim and R. M. Boynton (1966), in which the subjects are asked to describe colors by subjectively estimating percentages of their chromatic components – for instance, some orange might be characterized as 40% red and 60% yellow. See Byrne and Hilbert (2003) for more details on such an approach.

[24] Byrne and Hilbert (2003, sec. 3.2.3) introduce the notion of L-intensity, M-intensity, and S-intensity. For a given surface S under illuminant I, S's L-intensity is the degree to which it stimulates the L cones (analogously for M- and S-intensities). Then a reddish surface gets characterized not as *reflecting more light in the sensitivity range of the L cones than that of the M cones* but as *reflecting light in such a way that that reflection pattern would in turn excite L cones more than the M cones.*

and yellowishness. On the other hand, if a surface exhibits an M > L imbalance, but an S = (L + M) balance (add whatever "qualifications" are needed, Tye would say), then that object is unique green (with some saturation and lightness).

3.2 *From a Philosophical Point of View*
The reflectance theory is a particularly strong version of color realism. A list of its key assumptions follows.

1. All and only those surfaces that look the same in color share a common physical property that can be characterized in scientific terms and that is specifically causally responsible for evoking the corresponding experience of color.[25] Colors have to be *observer-independent* properties: properties that can be physically realized in the absence of human or other observers. These assumptions amount to *type physicalism* about color.[26] Type physicalism is one version of color realism.

2. Reflectance theorists also prefer the view that object colors are in some way local, or "inherent," properties of surfaces. In other words, the color of a surface does not depend on what light it is illuminated with, or what surround it is placed in. This assumption is not necessarily part of type physicalism, but those who defend the view find it intuitively plausible (Tye, 2000, pp. 150–162).

3. A key assumption is that colors are absolute, rather than perceiver-relative (Tye, 2000, note 4 on p. 167; McLaughlin, 2003; Jakab and McLaughlin, 2003). For instance, the [−7,−3,−1] chip of the OSA Uniform Color Scales, or samples at a paint store (perceptually determinate shades) like Real Red, Poinsettia, Royal Purple, or

---

[25] When we speak of looking the same in color, say, green, we can mean either a *perceptually determinate shade* like what we see on looking at a particular tree leaf, or a *broad color category* in which many different shades of green belong. The point, however, can be made at both levels of generality. In terms of broad color categories, the claim is that all and only green objects have some version of a more general surface property, call it greenness. In terms of perceptually determinate shades of green, the claim is that all and only surfaces with a particular perceptually determinate shade have a more specific objective color property that simply is the shade of green in question. (Metamers are perceptually indiscernible variants of such shades.)

[26] There are physicalist theories of color according to which particular colors should be identified with the disjunction of a variety of different physical properties (Jackson and Pargetter, 1997). Recently this view seems to have lost popularity among philosophers (see McLaughlin, 2003, for some critical considerations). Still, since this is a theoretical option (some have called it *disjunctive physicalism*), it is in contrast with the reflectance theory (see Tye, 2000, pp. 149–150, and note 4 on p. 167).

Ming Blue,[27] are such that each one of them is one and the same color for everyone – at least to every human perceiver.

4. The reflectance theory is also committed to the assumption that two perceived attributes of colors, namely, *unity* and the *unique-binary distinction,* are firmly grounded in the very stimulus properties that are the canonical causes of our color experience. For unity to be grounded in object color properties, object colors should exhibit the same measurable similarity metric as color space – the perceptual similarity space that is modeled in color science by (approximately) uniform color spaces like the CIELAB space (Wyszecki and Stiles, 1982). Similarly, for the unique-binary distinction to be grounded in object color, it should be the case that, say, the object color orange is some measurable "mixture," or compound, of the color properties red and yellow (mutatis mutandis for other colors). At this point, it might appear to the reader that the reflectance theory of color is not a philosophical theory after all, but rather a testable empirical proposal (which has certain philosophical importance). I think this is correct.

3.3 A list of Problems

1. Despite alleged solutions to the problem of metamerism, it still seems very difficult to maintain that all and only surfaces with the same color share some common physical attribute. One additional problem is that there are many surfaces that are colored but their color is not due to their surface reflectance. An area of a color monitor may look very similar, even indistinguishable, in color to a summer tree leaf, yet one is green due to light emission, whereas the other is green due to light reflection. Byrne and Hilbert's reply to this objection is to introduce the notion of *productance,* which is supposed to be a generalized notion of which reflectance is a special case (Byrne and Hilbert, 2003, sec. 3.1.2.; see also Hilbert, 1987, pp. 132–134). Productance is the ratio of emitted plus reflected light (the numerator) and illuminating light (the denominator), at all visible wavelengths. For nonemitting surfaces, productance is equivalent to reflectance. However, as Jakab and McLaughlin (2003) point out, the notion of productance is badly mistaken: The formula yields infinite productance values for emitting surfaces in the total absence of external illumination (think of a firefly or

---

[27] I found these colors on the following commercial website: http://www.plascon.co.za.

candlelight on a very dark night). So the problem of common ground (as McLaughlin [2003] calls it) remains unsolved (see also Jakab, 2003).

2. Byrne, Hilbert, and Tye like to cite color constancy in defense of their view that colors are local, inherent properties of objects (Tye, 2000, pp. 147–148; Byrne and Hilbert, 2003, sec. 3.1.). However, they do not seem to pay much attention to the fact that color constancy is only approximate (Wandell, 1995, pp. 314–315; Fairchild, 1998, pp. 156–157). This means, for instance, that when the afternoon daylight turns into sunset, or when we move from daylight to incandescent light, the illumination shift does actually cause a change in the color appearance of objects (as a striking case of this, think of the shopping nightmare when an expensive dress looks one color in the store and quite another in the street). True, such changes in perceived color due to illuminant change are most often diminished by mechanisms of color constancy; still, they are not completely abolished. What is left of them we tend not to notice because the weakness of our memory for perceptually determinate shades effectively masks them (Raffman, 1995, pp. 294–295; Tye, 2000, pp. 11–13). Still, such changes can easily be demonstrated in the laboratory, and sometimes they are noticeable in everyday life. This poses a problem for specifying perceptually determinate shades (see note 25) in terms of reflectance, abstracting away from the illuminant. Despite approximate color constancy, colors seem to some extent to be illuminant-relative.

Simultaneous color contrast is another problem case for color localism. For how could color be a local property of surfaces if the perceived color of surfaces depends on the colors in the surround? Tye's favored answer is that changes in perceived color due to surround effects are subtle illusions, or cases of *normal error* (Haugeland, 1981, p. 18; Matthen and Levy, 1984; Matthen, 1988; Dretske, 1995, pp. 91–92), just like the Müller-Lyer illusion, figural aftereffects, or simultaneous contrast in shape (Fairchild, 1998, p. 152; Tye, 2000, pp. 153–155). Some shape illusions do indeed seem to create perceptual error; Tye claims the same about simultaneous color contrast.

The problem with this reply is that color contrast is present in virtually every perceptual circumstance, not just in special settings like shape illusions. Any particular reflecting surface under constant illumination can look a whole variety of different shades (sometimes very different ones), depending on its surround. Which of the surrounds reveals *the true color* of the sample? Perhaps some neutral midgray background? Any such choice seems entirely arbitrary (McLaughlin, 2003), and it implies that in everyday life we very rarely see the true shades of objects – perhaps we do so only in the color scientist's laboratory. Along these lines, we are bound to conclude that slight illusions are ubiquitous in color perception – a conclusion severely at odds with representational externalism about color experience.

3. Color absolutism is already undermined by the illuminant and surround-dependence of color perception; however, the fatal blow to this idea comes from individual differences in color perception (Byrne and Hilbert, 1997, pp. 272–274; McLaughlin, 2003). It is a well-known fact that there are substantial differences in the color perception of normal trichromat humans.[28] Perceptually determinate shades like Poinsettia (a shade of red), Bridesmaid (a shade of pink), or Banff Spring (unsaturated bluish green) can look pretty different in color to two different, but equally normal, trichromats in the same circumstance of observation. These differences can be demonstrated in unique hue location experiments (Ayama et al., 1987; Laxar et al., 1988; Ikeda and Uehira, 1989; Kuehni, 2001; see also Hardin, 1988, pp. 79–80; Byrne and Hilbert, 1997, p. 272; Block, 1999). Given such individual variation in normal color vision, color absolutism seems challenged. For if a color sample looks unique green (with some lightness and saturation) to one color-normal subject Max in some standard circumstance of observation, whereas the same sample in the same circumstance looks slightly bluish green to another color-normal subject Samantha, then color absolutists face a strange consequence. If they insist that the sample is nevertheless exactly one color (say, unique green), then it becomes difficult to avoid the conclusion that while Max perceives its color *correctly*, Samantha perceives it *incorrectly*. This conclusion seems absurd, for why should we accept the idea that color-normal subjects in broadly normal, or even standard, circumstances of perception misperceive the colors? The alternative is to abandon color absolutism (that is, embracing color relativism) and to say that the sample is blue-green for

---

[28] Here I set aside color deficiencies, for they might be regarded as cases of *abnormal* color vision.

Samantha and unique green for Max (Jackson and Pargetter, 1997; McLaughlin, 2003).

Tye agrees that it would indeed be absurd to conclude that, in the example, either Max or Samantha misperceives the color of the sample. But he contends (Tye, 2000, pp. 89–93, note 23 on p. 169) that such individual differences in color perception are simply the result of differences in color discrimination. Samantha is better at color discrimination: She can discern the tinge of bluishness that is, as a matter of fact, present in the sample, whereas Max cannot. The difference between Max's and Samantha's color vision is like the difference between two gauges, one more finely calibrated than the other. A blunt ruler may say of a steel rod that it is roughly 19 inches long; a better ruler may say of the same rod that it is 19.35 inches long. Neither ruler is mistaken – and the same moral applies to individual differences in color vision.

This proposal does not seem enough to fend off the critique, however. It is likely that variation in trichromat color perception can take forms that cannot be accounted for by Tye's calibration approach. If the green sample looks slightly yellowish green to Max, and slightly bluish green to Samantha, this cannot be a mere difference in their ability to discriminate colors. Where Max discerns a tinge of yellowishness, Samantha discerns a tinge of bluishness; now the question of which one of them is right arises in a nastier way for the color absolutist. The sample is either yellowish or bluish; there is strong reason to assume, within the reflectance theory, that it cannot simultaneously be both (Jakab, 2001, ch. 6). Moreover, if, by assumption, the sample is objectively bluish green, whereas it looks to Max yellowish green, then it seems very hard to avoid the conclusion that Max, a perfectly normal color perceiver, misperceives the sample.

Byrne and Hilbert (2003, sec. 3.4) bite the bullet and claim that it is not absurd to think that normal color perceivers misperceive the colors slightly. In support, they mention that it is not uncommon for humans to have a slightly distorted shape perception due to certain ophthalmologic conditions (their example is aniseikonia). However, applying this principle to color perception, they quickly reach the conclusion (op. cit., note 50) that it is in principle unknowable which narrow reflectance range is unique green (mutatis mutandis for all perceptually determinate shades). For if different trichromat subjects single out different (in fact, often quite different) stimuli as unique green (unique yellow, unique red, etc.), then who has the authority to decide who is right and who is wrong? (See Jakab and McLaughlin, 2003, for further detail.)

4. Unity and the unique-binary distinction is another problem case for the reflectance theory, despite efforts to accommodate it. I shall address this issue in section 5.

4 Color Constancy and the Linear Models Framework

Recently, Laurence Maloney and Brian Wandell developed a theory of color perception to explain the phenomenon of color constancy and to help in understanding how information about surface reflectance is handled by color vision (Maloney, 1986; Maloney and Wandell, 1986; Wandell, 1995, ch 9; Maloney, 1999, 2002, 2003).[29] Color constancy is the phenomenon that despite changes in illumination, we perceive the colors of surfaces as approximately constant. Perceived color correlates better with the spectral reflectance of surfaces than with the pattern of light that comes to our eyes from the surfaces. This feature of color vision is remarkable since the only information color vision has as input is the pattern of light arriving at different points of our retinas. As Maloney and Wandell point out, no color vision could ever be color constant under *any* illuminant. Human color vision, for instance, is color constant only under a limited range of illuminants that are relatively close in characteristics to daylight (Shepard, 1997). That is, to achieve color constancy, color vision has to make "built-in" assumptions about the illuminant, and also about the reflectance characteristics of surfaces. In addition to constraints on the illuminant characteristics that obtain in the natural environment, characteristics of natural surface reflectances are also constrained (hence, such constraints can be assumed by color vision). For instance, surface reflectances in our environment are always smooth, continuous, slowly varying functions of wavelength (Maloney, 1986; Westland and Thomson, 1999). This characteristic derives from some general physical and chemical properties of terrestrial surfaces (Maloney, 1986, pp. 1677–1678). Color processing builds on these assumptions, and if they do not obtain

[29] An alternative account of color constancy called the retinex theory is developed by Edwin Land (McCann et al., 1976; Land, 1977); however, this account has some problematic features and predictions that the Maloney-Wandell model overcomes (Brainard and Wandell, 1986; Shepard, 1997, pp. 318–325).

in particular situations, color constancy in perception breaks down. (This happens in the light of sodium vapor lamps at night.)

The constraints on reflectances and illuminants are analyzed mathematically in a linear-models framework. This means that both illuminants and surface reflectances are represented by a linear combination (weighted sum) of a small number of basis functions. The basis functions are derived by linear algebraic methods (principal component analysis) from large sets of surface reflectances or illuminants, and so they represent generalizable properties of the whole set – reflectance characteristics of surfaces and relative energy distributions, or *spectral power distributions,* of lights. Given the basis functions, any particular reflectance *R* is characterized simply by the set of weights of the linear combination that best approximates *R.*

From a psychological point of view, the idea is that color vision represents particular reflectances by linear combinations of a small set of basis functions. The basis functions themselves embody the assumptions of color vision about environmental surface reflectances and illuminants. Only a limited range of target functions (surface reflectances or illuminant energy distributions) can be approximated by linear combinations of a given set of basis functions. Other target functions that do not share the characteristics represented by the basis will be poorly approximated, if at all. There is a general agreement among the previously cited authors that surface reflectances in our terrestrial environment can be very well approximated by five to seven basis functions, and three-dimensional linear models already yield a good approximation of large sets of natural reflectances. Within the linear-models framework, some linear algebraic transformations based on how photoreceptors transform color signals constitute a method of discounting the effect of illuminant changes (see Wandell, 1995, pp. 301–308, and Maloney, 2003, for an introduction; for a thorough review including mathematical details, see Maloney, 1999).

The color signal is the light actually arriving at the retina from particular surface areas. Mathematically, the color signal is the product of illuminant spectral power distribution and the reflectance of the distal surface in question. Since the problem for Maloney and Wandell's theory is to recover surface reflectance from the color signal, it is crucial that the visual system obtain an independent estimate of the actual illuminant. Generalized features of the "normal" range of illuminants are captured by the linear-model basis functions for illuminants (see, e.g., Wandell, 1995, pp. 306–308), but which particular illuminant obtains at any moment needs to be estimated. Then, from the color signal (the

product) and the illuminant characteristics (one of the factors), it becomes possible to estimate the surface reflectance of the distal object (the
other factor). There are different hypotheses regarding how the visual
system estimates the illuminant independently. In some cases, it might
happen by casting a glance at the illuminant (e.g., the sky or the sun).
Other cues may also be informative: specular highlights,[30] or the color
of the uniform background. Another strategy for illuminant estimation
is to make further assumptions about the particular scenes encountered,
for instance, that the average of all surface reflectances within a scene
is achromatic gray, or that the brightest surface in the scene viewed is a
uniform white reflector (Maloney, 1999). One advantage of such assumptions is that they make possible illuminant estimation in the absence of
the specific cues mentioned; however, it is a question of whether these
assumptions are actually correct for most of the scenes we might encounter in our environment. Given the different sources of illuminant
estimation, none of which is absolutely reliable (except, perhaps, looking at the illuminant), Maloney (2002) proposes that illuminant cues
are independently assessed, and then combined to obtain an estimate of
illuminant chromaticity. Just as depth is estimated via different modules
using different cues, illuminant estimation too builds on multiple sources
in a flexible way. The challenge then is to find an algorithm to describe
cue combination – in other words, to assess the relative importance of
each cue in a given scene and assign appropriate weights (Maloney, 2002,
p. 495).

5 Color Space and Color Similarity

5.1 *Representational Externalist Proposals to Accommodate Unity and the Unique-Binary Division*

Unity and the unique-binary division together amount to the internal
structure of color space, that is, the perceptual similarity space for colors.

---

[30] Shiny objects have a mirrorlike component of their reflectance whose spectral distribution is typically even: It reflects approximately the same percentage of light at any
visible wavelength. That is, the color of the specular highlight closely approximates
that of the illuminant. Think of a shiny red ball in sunlight: The ball looks red except for the white spot arising from the sun's reflection. There are exceptions to this
rule, though. The characteristic color appearance of gold is the result of a non-even
reflectance distribution of both the specular component (interface reflection) and the
scattered, nonspecular component (body reflection). For this reason, gold in white light
produces yellow specular highlights, and so it would be a misleading cue for illuminant
estimation. On the two mechanisms of reflectance, see Wandell, 1995, pp. 292–293.

The two features are very closely related; however, sometimes they are discussed separately (see esp. Tye, 2000, pp. 162–165; Bradley and Tye, 2001).[31] In this section I discuss them together, focusing on Byrne and Hilbert's work (Hilbert, 1987; Byrne and Hilbert, 1997, 2003; Hilbert and Kalderon, 2000), but also including Tye and Bradley's proposals.

An important and much-debated observation about the structure of color space is that perceived color similarity does not seem to correspond, in any interesting way, to the measurable physical similarity relations that characterize object colors. As some authors claim (e.g., Thompson, 1995, pp. 122–133, and 2000; Matthen 1999), there is no interesting isomorphism between the similarity space for surface reflectances and color space. M. Matthen (1999, pp. 64–69, 73–76, 82–83) argues that color experience distorts the similarity relations of the (physical) colors. Of course, a lot might depend on how we set up the similarity space for reflectances. The reflectance theory of color offers us some guidelines in this respect. Hilbert (1987, p. 111) argues that the triplets-of-integrated-reflectances-based similarity space[32] (TIR-space, for short) orders reflecting surfaces in a way that is close in similarity structure to color space only if reflectances are weighted by cone sensitivity profiles before integrating.[33] As a result of E. Thompson's critique (1995, pp. 122–133), Byrne and Hilbert (1997, p. 285, note 32) abandon the idea that the similarity metric of the TIR similarity space and color space are isomorphic in any interesting way. Instead, they try out a different solution to accommodate perceptual color similarity within the reflectance theory (Byrne and Hilbert, 1997, pp. 274–279). They argue that perceived similarity

---

[31] Tye and Bradley discuss only the unique-binary distinction in their paper.

[32] The TIR-space is a three-dimensional similarity space whose dimensions are reflectances integrated above the sensitivity ranges of the short-, middle-, and long-wave cones, respectively. In reality, what Hilbert (1987, p. 111) has in mind (following McCann et al., 1976, Part II) is reflectance, weighted separately by the three spectral-sensitivity-profiles retinal cone types and then integrated, yielding three scalars that characterize the photon absorptions (or the resulting excitations) of the three cone types (Maloney and Wandell, 1986, p. 29; Wandell, 1995, Figure 4.18, p. 92).

[33] In fact, McCann et al.'s observation (1976, pp. 448–451) was that cone outputs (TIRs weighted by cone spectral sensitivities) need to be further transformed by a power function in order to achieve a good fit between perceived colors and TIRs. This means that the authors introduce a sensory function whose exponent characterizes the sensitivity of the perceiver to color stimuli, in order to achieve the desired fit. As they summarize: "Our results show that the color sensations are very highly correlated with the triplets of reflectance" (McCann et al., 1976, p. 446). The authors conducted their study to seek empirical support for Land's retinex theory; they did not frame their findings in terms of the opponent-processing theory.

is the result of a process of categorization implicit in color perception. For example, red looks more similar to orange than to green. This observation is explained by saying that red and orange surfaces are both represented as reddish, whereas green surfaces aren't. Similar account can be given of observations that purples look more similar to blues than to either yellows or greens. Perceptually, purples are bluish and reddish (they are *perceptually represented as* both bluish and reddish). Therefore, purples are perceptually more similar to blues than to yellows or greens – because neither greens nor yellows are represented as bluish (or reddish). Byrne and Hilbert's arguement that "red looks more similar to orange than to green" should be taken to mean that red surfaces are represented as having more properties in common with orange surfaces than with green ones – that is, there is something like a counting of shared versus distinct categories implicit in color perception.

Even if this claim is true, it provides only a pretty coarse-grained account of color similarity, at least without the introduction of a whole host of subordinate color categories (in addition to reddish, greenish, yellowish, and bluish, which are explicitly mentioned by Byrne and Hilbert). For consider the following claim: Magenta looks more similar to purple than to purplish blue. Perceptually, magenta is slightly bluish red; purple is reddish and bluish approximately to the same extent, whereas purplish blue is slightly reddish and predominantly blue, so the claim is true – it would likely come out as a result from suitable experiments. However, all these colors are both reddish and bluish, so the just-presented "digital" schema cannot account for this datum without further refinement. This was probably a reason why Byrne and Hilbert (2003, sec. 3.2.) turned to a different representationalist solution that, instead of discrete category counting, appeals to an analog representation of color similarity. In effect, the two authors seem to have returned to the idea that they earlier abandoned as a result of Thompson's critique.

The new representationalist model of color similarity was inspired by C. Sternheim and R. Boynton's and J. Werner and B. Wooten's experiments (Sternheim and Boynton 1966; Werner and Wooten, 1979; Byrne and Hilbert, 2003, sec. 3.2.1.), in which subjects were asked to estimate the relative amount of hues in colors presented to them (for example, 30% green and 70% yellow). Participants both understood the instruction and gave similar answers. A matrix of perceived color similarities can straightforwardly be derived from such data.

To account for these findings, Byrne and Hilbert introduce four hue-magnitudes: **R** (for reddishness), **G** (greenishness), **Y** (yellowishness), and **B** (bluishness). They suggest that objects that look a binary hue (e.g., purple) can be seen as having proportions of two of these four hue-magnitudes, while objects that look a unique hue (e.g., unique red) can be seen as having 100% of one of these hue-magnitudes. Thus, for example, objects that look unique red have 100% of **R**. As concerns objects that look purple, they have "**R** and **B** in a similar proportion, say a 55% proportion of **R** and a 45% proportion of **B**" (op. cit., sec. 3.2.1.).

The hue-magnitudes in turn are spelled out in terms of the notions of L-intensity, M-intensity, and S-intensity of color signals. The L-intensity of a color signal is the degree to which it would stimulate the L cones; M- and S-intensities are interpreted analogously. Byrne and Hilbert suggest that "an object has some value of **R** if and only if, under equal energy illuminant, it would reflect light with a greater L-intensity than M-intensity – the greater the difference, the higher the value of **R**" (2003, sec. 3.2.3.). The M > L relation gives rise to a value of **G**. If a color signal has an S-intensity that is greater than the sum of its L- and M-intensities, then it has a value of **B**; If, however, the sum of L- and M-intensities exceeds the S-intensity, that will result in a value of **Y**.

In sum, Byrne and Hilbert (1997, sec. 3.2.2.) claim that for unique red surfaces, the L- , M- , and S-intensities are such that L > M and (L + M) = S; that is, the L cone excitation is greater than the M cone excitation, and the S cone excitations equal the sum of M and L cone excitations. They also mention that LMS-intensities can be derived from the color signal and the cone sensitivities.[34]

This notion of hue-magnitudes was proposed in order to provide a representational externalist account of unity: to demonstrate that object colors do indeed exhibit, perceiver-independently, the very similarity relations that color experience presents them as having. In another place, Hilbert and Kalderon (2000, p. 198; their italics) formulate this claim nicely:

[T]he relations of similarity in hue, saturation, and brightness supervene on color properties that exist independently of viewing subjects. In speaking of the visual system of *selecting* these relations, we are not claiming that visual experience *brings them into being*. We are merely claiming that a pre-existing relation only counts as similarity in hue, saturation, or brightness in virtue of an antecedent classificatory function of the visual system. Facts about visual experience fix which of the similarities among certain objective properties *count as* similarities with respect to

---

[34] Byrne and Hilbert 2003, note 44. The natural suggestion here is that LMS-intensities can be so derived by taking the product of the color signal and the cone sensitivity functions.

color. But that is consistent with higher-order color similarities, so understood, supervening on objective color properties. The properties represented by color experience are ontologically independent of those experiences, as are the relations of similarity and difference among them.

As I shall argue in a moment, this claim is mistaken. The visual system does make a selection from the innumerable available surface reflectance properties, but it does other things as well. As a result, the arising perceptual similarity relations of the colors cannot be counted as any sort of veridical representation of selected (derivative, anthropocentric: see Hilbert, 1987, pp. 13–15, 115, 119–120; Tye, 2000, p. 161), but still perceiver-independent, similarities of the corresponding stimuli.

5.2 Why the Proposals Don't Work

Here are some preliminary remarks. The problem of unity can be thought of as a mapping between two similarity spaces: some stimulus similarity space, and color space – the perceptual similarity space for colors. Points in the former space correspond to measurable characteristics of color stimuli; points in the latter correspond to color experiences characterized by three perceptual dimensions (hue, lightness, and saturation; or lightness, reddishness-greenishness, and yellowishness-bluishness). Similarities between color stimuli are expressed as distances between points in the stimulus space. Perceptual similarities of the colors are expressed by distances between points of color space.[35] Unity is a veridical aspect of color representation if and only if there exists an isomorphic mapping between the distance graphs of the two spaces. The hue-magnitudes proposal easily translates into this formulation of unity. The Natural Color System (NCS, as explained previously) is derived from subjects' judgments about the proportions of hue components in perceived colors. The NCS color space has three dimensions: white-black, red-green, and yellow-blue that arise from the judgments of hue proportions. Points in this space correspond to color percepts. Distances between two points correspond to similarities/differences between the corresponding color percepts (Sivik, 1997). If this structure is isomorphic

---

[35] There are different methods for modeling color space. Most of these methods result in similar overall patterns that conform well with the structure of Munsell color space (Indow and Aoki, 1983; Indow, 1988; Jameson and D'Andrade, 1997, p. 299). There may be exceptions to this rule: Using just-noticeable color differences to derive the metric structure of the color space (Wyszecki and Stiles, 1982) does not seem to coincide with direct estimates of the subjective differences between the colors (Judd, 1967; Izmailov and Sokolov, 1991).

with another similarity space that characterizes the hue-magnitudes in strictly the terms of stimulus properties, then the idea of unity as an aspect of veridical representation is defended. Otherwise it has to go.

Now a word about isomorphism. Philosophers like to emphasize that isomorphism is a cheap notion in the sense that everything is isomorphic with everything else. Take any two similarity spaces and there will surely be some mapping that takes one into the other. Just point out the mapping – or simply say that there must be one – and the isomorphism claim is defended.[36] Obviously, this is an empty sense of isomorphism. But it can be improved. In particular, it is possible to impose limitations on (1) the relations or properties in terms of which similarity should obtain, and (2) the type of similarity metric that defines similarity in both spaces, and (3) the type of mapping that has to obtain between the two spaces. Constrained this way, isomorphism is no longer so cheap a notion. Reflectance theorists propose to characterize the relevant stimulus similarity space as a TIR-space, or a space whose dimensions are certain proportions of TIRs (Tye, 2000, pp. 159–161; Matthen, 2001), or, as Byrne and Hilbert do, a proportions-of-LMS-intensities space.[37] Color space is assessed via subjects' responses. Regarding the similarity metric, one idea could be to use the same metric in both similarity spaces. However, this may not be the right thing to do, since there is evidence that color space is non-Euclidean (Izmailov and Sokolov, 1991; Wuerger et al., 1995), whereas the corresponding stimulus space (LMS-space) is reasonably defined as Euclidean. This problem has not yet been considered by proponents of the reflectance theory. In any event, there might be independent evidence that helps to constrain the choice of similarity metric for both spaces. Using the relevant metrics, one can calculate interpoint distances in the two spaces, and endorse reasonable criteria of how these distances should be related in order for isomorphism to obtain. In Byrne and Hilbert's spirit, such a criterion could be that proportions of the hue-magnitudes, calculated from their physical characterization, should sufficiently well approximate the same proportions arising from subjects' estimates.

---

[36] As Laurence Maloney remarked (electronic communication), from a mathematician's point of view this idea is near nonsense. In mathematics, isomorphism is precisely defined, and it applies to certain mappings but not to others.

[37] Color stimuli are often characterized by the excitations of the three cone types they produce, or by simple transforms of those excitations. Chromaticity diagrams are such representations (MacLeod and Boynton, 1979; Wyszecki and Stiles, 2000). In what follows, I shall call such stimulus similarity spaces LMS-spaces.

For the hue-magnitudes proposal to become a representational externalist account of unity, it should be the case that if, say, an orange surface S_O in daylight has an **R** value that is twice its **Y** value (calculated using Byrne and Hilbert's physicalistic definition), then trichromat subjects looking at the surface in daylight should judge (on average at least) that the surface is 66% reddish and 33% yellowish (setting aside saturation, e.g., assuming that S_O is highly saturated). This would justify the key claim that color experience veridically represents, or conveys without distortion, the proportions of hue-magnitudes as defined by the physicalistic account.

If, however, the magnitude ratios arising from calculations of the hue-magnitudes, physicalistically defined, and the corresponding ratios arising from subjects' reports on their color experience differ substantially, then the proportions of hue-magnitudes physicalistically defined are not veridically represented in color experience, but rather they are distorted. For instance, it might happen that, for a given color signal CS, calculation based on Byrne and Hilbert's physicalistic definition outputs 35% **Y** and 65% **G**, whereas subjects looking at the surface producing CS would report, on average, 55% **Y** and 45% **G**. In such a case, we'd have to say that the subjects' color experience does not veridically represent CS's proportion of hue-magnitudes, physicalistically defined. It misrepresents that proportion – perhaps it is not even the job of color perception to represent that proportion. At any rate, there is a serious empirical question here, something testable that may turn out right or wrong. Moreover, the refinements by R. Hunt (1982) and Werner and Wooten (1979) reviewed previously suggest that the simplified opponent-processing theory is really just a caricature of the more complicated transformation that takes place between cone signals (LMS-space) and color space. So if Byrne and Hilbert use the simplified opponent-processing theory to calculate physical hue-magnitudes (they do so in their 2003 piece), then the odds are high that there will be no match between the proportions of hue-magnitudes (calculated using the simplified model) and subjects' reports of proportions of hue-magnitudes based on their color experience (since color experience arises from a transformation more complex than the simplified opponent-processing model).[38]

One armchair move to avoid this problem would be to define the hue-magnitudes **R, G, Y, B** such that they are the outputs of the whole

---

[38] In fact, some predictions of the hue-magnitudes account, as it is presented in Byrne and Hilbert (2003), are obviously mistaken (Jakab and McLaughlin, 2003).

Werner-Wooten or Hunt algorithm that takes the LMS-intensities as input (see Bradley and Tye, 2001, p. 482). In this case – assuming that the Werner-Wooten or Hunt model is correct – the calculations of hue-magnitudes would probably be in pretty good agreement with trichromat subjects' average estimates of the hue-magnitudes. This is simply because it was the goal of those models to provide a function that takes the LMS-space into color space.

But the question now becomes: What do the hue-magnitudes, thus calculated, have to do with the distal surface reflectance properties (or even with the color signals) that are supposedly veridically represented in color experience? After hitting the retina, color signals undergo a wavelength-specific weighting (by the cones), then follow some process of opponent recombination, including raising the cone signals to some power, multiplicative scaling, and additive shifts, and the output is what one would call the hue-magnitudes. If Byrne and Hilbert took this line, they'd have to tell us in what sense this is a *physicalistic definition* of the hue-magnitudes – in what sense are the thus-defined hue-magnitudes and their proportions generic-surface reflectance properties that the surfaces actually have, perceiver-independently?

By my lights, a property is perceiver-independent if it can be physically instantiated in the absence of perceivers. If we characterize the hue-magnitudes by appeal to Hunt's or Werner and Wooten's model, we make tacit reference to functions that are realized by processes in the visual system of the trichromat observer. Therefore, the hue-magnitudes are not observer-independent in the sense I just specified. Where there are no perceivers, the functions that output the hue-magnitudes are not calculated, and so the hue-magnitudes are not physically instantiated in any way. Under this notion, and in the absence of human perceivers, surfaces or color signals would have their hue-magnitudes only as certain dispositional properties (they would elicit such-and-such a response from trichromat human observers, did such observers come around to look) or abstract properties like mathematical objects (in the sense in which the 527th power of my body weight in milligrams is an existing property of me, even though no one ever calculated it). Furthermore, I take it that dispositions of stimuli to elicit perceptual reactions cannot be counted as causes of the perceptual reactions in question.[39] Nor can uninstantiated

---

[39] A disposition is not the cause of its own manifestation – it is the basis of the disposition that is the cause of the manifestation (McLaughlin, 1995, p. 123). Dispositions can figure in causal explanations because they correspond to (or, simply, *are*) lawlike causal links

abstract objects be the causes of color experience. So the hue-magnitudes, thus specified, cannot be the observer-independent, causally effective object color properties that are assumed in Byrne and Hilbert's theory to be veridically represented by color experience.[40]

As we now see, this way out for Byrne and Hilbert is in reality a dead end. In order to save their theory, they are supposed to provide an account of the hue-magnitudes in terms of perceiver-independent and causally effective distal surface properties (like surface reflectance) that are specifically causally responsible for eliciting trichromat humans' experiences of color. It is the proportions of hue-magnitudes defined in this way that have to match subjects' reports on their color experience, in order to support the claim that unity is an aspect of veridical perceptual representation of color.[41]

Another armchair move to save the externalist account of unity is implicit in Bradley and Tye's response to the problem of unique-binary division (Bradley and Tye, 2001, p. 482). They say that no matter what transformations color vision performs on color signals, there is always

---

between the bases of dispositions and their manifestation. But to say that it is the lawlike causal link itself that causes something would be a confusion. The causal link is a specific, predictable interaction (i.e., the event of causation; the evoking of the manifestation by the basis) in specific circumstances.

[40] Dispositions of stimuli to elicit perceptual reactions are not perceiver-independent in the way surface reflectance is. Surface reflectance is a disposition whose manifestation does not include any response of perceiving organisms, and so this disposition can be manifested in the absence of perceivers. However, a surface's disposition to evoke color experience of a certain kind in humans is evidently a disposition whose manifestation includes (i.e., entirely consists of) perceptual responses. Hence, such a disposition cannot manifest itself in the absence of perceivers.

[41] By the way, there is a problem behind the Sternheim-Boynton method that Byrne and Hilbert appeal to in their account of color content. What does it mean when the subject says that the sample looks 35% reddish? Is there any way of checking the correctness of such a report? In other words, how can we make sense, in this particular case, of a *function* from the subjects' experience to the verbal report on their experience? Since the only third-person access to subjects' color experience is via their report, how could we treat this case as a function (from experiences to reports), similarly as to sensory functions? It looks as though there is a Stalinesque-Orwellian trap (Dennett, 1991) behind the Sternheim-Boynton method. Two quick replies follow. First, other methods for behaviorally indicating the quality of one's color experience seem to work better. An example is simply asking the subjects to pick, or adjust by a knob, the green that looks neither yellowish nor bluish to them. In this case, at least there is no problem of translating perceptual experience into percentages. Second, very similar philosophical problems attend the study of sensory functions. The domains of sensory functions are sets of stimuli, but their range is sensory experiences that we can only access via behavior. Still, we continue to assume that behavior in psychophysical experiments is an indicator of something else – the subject's sensory experience, or some aspect of it.

some surface reflectance property S_{R35} such that, when appropriately transformated in the visual system, it will give rise to a certain degree of reddish look (say, it will "look 35% reddish," according to subjects' reports). Similarly, there is some other reflectance property S_{Y65} such that its color signal, when processed by color vision, will give rise to a certain degree of yellowish look (say, 65%). No matter what that property is, there is some such *stimulus* property, claim Tye and Bradley. Moreover, no matter what transformations color vision performs on color signals, when a surface has both S_{R35} and S_{Y65}, it will look reddish and yellowish to the specified degrees.[42] Therefore, S_{Y65} will be *veridically represented* by the degree of yellowish look it gives rise to, and the same applies to S_{R35}.[43]

Now for a reply. Bradley and Tye (2001) do not mention unity in their paper, and I think their account of the unique-binary division cannot be extended to accommodate it. It is true, of course, that the appropriate transforms of S_{R35} and S_{Y65} exhibit the required ratio of hue-magnitudes. S_{R35} and S_{Y65}, when measured by a colorimeter that outputs, say, CIELAB coordinates,[44] will output a* and b* values whose proportion will likely be similar to the average judgment of proportions of hue-magnitudes made by trichromat subjects. But colorimeters are funny measuring instruments, compared to measuring tapes, speedometers, and the like, because in important respects, they simulate the functioning of our color vision. In doing so, they apply wavelength-specific weighting, power transformations, multiplicative scaling, and additive shifts to their physical

[42] This is again vastly oversimplified, the effect of the illuminant and simultaneous contrast being the two key factors ignored. My point is, even if, for the sake of argument, we agree to ignore these factors, Bradley and Tye's proposal is still mistaken. Note also that S_{R35} and S_{Y65} can be relative to each other: The same amount of reddishness (L−M, or some function of it) can give rise to different degrees of reddish look depending on the blue-yellow imbalance present at the same time (see Byrne and Hilbert, 2003).

[43] At this point, Bradley and Tye's move is really a combination of the two ideas I attempt to separate here. They say (Bradley and Tye, 2001, pp. 482–483; their italics): "So long as there is, for each channel, a well-defined function that connects the values of S, M, and L with the resultant color experience, that function, *linear or not*, will have an objectivist counterpart function connecting the relevant color with S*, M*, and L*." For S*, M*, and L* see note 19. They continue: "Thus, as before, given how the human visual receptors collectively operate, for each color experience, there is a complex surface disposition normally tracked by the experience – a complex disposition with which the color represented by the experience may be identified."

[44] Color coordinates in the CIELAB system are L* (lightness), a* (red-green dimension: positive values mean reddish hues, negative ones greenish hues), and b* (yellow-blue dimension: positive values mean yellowish hues, negative ones bluish hues). The CIELAB space is a uniform color space (see section 3.2).

input. Therefore, to infer from the colorimeter reading that S_{R35} and S_{Y65} *themselves* (and not just their appropriate transforms) exhibit the same hue-magnitude ratios as reported by trichromat perceivers would be a textbook example of how one can beg the question. Alternatively, use an ordinary luminance or reflectance-measuring instrument that does not simulate human color vision, and it will turn out that S_{R35} and S_{Y65} do not stand in the required proportion. At this point, a reasonable conclusion seems to be that if we endorse our stricter criterion for isomorphism, there simply is no perceiver-independent property that would fill in the externalist bill. The perceiver-independent causal antecedents of color experience do not exhibit the similarity relations they look to us to exhibit.[45]

Finally, note an important difference between measurement and sensation. In general, artificial measuring instruments in science and everyday life work in such a way that either their output correlates linearly with the phenomenon to be measured[46] or, if a nonlinear transformation occurs in the process of measurement, that is clearly indicated. Measuring instruments typically do not conceal any nonlinear transformation that occurs in the process of deriving their output from the input. And this is in sharp contrast with almost every type of sensation. First, sensation, in the majority of cases, is nonlinear. Its models reflect this fact. Fechnerian sensory functions are logarithmic; Stevensian sensory functions are power functions.[47] For this reason, sensation most often does not preserve any sort of linear isomorphism that could justify externalist

---

45 In citing Werner and Wooten, Bradley and Tye (2001, p. 482) are completely oblivious about those authors' observation about individual differences (see section 2), let alone the devastating implications those observations have for color absolutism. If the sensory function that transforms color signals into points (vectors) of color space is indeed so different from one color-normal individual to the other, then this helps to explain the individual differences in unique hue location, and the corresponding differences in hue naming (Werner and Wooten, 1979, esp. p. 425; Kuehni, 2001). It also raises the question for the color absolutist, whose sensory function is "the correct" one, when different subjects with different chromatic response curves describe the same color stimulus in the same circumstance in substantially different ways. Finally, Werner and Wooten's observations imply that Tye's absolutist solution to the individual differences problem, namely, that individual differences in trichromatic color perception are merely due to individual differences in color discrimination (Tye, 2000, pp. 89–93), is incorrect.

46 Of course, this can only be checked by another measurement, but it seems to be an important assumption anyway. In simple cases like measuring sticks (or even speedometers), it also seems obviously true.

47 That is, linear sensory functions can arise as a special case of Stevensian sensory functions: power functions with exponent 1. Psychophysical experiments showed that perceived length is quite close to being a linear function of actual length (e.g., Sekuler and Blake, 2002, p. 606).

claims about represented stimulus similarity – let it be similarity between colors or temperatures (Akins, 1996; for contrast, see Dretske, 1995, ch. 3). Second, the sensory systems that implement such functions do not in any way inform the rest of the cognitive system about the fact that there is a nonlinear transformation going on. It is psychophysics that informs us about the nonlinear features of sensation. My summary point so far is that Matthen (1999) and earlier Hardin (1988) are correct about color similarity and the unique-binary distinction. This, however, need not lead us to abandon physicalism about color (McLaughlin, 2003).

5.3 Opponent-Processing and Linear Models: Two Steps of the Same Ladder

L. Maloney (1999, pp. 409–414) discusses how the linear models framework relates to the opponent-processing model of color perception. Briefly, the idea is that, following W. Stiles (1961, p. 264; Maloney, 1999, p. 410), for purposes of theoretical analysis, color vision can be divided into two very general stages: (i) adaptational states of the pathways of chromatic processing, and (ii) the processes that adjust and modify these adaptational states. Color processing consists of a number of transformations of retinal signals, including multiplicative scaling, additive shifts, and opponent recombination. The outcome of all these transformations is color appearance. These transformations contain certain parameters (coefficients for multiplicative scaling, constants for additive shift, and so on) that are systematically modified by some characteristics of visual stimulation.

The general schema is that transformations on receptor inputs at a given retinal location are influenced by previous retinal input and simultaneous input at other parts of the retina. This information about retinal surround determines the parameters for transformation of the cone signals at the retinal point under consideration. Now, the linear-models-based algorithms of surface reflectance estimation figure in adaptational control: They are *part* of the transformations by which color appearance is reached from retinal input (Maloney, 1999, p. 413). The first transformation of photoreceptor excitations is their multiplication by the so-called *lighting matrix* Λ_ε^{-1} (Wandell, 1995, p. 307; Maloney, 1999, p. 413, and 2003). The lighting matrix is illumination-dependent, and this transformation has the function of discounting the effect of illuminant changes, thereby achieving (approximate) color constancy. The result of this transformation is the visual representation of surface reflectance by linear-models weights. This representation then undergoes a further transformation that determines color appearance. This further transformation (function F in Maloney, 1999, p. 413) is arbitrary in the sense that, in principle, some species with trichromatic color vision and photoreceptors of the same kind as ours could discriminate the same reflectance types as trichromat humans can, form the same linear-models-weights representations of them, yet still apply some different F function (second-site multiplicative attenuation, opponent recombination: Maloney, 1999, p. 410) to them so that despite the fact that such organisms discriminate the same reflectance ranges by their color experiences as we do, their color space (unique-binary division, similarity metrics) would be substantially different from ours. As Maloney says (1999, p. 413), in principle, any one-to-one transformation of the linear-models-weights representation would equally well serve to determine color appearance; constraints on this transformation should come from further assumptions about how this second stage of color processing operates in humans.

What I wish to conclude from Maloney's picture is that particularities of surface reflectance estimation by color vision do not alone determine color appearance. Color appearance crucially depends on further transformations in the visual system that are independent of information about surface reflectance but that play a key role in shaping our color space. There is no necessary connection between information represented about surface reflectance, on the one hand, and color appearance, on the other.

5.4 Is Color Space Euclidean?

In a fascinating study of the structure of color space, S. M. Wuerger, L. T. Maloney, and J. Krauskopf (1995) argue that imposing a Euclidean geometry on LMS-space (the space of the photoreceptor excitations L, M, and S) by specifying three coordinate axes in it that are linear combinations of the photoreceptor excitations (e.g., $L + M + S, L - M, L + M - S$) is not consistent with the proximity judgments of colors made by human subjects. Their results show that there is no Euclidean metric on color space that accounts for the proximity judgments – assuming that color space is a linear transformation of the cone excitations L, M, and S. In other words, color space (the perceptual similarity space for colors arising from the similarity judgments for suprathreshold differences) cannot have both a linear structure and a Euclidean metric.[48] Assuming

---

[48] By "linear structure" the authors mean that color space is a linear transformation of L, M, and S. The key assumption for the authors' test is that the transformation from LMS-space to color space is such that it preserves the length ratios of collinear line segments (Wuerger et al., 1995, pp. 828–829). This is true of all linear transformations. Linear

that the transformation Ψ between LMS-space and color space is linear, the Euclidean hypothesis (the hypothesis that there is a Euclidean metric on color space that accounts for proximity judgments) fails on empirical tests.

Wuerger et al. constructed their own similarity judgment tasks to test the Euclidean hypothesis. Their subjects were asked to compare a test stimulus (say, **a**) to a standard (**b**) and a comparison (**c**) and to say which of **b** and **c** were more similar to **a**. In a single series of judgments, nine different but equally spaced (in LMS-space) comparison stimuli were used, one of which was identical to the standard. Three variants of this task were used that differed in the ways **a**, **b**, and **c** were arranged on the screen, and in the corresponding instructions (for the details, see Wuerger et al., 1995, pp. 829–830).

Via these tasks, two features of the Euclidean hypothesis were examined: (i) the additivity of angles and (ii) the assumption that the variability of similarity judgments increases with increase in the distance between the stimuli to be compared. The latter criterion implies a distinct psychological assumption, namely, Weber's law, according to which the error in estimating distances grows with the distances themselves. The link between the Weberian notion and the Euclidean hypothesis was simply that greater variability in the proximity judgments indicates that subjects' color space assigns greater differences (interpoint distances) to stimuli with greater separation in LMS-space.

The additivity-of-angles task was based on the geometric idea in Figure 10.1. Take a standard stimulus a_1 in LMS-space, and its image A_1 in color space ("proximity-judgment space"). Take a series of comparison stimuli c_1–c_9 that lie on the same line (call it λ_2) in LMS-space. Assuming that color space is a linear transform of LMS-space, the images of the comparison stimuli, C_1–C_9, will lie on one straight line in color space as well (call this line Λ_2). By repeatedly choosing from stimuli c_1–c_9 the one that looks most similar to a_1, the closest point to A_1 on Λ_2 (i.e., in color space) can be determined. Let the closest point be called B_2. Then the assumption is, the A_1–B_2 line is perpendicular to Λ_2,[49] because B_2 is the

transformations are those in which the output space's dimensions are linear combinations of the dimensions of the input space. For example, given a vector space **A** with three dimensions x, y, and z, **A**'s linear transforms will be the spaces with dimensions $p = w_1 x + w_2 y + w_3 z + c_1$, $q = w_4 x + w_5 y + w_6 z + c_2$, $r = w_7 x + w_8 y + w_9 z + c_3$.

[49] In color space, the a_1 – b_2 line in LMS-space need not be perpendicular to c_1 – c_9. See Figure 10.1. The authors used linear interpolation, based on the choices from c_1 – c_9, to find the closest point to A_1 on Λ_2 (and to A_2 on Λ_1, as we will discuss).

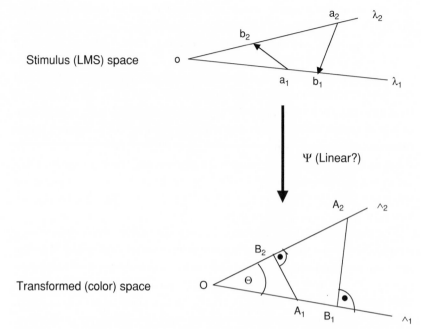

FIGURE 10.1. Schema for the additivity-of-angles experiment. Black dots in the arcs indicate right angles. *Source*: Redrawn from Figure 1 in Wuerger et al., 1995, p. 828.

closest point to $\mathbf{A}_1$ on Λ_2. Now take another test stimulus $\mathbf{a}_2$ on line Λ_2, and its image $\mathbf{A}_2$ on Λ_2 in color space. Take another series of comparison stimuli ($\mathbf{c}_{11}$–$\mathbf{c}_{19}$) that lie on the same line (call it λ_1) as the first test stimulus, $\mathbf{a}_1$ in LMS-space. Again, the images of these points, $\mathbf{C}_{11}$–$\mathbf{C}_{19}$, will lie on the same straight line Λ_1 in color space. By repeatedly choosing from $\mathbf{c}_{11}$–$\mathbf{c}_{19}$ the one looking most similar to $\mathbf{a}_2$, we can determine the closest point ($\mathbf{B}_1$) to $\mathbf{A}_2$ on Λ_1 in color space. Then the line $\mathbf{A}_2$–$\mathbf{B}_1$ will be perpendicular to Λ_1. From this constellation, it is possible to calculate the angle Θ between the lines Λ_1 and Λ_2. Given this procedure, the additivity of angles in color space can be tested by taking three lines Λ_1, Λ_2, and Λ_3, estimating the angles between all three pairs of them ($[\Lambda_1, \Lambda_2]$, $[\Lambda_2, \Lambda_3]$, and $[\Lambda_1, \Lambda_3]$), and seeing if two of these angles add up to yield the third.

The Euclidean hypothesis failed this test. Angles proved nonadditive in each of the three task variants used. Tested by a similar method, the variability of the proximity judgment responses failed to increase with the distance between the stimuli to be compared.

Thus, it seems that color space is not both linear in structure and Euclidean. It can be linear such that some non-Euclidean distance metric accounts for similarity data. The authors found that their data from the angle-additivity experiment are consistent with the assumption that the observer employs a city-block metric to judge the proximity of colored lights.[50] Alternatively, color space may be Euclidean but nonlinear in structure, in which case the lines and planes in LMS-space are nonlinear curves in the Euclidean color space. C. Izmailov and E. Sokolov (1991) suggest that from multidimensional scaling data a nonlinear structure of color space arises. They found that when equibright spectral lights were scaled for similarity, the data could be accounted for only in a three-dimensional Euclidean space. That is, equally bright lights do not lie in the same plane of a Euclidean space constructed to accommodate similarity judgments (see also Wuerger et al., 1995, p. 834).

A question that these studies raise is whether the transformation Ψ between LMS-space and color space is indeed linear. Wuerger et al. tested this within their own experimental paradigm (p. 834) and found that their data support the linearity of Ψ. However, as we saw, it is not generally agreed that the cone-space to color-space transformation is linear. Werner and Wooten (1979) and Hunt (1982) all argued that it is not. However, Kuehni (2000) proposed linear transformations that take cone responses into uniform color spaces like the Munsell and OSA-UCS system. If Ψ is nonlinear, then Wuerger et al.'s method to estimate the angle (Θ) between two lines in color space needs to be modified. In this case, the additivity-of-angles test may even be satisfied for color similarity judgments. On the other hand, if Ψ is linear, then the Euclidean hypothesis has to go – in this case, we do not seem to have the alternative option (Euclidean metric on a nonlinear color space).

A more recent conjecture[51] is that nonadditivity of angles may be a result of the phenomenon of *crispening* (Takasaki, 1966; Whittle, 1992; Maloney and Yang, 2003), namely, that unsaturated colors tend to look more different than saturated ones. Two points near the neutral (white) point in LMS-space, separated by a distance **d**, will look more different than two other points, far away from whites (in some saturated-colors

---

[50] Euclidean distance is computed by the formula $(|x_2 - x_1|^2 + |y_2-y_1|^2)^{1/2}$ where x_1 and y_1 are coordinates of point P_1, and x_2, y_2 are coordinates of point P_2 (using a two-dimensional space just for simplicity). City-block metric uses power 1 instead of power 2: According to it, the distance of P_1 and P_2 will be $(|x_2 - x_1| + |y_2 - y_1|)$ – the sum of the distances between the corresponding coordinates.

[51] Maloney, personal communication, September 10, 2003.

range), separated by **d**. That is, equal distances in LMS-space do not always correspond to equal distances in color space – a case of nonlinearity
arising in the transition from LMS-space to color space. It may be the
case that if we performed a transformation on color space (i.e., the proximity judgment data supposed to reveal the structure of color space) that
compensates for, or "undoes" the effect of, crispening, then, in the resulting "straightened," or linearized, version of color space, additivity of
angles would obtain. Such a result would demonstrate that the source of
nonlinearity in color space is the crispening effect. This is certainly a question for future study. Note also that farther away from the neutral point,
color space appears more linear. That is, the relation between separation
of points in LMS-space and the corresponding similarity judgments is
closer to a linear one.

In presenting these issues, my aim was to further illustrate the idea
that human color vision transforms an impoverished realm of stimuli
(reflectance, or light radiation, in three overlapping, broad ranges of
wavelength) in intricate ways, thereby essentially constructing our world
of perceived color. Colors are physical properties of environmental objects, but perceived colors (i.e., color appearances; perceptual modes of
presentation of object colors) are products of our mind.

6 Outlook: Systematic Misperception and the Modularity of Mind

J. Fodor (1990) argues that the modular, encapsulated organization of
perception assures that certain lower levels of perceptual processing are
free of top-down influences, that is, the influence of previously acquired
experience, knowledge, motivational states, values, and so on. Not all levels of perceptual processing are permeated by concept-driven processes.
Moreover, this kind of cognitive organization is advantageous since this
way, the cognitive system receives "pure" perceptual information about
the environment – input that is unbiased, uninfluenced by higher cognitive processes. Moreover, an unbiased perceptual representation can be a
solid foundation for the rest of cognition: something that can be trusted
as information about *the world as it is,* independently of how we conceive
of it. It eventually gets influenced by top-down processes, but in the first
place it exists separately from those influences.

If we generalize the view that I have proposed in the present chapter, we reach the idea that perception, in certain respects, systematically
misrepresents the world (or even *constructs* a world of perceived color). I
explained this idea using color vision as an example. However, the moral

seems generalizable to other perceptual modalities. K. Akins (1996) argues that heat sensation also fails to exhibit anything like a systematic mapping between external temperatures and inner sensory states – nor is this the evolutionary function of heat sensation. Akins suggests (p. 364) that two aspects of perception are straightforwardly compatible. First, perception makes it possible that the organism's behavior is directed toward certain salient objects and properties of its environment. Second, to achieve this, perception need not use encoding that is veridical in every respect. In other words – now I am continuing Akins's line of thought – perception need not be veridical in every respect, in order to be adaptive. For an organism with color vision, the important thing is only that color perception enhances visual surface discrimination, that it helps to pick out red or orange berries from among green leaves and the like. The exact similarity relations of perceived colors, or whether these relations map, in an isomorphic way, the similarities in surface reflectance, may be irrelevant for survival. Color perception in this case veridically informs the organism about the difference between the surface of the berries and that of the leaves. It may well be nonveridical with respect to such subtle matters as the exact relations of similarity in terms of the relevant stimuli, but that need not be a disadvantage for the organism. Distortion may in some cases be indifferent. In other cases it may even be advantageous, if it serves to enhance the detection of small physical differences that signal events of high importance to the organism. As an example, think of the classical experiments in speech perception according to which equal physical differences between speech sounds are detected with a greater probability when they fall on a category border than when they fall within a perceptual category (Pisoni, 1973; Pisoni and Tash, 1974; Pisoni and Lazarus, 1974).

I think Fodor's view of the function of perception and (my view of) Akins's view are compatible: Even if perception is not veridical in every respect, it is veridical in certain respects, and this is enough to provide a solid foundation for cognition. Partially distorted but stable input is still better than input that is forever changing not due to changes in stimulation but as a result of top-down influence.

Acknowledgments

While preparing this chapter, the author was supported by the Natural Sciences and Engineering Research Council of Canada (PDF-242003-2001). The author is grateful to Laurence Maloney for his comments and observations on earlier versions.

References

Akins, K. (1996). Of Sensory Systems and the "Aboutness" of Mental States. *Journal of Philosophy*, 93, 337–372.

Akins, K., and Hahn, M. (2000). The Peculiarity of Color. In Steven Davis (Ed.): *Color Perception: Philosophical, Psychological, Artistic and Computational Perspectives*. New York: Oxford University Press.

Ayama, M., Nakatsue, T., and Kaiser, P. K. (1987). Constant hue loci of unique and binary balanced hues at 10, 100, and 1000 Td. *Journal of the Optical Society of America A*, 4, 1136–1144.

Block, N. (1997). Inverted Earth. In N. Block, O. Flanagan, and G. Güzeldere (Eds.): *The Nature of Consciousness*. Cambridge, Mass.: MIT Press.

Block, N. (1999). Sexism, racism, ageism, and the nature of consciousness. *Philosophical Topics*, 26 (1–2), 39–70.

Block, N., Flanagan, O., and Güzeldere, G. (Eds.) (1997). *The Nature of Consciousness*. Cambridge, Mass.: MIT Press.

Bradley, P., and Tye, M. (2001). Of colors, kestrels, caterpillars and leaves. *Journal of Philosophy*, 98, 469–487.

Brainard, D. H., and Wandell, B. A. (1986). Analysis of the retinex theory of color vision. *Journal of the Optical Society of America A*, 3, 1651–1661.

Byrne, A., and Hilbert, D. R. (2003). Color realism and color science. *Behavioral and Brain Sciences*, 26, 3–64.

Byrne, A., and Hilbert, D. R. (Eds.) (1997). *Readings on Color, Vol. 1. The Philosophy of Color*. Cambridge, Mass.: MIT Press.

Campbell, J. (1997). A Simple View of Color. In Alex Byrne and David R. Hilbert (Eds.): *Readings on Color, Vol. 1. The Philosophy of Color*. Cambridge, Mass.: MIT Press, 177–190.

Chalmers, D. (1996). *The Conscious Mind: In Search of a Fundamental Theory*. New York: Oxford University Press.

Davies, M. (1997). Externalism and Experience. In Ned Block, Owen Flanagan, and Güven Güzeldere (Eds.): *The Nature of Consciousness*. Cambridge, Mass.: MIT Press.

Dennett, D. C. (1991). *Consciousness Explained*. Boston: Little, Brown.

DeValois, R. L., and DeValois K. K. (1997). Neural Coding of Color. In Alex Byrne and David R. Hilbert (Eds.): *Readings on Color, Vol. 2. The Science of Color*. Cambridge, Mass.: MIT Press, 93–140.

DeValois, R. L., Smith C. J., Kitai, S. T., and Karoly, S. J. (1958). Responses of single cells in different layers of the primate lateral geniculate nucleus to monochromatic light. *Science*, 127, 238–239.

Diamond, S. (Ed.) (1974). The Roots of Psychology. New York: Basic Books.

Dretske, F. (1995). *Naturalizing the Mind*. Cambridge, Mass.: MIT Press.

Fairchild, M. D. (1998). *Color Appearance Models*. Reading, Mass.: Addison-Wesley.

Flanagan, O. (1992). *Consciousness Reconsidered*. Cambridge, Mass.: MIT Press.

Fodor, J. A. (1990). Why Should the Mind Be Modular? In J. Fodor, *A Theory of Content and Other Essays*. Cambridge, Mass.: MIT Press.

Hardin, C. L. (1988). *Color for Philosophers: Unweaving the Rainbow*. Indianapolis: Hackett.

Haugeland, J. (1978). The nature and plausibility of cognitivism. *Behavioral and Brain Sciences*, 2, 215–260. Also in John Haugeland (Ed.): *Mind Design*. Cambridge, Mass.: MIT Press, 1981, 1–34.

Helmholtz, H. v. (1911). *Physiological Optics*, 3d ed., vol. 2. Ed. J. P. C. Southall, Optical Society of America, Rochester (1924). Reprinted, New York: Dover, 1962.

Hilbert, D. R. (1987). *Color and Color Perception: A Study in Anthropocentric Realism*. Stanford, Calif.: Center for the Study of Language and Information.

Hilbert, D. R. (1992). What is color vision? *Philosophical Studies*, 68, 351–370.

Hilbert, D. R., and Kalderon M. E. (2000). Color and the Inverted Spectrum. In Steven Davis (Ed.): *Color Perception: Philosophical, Psychological, Artistic and Computational Perspectives*. New York: Oxford University Press.

Hunt, R. G. W. (1982). A model of colour vision for predicting colour appearance. *Color Research and Application*, 7 (2), 95–112.

Hurvich, L. M. (1981). *Color Vision*. Sunderland, Mass.: Sinauer Associates.

Hurvich, L. M. (1997). Chromatic and Achromatic Response Functions. In Alex Byrne and David R. Hilbert (Eds.): *Readings on Color*, Vol. 2. *The Science of Color*. Cambridge, Mass.: MIT Press, 67–91.

Hurvich, L. M., and Jameson, D. (1955). Some quantitative aspects of an opponent-colors theory. II. Brightness, saturation, and hue in normal and dichromatic vision. *Journal of the Optical Society of America*, 45, 602–616.

Ikeda, M., and Uehira, I. (1989). Unique hue loci and implications. *Color Research and Application*, 14, 318–324.

Indow, T. (1988). Multidimensional studies of Munsell solid. *Psychological Review*, 95 (4), 456–470.

Indow, T., and Aoki, N. (1983). Multidimensional mapping 178 Munsell colors. *Color Research and Application*, 5 (3), 145–152.

Izmailov, C. A., and Sokolov, E. N. (1991). Spherical model of color and brightness discrimination. *Psychological Science*, 2 (4), 249–259.

Jackson, F. (2000). Philosophizing About Color. In Steven Davis (Ed.): *Color Perception: Philosophical, Psychological, Artistic and Computational Perspectives*. New York: Oxford University Press.

Jackson, F., and Pargetter, R. (1997). An Objectivist's Guide to Subjectivism About Color. In Alex Byrne and David R. Hilbert (Eds.): *Readings on Color*, Vol. 1. *The Philosophy of Color*. Cambridge, Mass.: MIT Press, 67–79.

Jakab, Z. (2001). Color experience: Empirical evidence against representational externalism. Ph.D. thesis, Carleton University, Ottawa. Available at: http://www.carleton.ca/iis/TechReports.

Jakab, Z. (2003). For a truly colored world: Review of Michael Tye: *Consciousness, Color, and Content* (MIT Press, 2000). *Color Research and Application*, 28 (5), 384–391.

Jakab, Z., and McLaughlin, B. (2003). Why not color physicalism without color absolutism? *Behavioral and Brain Sciences*, 26 (1), 34–35.

Jameson, D., and Hurvich, L. M. (1955). Some quantitative aspects of an opponent-colors theory: I. Chromatic responses and spectral saturation. *Journal of the Optical Society of America*, 45, 546–552.

Jameson, D., and Hurvich, L. M. (1968). Opponent-response functions related to measured cone photopigments. *Journal of the Optical Society of America*, 58, 429–430.

Jameson, K., and D'Andrade, R. (1997). It's Not Really Red, Green, Yellow, Blue: An Inquiry into Perceptual Color Space. In C. L. Hardin and Luisa Maffi (Eds.): *Color Categories in Thought and Language*. New York: Cambridge University Press, 295–319.

Johnston, M. (1992). How to speak of the colors. *Philosophical Studies*, 68, 221–263.

Judd, D. B. (1967). Interval scales, ratio scales, and additive for the sizes of differences perceived between members of geodesic series. *Journal of the Optical Society of America*, 57, 380–386.

Kirk, R. (1994). *Raw Feeling*. Oxford: Clarendon Press.

Kuehni, R. G. (2000). Uniform color space modeled with cone responses. *Color Research and Application*, 25 (1), 56–63.

Kuehni, R. G. (2001). Determination of unique hues using Munsell color chips. *Color Research and Application*, 26 (1), 61–66.

Land, E. H. (1977). The retinex theory of color vision. *Scientific American*, 237 (6), 108–128.

Laxar, K., Miller, D. L., and Wooten, B. R. (1988). Long-term variability in the spectral loci of unique blue and unique yellow. *Journal of the Optical Society of America A*, 5, 1983–1985.

Levine, J. (1983). Materialism and qualia: The explanatory gap. *Pacific Philosophical Quarterly*, 64, 354–361.

MacAdam, D. L. (1997). The Physical Basis of Color Specification. In Alex Byrne and David R. Hilbert (Eds.): *Readings on Color*, Vol. 2. *The Science of Color*. Cambridge, Mass.: MIT Press, 33–63.

Maloney, L. T. (1986). Evaluation of linear models of surface spectral reflectance with small numbers of parameters. *Journal of the Optical Society of America A*, 3 (10), 1673–1683.

Maloney, L. T. (1999). Physics-based models of surface color perception. In K. R. Gegenfurtner and L. T. Sharpe (Eds.): *Color Vision: From Genes to Perception*. Cambridge: Cambridge University Press, 387–418.

Maloney, L. T. (2002). Illuminant estimation as cue combination. *Journal of Vision*, 2, 493–504.

Maloney, L. T. (2003). Surface Color Perception and Environmental Constraints. In R. Mausfeld and D. Heyer (Eds.): *Colour Vision: From Light to Object*. Oxford: Oxford University Press.

Maloney, L. T., and Wandell, B. A. (1986). Color constancy: A method for recovering surface reflectance. *Journal of the Optical Society of America A*, 3 (1, January), 29–33.

Maloney, L. T., and Yang, J. N. (2003). Maximum likelihood difference scaling. *Journal of Vision*, 3 (8), 573–585.

Matthen, M. (1988). Biological functions and perceptual content. *Journal of Philosophy*, 85 (1), 5–27.

Matthen, M. (1999). The Disunity of Color. *Philosophical Review*, 108 (1), 47–84.

Matthen, M. (2001). What colors? Whose colors? *Consciousness and Cognition*, 10 (1), 117–124.

Matthen, M., and Levy, E. (1984). Teleology, error, and the human immune system. *Journal of Philosophy*, 81 (7), 351–372.

McCann, J. J., McKee, S. P., and Taylor, T. H. (1976). Quantitative studies in retinex theory: A comparison between theoretical predictions and observer responses to the "Color Mondrian" experiments. *Vision Research*, 16, 445–458.

McLaughlin, B. (1995). Disposition. In Jaegwon Kim and Ernest Sosa (Eds.): *A Companion to Metaphysics*. Oxford and Cambridge, Mass.: Blackwell, 121–124.

McLaughlin, B. P. (2003). Color, Consciousness, and Color Consciousness. In Quentin Smith (Ed.): *New Essays on Consciousness*. Oxford: Oxford University Press, 97–152.

MacLeod, D. I. A., and Boynton, R. M. (1979). Chromaticity diagram showing cone excitations by stimuli of equal luminance. *Journal of the Optical Society of America A*, 69, 1183–1186.

Meyer, G. W. (1988). Wavelength selection for synthetic image generation. *Computer Vision, Graphics, and Image Processing*, 41, 57–79.

Müller, J. (1826a). *Zur vergleichenden Physiologie des Geichtssinnes des Menschen und der Thiere*. Leipzig: Cnobloch.

Müller, J. (1826b). *Über die phantastischen Geichtserscheinungen* (On Visual Fantasy Phenomena). Coblenz (pp iii–iv; 4–7; 33–35).

Müller, J. (1833–1837). *Handbuch der Physiologie des Menschen*, 2 vols. Coblenz: Holscher. (Translated by William Baly as *Elements of Physiology*, 2 vols. London, 1838–1842.)

Müller, J. (1843). *Elements of Physiology*, Vol. 2. Philadelphia: Lea and Blanchard, Book 5, parts 1–8.

Pisoni, D. B. (1973). Auditory and phonetic memory codes in the discrimination of consonants and vowels. *Perception and Psychophysics*, 13 (2), 253–260.

Pisoni, D. B., and Lazarus, J. H. (1974). Categorical and noncategorical modes of speech perception along the voicing continuum. *Journal of the Acoustical Society of America*, 55 (2), 328–333.

Pisoni, D. B., and Tash, J. (1974). Reaction times to comparisons within and across phonetic categories. *Perception and Psychophysics*, 15 (2), 285–290.

Raffman, D. (1995). On the Persistence of Phenomenology. In Thomas Metzinger (Ed.): *Conscious Experience*. Thoverton, UK: Imprint Academic, 293–308.

Sekuler, R., and Blake, R. (2002). *Perception*. New York: McGraw-Hill.

Shepard, R. N. (1997). The Perceptual Organization of Colors: An Adaptation to Regularities of the Terrestrial World? In Alex Byrne and David R. Hilbert (Eds.): *Readings on Color*, Vol. 2. *The Science of Color*. Cambridge, Mass.: MIT Press, 311–356.

Sivik, L. (1997). Color Systems for Cognitive Research. In C. L. Hardin and Luisa Maffi (Eds.): *Color Categories in Thought and Language*. Cambridge: Cambridge University Press, 163–193.

Sternheim, C. E., and Boynton, R. M. (1966). Uniqueness of perceived hues investigated with a continuous judgmental technique. *Journal of Experimental Psychology*, 72, 770–776.

Stiles, W. S. (1961). Adaptation, chromatic adaptation, colour transformation. *Anales Real Soc. Espan. Fis. Quim.*, Series A, 57, 149–175.

Stroud, B. (2000). *The Quest for Reality – Subjectivism and the Metaphysics of Color.* New York: Oxford University Press.

Takasaki, H. (1966). Lightness change of grays induced by change in reflectance of gray background. *Journal of the Optical Society of America*, 56, 504–509.

Thompson, E. (1995). *Colour Vision: A Study in Cognitive Science and the Philosophy of Perception.* London and New York: Routledge.

Thompson, E. (2000). Comparative Color Vision: Quality Space and Visual Ecology. In Steven Davis (Ed.): *Color Perception: Philosophical, Psychological, Artistic and Computational Perspectives.* New York: Oxford University Press.

Thompson, E., Palacios A. G., and Varela F. J. (1992). Ways of coloring: Comparative color vision as a case study for cognitive science. *Behavioral and Brain Sciences*, 15, 1–74.

Tye, M. (1995). *Ten Problems of Consciousness.* Cambridge, Mass.: MIT Press.

Tye, M. (2000). *Consciousness, Color, and Content.* Cambridge, Mass.: MIT Press.

Wandell, B. A. (1995). *Foundations of Vision.* Sunderland, Mass.: Sinauer Associates.

Werner, J. S., and Wooten, B. R. (1979). Opponent chromatic mechanisms: Relation to photopigments and hue naming. *Journal of the Optical Society of America*, 69, 422–434.

Westland, S., and Thomson, M. (1999). Spectral Colour Statistics of Surfaces: Recovery and Representation. Derby: Colour & Imaging Institute, Kingsway House East, Derby University.

Whittle, P. (1992). Brightness, discriminability, and the "Crispening Effect." *Vision Research*, 32, 1493–1507.

Wuerger, S. M., Maloney, L. T., and Krauskopf, J. (1995). Proximity judgments in color space: Tests of a Euclidean color geometry. *Vision Research*, 35 (6), 827–835.

Wyszecki, G., and Stiles, W. S. (1982). *Color Science: Concepts and Methods, Quantitative Data and Formulae.* New York: Wiley.

Wyszecki, G., and Stiles, W. S. (2000). *Color Science: Concepts and Methods, Quantitative Data and Formulae.* 2d ed. New York: Wiley.

Young, T. (1802). On the theory of light and colors. *Phil. Trans. R. Soc. Lond.*, 92, 12–48.

CONSCIOUSNESS

11

A Neurofunctional Theory of Consciousness

Jesse J. Prinz

Introduction: The Problems of Consciousness

Reading the philosophical literature on consciousness, one might get the idea that there is just one problem in consciousness studies, the hard problem. That would be a mistake. There are other problems; some are more tractable, but none is easy, and all interesting. The literature on the hard problem gives the impression that we have made little progress. Consciousness is just an excuse to work and rework familiar positions on the mind–body problem. But progress is being made elsewhere. Researchers are moving toward increasingly specific accounts of the neural basis of conscious experience. These efforts will leave some questions unanswered, but they are no less significant for that.

To move beyond the hard problem, I would like to consider some real problems facing consciousness researchers. First, there is a What Problem. This is the problem of figuring out what we are conscious of. What are the contents of conscious experience? For those looking at the brain, it is closely tied to a Where Problem. Where in the brain does consciousness arise? Locating consciousness may not be enough. We need to address a How Problem. How do certain states come to be conscious? This can be construed as a version of the hard problem, by asking it with right intonation: How could certain physical states possibly be experienced? But there is another reading that is also worth investigating. The How Problem can be interpreted as asking, What are the psychological or neuronal mechanisms or processes that distinguish conscious states from unconscious states? The mechanisms and processes can be pinpointed by pursuing a question closely related to a When Problem: Under what conditions do the states that are potentially conscious become conscious? Once we know what we are conscious of and when we are conscious, we can begin to address a Why Problem: Why do we have conscious states? And finally, there is a Who Problem: Who is conscious? Nonhuman animals? Human infants? Machines?

I will discuss these problems in turn. Some already have answers, others have answers on the way, and at least one may never be answered in full. What I offer here is a progress report on a theory of consciousness that I have presented elsewhere (Prinz, 2000, 2001). I hope to show that progress is indeed being made.

1 What Are We Conscious Of?

I think the best answer to the What Problem was given by Ray Jackendoff in his 1987 book, *Consciousness and the Computational Mind*. Jackendoff began with the observation that perceptual systems are organized hierarchically, with different subsystems representing features at varying degrees of abstraction. There is a movement from very fine-grained local features, with minimal global integration, to very abstract categorical representations that are especially useful for capturing invariance across perceptual vantage points. Between these extremes, the disjointed local and the abstract categorical, there are postulated to be intermediate-level subsystems. These are vantage point specific, rather than invariant, and they also have global organization – the parts are bound together coherently. Jackendoff based this picture on perceptual psychology from the 1970s and 1980s. He was particularly inspired by David Marr's theory of vision and by models of categorical speech perception in phonology. These models may not have held up in detail, but the overall approach remains current. Perception is still widely believed to be hierarchical. The major difference between the state of play now and back in 1987 is that we have a much better understanding of the perceptual hierarchies, and that understanding has been fueled by advances in neuroscience. Jackendoff had virtually nothing to say about the brain.

To assess Jackendoff's conjecture, we can start with the neuroscience of vision. Is there evidence that an intermediate level of processing in the visual brain is privileged with respect to consciousness? I think the answer is yes. First consider low-level vision, which is associated with activity in the primary visual cortex (V1). Destruction of V1 ordinarily eliminates visual consciousness, but it may do so by preventing higher areas from receiving visual signals. People with V1 damage sometimes experience visual hallucinations, and some also experience residual visual experience when presented with rapidly moving, high-contrast stimuli (Seguin, 1886; Sahraie et al., 1997). F. Crick and C. Koch (1995) have argued that V1 cannot be the seat of consciousness because it encodes the wrong information. Some research suggests that V1 does not show context effects that are typical of conscious color vision, and V1 may not be responsive to certain imagined contours that we experience in certain optical illusions. Rees et al. (2002) cite evidence that V1 activity drops during eye blinks, despite the fact that we are not consciously aware of visual interruptions while blinking. Some of these data are controversial. For example, M. Seghier et al. (2000) and other groups have found evidence for the processing of illusory contours in V1. It is not yet clear what role that processing plays, however. Higher visual areas are much more consistently responsive (Mendola et al., 1999). Moreover, B. M. Ramsden et al. (2001) found that the activity in V1 during the perception of illusory contours was the inverse of activity during perception of real contours. In higher visual areas, activity for real and illusory contours is alike. Also, receptive fields in V1 may be too small to respond to illusory contours (or color context effects) over large gaps. Higher visual areas may be needed to perceive these. It is premature to rule conclusively on whether V1 is a locus of visual consciousness, but current evidence provides little reason to think that it is.

Now consider high-level vision, which is associated with activity in areas of inferotemporal cortex. N. Logothetis and colleagues have been arguing that these areas are crucial for visual consciousness. When we look at two distinct images simultaneously, one in each eye, we experience only one or the other. Logothetis and colleagues presented such stimuli to monkeys and trained them to indicate which of two stimuli they were seeing. They measured cellular response in various visual areas as the monkeys made their reports. It turned out the 90% of the measured cells in high-level visual areas correlated with indicated percept, compared to 38% of the measured cells in intermediate level areas (Leopold and Logothetis, 1996; Sheinberg and Logothetis, 1997). Logothetis (1998, p. 541) provisionally concludes that activity in inferotemporal cortex (IT) "may indeed be the neural correlate of conscious perception."

There are three difficulties with this conclusion. First, the fact that fewer than half of the intermediate cells correspond to the perceived object does not show that conscious perception is not located at the intermediate level. The cells in that 38% may be responding in a distinctive way, and that distinctive way of responding may mark the difference between conscious and unconscious vision. In fact, the intermediate cells that correspond to the perceived object are almost certainly responding distinctively. After all, they seem to be the only cells that are allowing information to propagate forward to higher areas. The 90% in IT reflects a selection process at the prior stage of processing. Second, there is independent reason to think that IT is not the locus of visual awareness. Cells in IT encode the wrong information. They tend to abstract away from size and specific orientation (Vuilleumier et al., 2002). Indeed, many IT cells respond the same way to a shape oriented to the left and the same shape oriented to the right (Baylis and Driver, 2001). Third, damage in high-level visual areas tends to cause deficits in recognition of images, but not visual experience as such. Damage to intermediate-level visual areas impairs both. This is the difference between associative and apperceptive agnosia (Farah, 1990).

In contrast to cells in IT and in V1, cells in intermediate-level visual areas do correspond to the contents of experience. They represent illusory contours and color constancy, and they represent objects from a specific orientation, as they appear in consciousness. Damage to intermediate areas (V2, V3, V4, and V5) eliminates consciousness of the stimulus features processed in these areas (Zeki, 1993), and these areas are active during visual experiences and visual hallucinations (Ffytche et al., 1998). This confirms Jackendoff's conjecture that the intermediate level is the level of consciousness.

Jackendoff speculates, in his book, that the intermediate level is the locus of consciousness in all perceptual modalities, not just vision. Though hardly confirmed, that generalization is consistent with current data from neuroscience. All sense modalities seem to be hierarchical. In audition, there is a hierarchy extending from the superior temporal plane into the superior temporal gyrus. Researchers have divided this pathway up into core, belt, and parabelt regions, which correspond to high, intermediate, and low levels of processing (Kaas and Hackett, 2000). In touch, there is a hierarchy that extends from Brodmann areas 3a, 3b, 1, and 2 into areas 5 and 7 (Friedman et al., 1986; Kaas, 1993). There is a taste hierarchy as well, including the insula and moving into orbitofrontal cortex (Rolls, 1998). Even smell, the ancient sense, has hierarchical organization, which extends from piriform cortex into orbitofrontal and prefrontal areas (Savic et al., 2000). I have argued elsewhere that a sensory hierarchy can also be identified for the perception of internal bodily states, which forms the basis of both interceptive experience and emotional experience, a special case of interoception (Prinz, 2004).

These hierarchies differ in organizational detail, of course, but there is a general pattern. Highly specific, local features are registered first, then combinations of those features are registered, and, finally relatively abstract or categorical stimulus properties are registered. It may be a long time before we have good evidence pinpointing consciousness in these hierarchies, but the functional profiles invite speculation. The smallest units of perception, whether they are isolated tones or tiny features of texture, are too discrete to map onto units of experience. More complex features, especially those that abstract away from the idiosyncratic features of a particular perceptual episode, seem more abstract than the contents of experience. Only in the middle, where tones merge together into the components of a melody and minute edges merge into textural patterns, do we seem to have conscious experience. Existing findings are consistent with these speculations. For example, in auditory processing, there is a distinction between disorders of recognition, where sounds can be heard intact, and disorders of perception, where sounds are disrupted or impoverished. This corresponds to the distinction between associative and apperceptive agnosia in vision, which supports the intermediate-level conjecture (Vignolo, 1982). Time will tell if the conjecture holds up. If it does, Jackendoff's answer to the What question will be upheld.

2 How Do We Become Conscious?

Jackendoff may have been right about the What Problem, but he does not offer a solution to the How Problem. He does not specify the conditions under which intermediate-level perceptual representations become conscious. The implied answer is that consciousness arises whenever intermediate-level representations are active. Consciousness is the tokening of intermediate-level representations. This would be a seductively simple theory, but it cannot be right. There is good reason to think that activity can occur at the intermediate level without conscious experience. Consider subliminal perception. One can extract information about the meaning, form, or identity of an object even if it is presented under conditions that prevent conscious awareness (e.g., Bar and Biederman, 1998; Dell'Acqua and Grainger, 1999). To extract such information, one must process the stimulus all the way through the perceptual hierarchy. Thus, intermediate-level activation is not sufficient for consciousness.

A more dramatic demonstration of this point comes from studies of patients with unilateral neglect. In subliminal perception studies, stimuli are presented very briefly. Such studies can be interpreted as demonstrating

memory effects, rather than unconscious perception. Subjects may consciously perceive the stimuli and then forget them. This interpretation is
unavailable for cases of neglect. In these cases, right inferior parietal injuries prevent patients from attending to the left side of their visual fields
or the left side of visually presented objects, or the left side of their own
bodies. They report no conscious experience of things on the left. This
effect has no temporal cap. For example, patients can stare at an object
for a long time and still be oblivious to its left side. Nevertheless, there
is evidence that patients with neglect are registering information about
the objects that they fail to experience (Marshall and Halligan, 1988;
Bisiach, 1992; Driver, 1996). This suggests that they are processing neglected objects through (and perhaps beyond) the intermediate level.
This conclusion has been supported by neuroimaging studies, which
demonstrate intermediate-level activity in the visual systems of unconsciously perceived objects in patients with right parietal injuries (Rees
et al., 2000; Vuilleumier et al., 2001).

   Neglect provides two important lessons for consciousness. First, apparent cases of unconscious perception are not always readily interpretable
as memory effects. There is every reason to think that patients with neglect
are capable of perceiving, to some degree, without conscious experience.
This shows that mere activity in a perceptual system is not sufficient for
consciousness. The second lesson comes from a moment's reflection on
the cause of the deficit in neglect. Conscious experience of the left side is
lost as a result of injuries in centers of the parietal cortex that have been
independently associated with attention. Neglect is an attention disorder. This suggests that consciousness requires attention. This hypothesis
gains experimental support from behavioral studies with normal subjects. Evidence suggests that people will fail to consciously perceive a
centrally presented object if their attention is occupied by another task.
The phenomenon was labeled inattentional blindness by A. Mack and
I. Rock (1998). In their studies, many subjects fail to notice a centrally
presented object that is briefly displayed, and unexpected. Inattentional
blindness can also be sustained for a relatively long time. Daniel Simons
and Christopher Chabris (1999) had subjects watch a video in which two
teams were tossing a basketball. Subjects were asked to count how many
times the ball was passed by a particular team – an attention-demanding
task. During the game, a person in a gorilla suit strolled across the center
of the screen. The gorilla is highly salient to passive viewers, but 66% of
the subjects who were counting passes failed to notice the gorilla. It is
possible that these subjects consciously perceived the gorilla and simply

didn't categorize it; but it is equally possible that they didn't perceive it consciously at all. The latter interpretation would be most consistent with the neglect findings. All these findings provide strong support for a link between attention and consciousness (see also Luck et al., 1996, on the attentional blink).

These findings present a solution to the How Problem. Consciousness seems to arise in intermediate-level perceptual subsystems when and only when activity in those systems is modulated by attention. When attention is allocated, perception becomes conscious. Attentional modulation of intermediate-level representations is both necessary and sufficient for consciousness. This also makes sense of the Logothetis studies. When presented with conflicting simultaneous stimuli, we may be able to attend to only one. The shifts in attention lead to shifts in consciousness.

Some researchers object to the thesis that attention is necessary. Christof Koch (personal communication) gives the example of staring at a solid expanse of color, spanning the visual field. Do we really need to attend in order to experience the hue? I think the answer is yes. Contrary intuitions stem from the fact that attention is often associated with effort and focus. In this case, attention is spread evenly across the field, or perhaps it bounces around, as if scanning fruitlessly for points of interest. In either case, the allocation of attention might not require effort. It does, however, involve selection. Attention is generally in the business of picking out salient objects. This is a limiting case. There are no objects, and so the whole field, or large portions, are selected. I don't have demonstrative proof of this conjecture. My point is that Koch's alleged counterexample is not decisive. There is a plausible interpretation that involves attention. If all else were equal, there would be no way to decide between these interpretations. But all else is not equal. Evidence from the inattentional blindness literature suggests that we are not conscious when attention is withdrawn. In the color case, we must be attending.

So far, I have said very little about what attention actually is. We need an answer to this question before we can make any serious progress. Fortunately, psychologists and neuroscientists have developed numerous models of attention. The general theme of these models is that attention is a selection process that allows information to be sent to outputs for further processing. It is well established that visual attention modulates activity through the visual hierarchy, and changes can include increase in activity for cells whose receptive fields include attended objects, and decrease in activity in cells responding to unattended objects (Fries et al., 2001; Corchs and Deco, 2002). There is some evidence that the relevant changes involve changes in rates of neuronal oscillation (Niebur et al., 1993; Fries et al., 2001). Whatever the specific mechanism, changes in perceptual processing seem to have an impact on what can propagate forward for further processing. Attention thus affects a virtual change in connectivity between neuronal populations (Olshausen et al., 1994). Attention is a routing gate.

This raises another question. When perceptual information can propagate forward, where does it go? The most plausible answer is that attention allows perceptual information to access working memory stores. "Working memory" refers to our ability to retain information for brief periods of time, and to manipulate it in various ways. In the brain, a working memory area is understood as a neuronal population that responds to features of a perceived stimulus but maintains activity when that stimulus is removed. Working memory areas have been identified in frontal and temporal cortex (e.g., Nakamura and Kubota, 1995; Miller et al., 1996). When we perceive, what we perceive is made available for temporary storage. Attention is gatekeeper to working memory.

If all of this is right, the solution to the How Problem can be stated as follows: Consciousness arises when intermediate-level perception representations are made available to working memory via attention. I call this the AIR theory of consciousness, for attended intermediate-level representations. The suggestion that working memory access is important to consciousness has many defenders (e.g., Crick and Koch, 1995; Baars, 1997; Sahraie et al., 1997). G. Rees (2001) has defended a view that is especially close to the AIR theory. He uses neuroimaging evidence to support the conjecture that consciousness involves a large-scale neural network that includes perception centers, along with a prefrontal working memory center and parietal attention centers. I take this convergence to be an encouraging sign that we are closing in on the How Problem.

3 Why Are We Conscious?

The AIR theory is a functionalist account of what consciousness is. It says that consciousness can be understood in terms of an information-processing role. As such, it lends itself to a straightforward answer to the Why Problem. In asking why we are conscious, we are really wondering what consciousness is for or what purpose it serves. Does it do any work for us? That depends on what it is. If consciousness is the property of inner representations of a certain sort of being made available in a certain way, then the answer depends on what these representations are and what their availability does for us.

It should be evident that consciousness is extremely valuable, if the AIR theory is right. Consciousness serves the crucial function of broadcasting viewpoint-specific information into working memory. Viewpoint-specific representations are important for making certain kinds of decisions. If we encounter a predator, for example, it is useful to know whether it is facing us or facing in another direction. Consciousness provides us with this information by making it available to working memory. Working memory is not just a transient storage bin. It is the store from which decisions are made and actions are chosen. It is crucial for decisions that have not been rehearsed much in the past. Without working memory, we would be reduced to reflexive response. There is also evidence that working memory is a gateway to episodic memory (Buckner et al., 1999). Information is sent to working memory before getting encoded in a long-term store. If that is right, then without consciousness, we would have no biographies, no memories of our past. In these ways, consciousness is what distinguishes us from paramecia and jellyfish.

It might be objected that the functions I have been describing could be achieved without consciousness. In another possible cognitive system, viewpoint-specific representations could be broadcast to working memory without any experience or phenomenology. This is perfectly true, but the "could" renders the point irrelevant. Flyswatters serve the function of killing flies. That function could also be achieved by insect-eradicating lamps. It would be a howler to infer from this that flyswatters have no function. The same function can be carried out in different ways. In us, the function of broadcasting viewpoint-specific information to working memory is achieved by mechanisms that give us conscious experiences. Those experiences have the function they serve.

The objector might press the point, however. If the function of broadcasting information can be realized without consciousness, then consciousness cannot be identified with that function. Rather, it must be identified with something that happens to realize that function in us. This objection is well taken, but it is not fatal to the AIR theory. It requires a clarification. Consciousness does not arise whenever intermediate-level representations are broadcast to working memory; it arises when that is done via the kinds of mechanisms found in us. Just as a flyswatter kills flies in a certain way, consciousness is broadcasting in a certain way. To figure out what way, we need to turn to the science of attention. We need to look at specific models of how attention routs information into working memory. Various models of attention have been worked out in physiological detail. Earlier, I mentioned excitation, inhibition, and changes in oscillation frequency as examples. The idea is that we will ultimately be able to give a true and complete neurocomputational model to go along with the AIR theory. The neurocomputation model will specify the mechanisms that matter for consciousness.

Specified in this way, the AIR theory is different from other functionalist theories in the philosophy of mind. It is a neurofunctional theory: Consciousness is identified with a functional role that is implemented by mechanisms specifiable in the language of computational neuroscience. Computational neuroscience is a functional theory; it describes what neurons and neural networks do. But it is also a neuroscientific theory, because it describes those functions at a level of architectural specificity that can be investigated using the tools of neuroscience. Such functions might be realizable by materials other than those found in the nervous system, but they cannot be realized by everything under the sun. The computations that our brains performed, with their specific temporal profiles, place structural and organizational constraints on the things that can perform them. In describing things at a fine level of neurocomputational detail, we should not forget, however, that all of these neuronal mechanisms are in the business of implementing functions that can be described at a psychological level of analysis. As stated previously, the AIR theory is a characterization of that psychological level. The psychological level implemented by the neurocomputational level specifies what the neurocomputations are achieving, and, thus, the psychological level tells us what consciousness is for.

4 Who Is Conscious?

Now only the Who Problem remains. On the face of it, the problem is easy. Here's a formula for answering it. Find out what the mechanisms of consciousness are in us. Then see what other creatures have those mechanisms. On the story I have been telling, the mechanisms of consciousness are the attentional mechanisms that make intermediate-level perceptual representations available to working memory. Now we can ask: What creatures have those?

Higher mammals, especially primates, probably have neural mechanisms that are very much like ours. Much of what we know about the cellular neurophysiology of human attention and working memory comes from studies of monkeys. There also seems to be great continuity between primates and rats. Rats are used as animal models for attention and working memory abnormality in schizophrenia and attention deficit/hyperactivity disorders. The functional neuroanatomy of a rat is similar to our own in many respects. For example, rat prefrontal cortex plays a key role in rat working memory (e.g., Izaki et al., 2001).

Creatures that are more distantly related resemble us in functional respects. For example, researchers have found working memory capacities in octopuses (Mather, 1991), pigeons (Diekamp et al., 2002), bees (Menzel, 1984), and slugs (Yamada et al., 1992). But such creatures have neuronal mechanisms that are quite different from ours, and this raises a question. Is there any reason to think they are conscious? This question is far more difficult than the questions I have so far considered, and I see no obvious way to answer it. In other words, I see no obvious way to know how similar a mechanism must be to those found in our own brains in order to support conscious experiences. The problem can be put in terms of levels of analysis. Let's suppose we find that octopuses have viewpoint-specific representations and attentionally gated working memory. At the psychological level, they are like us. But suppose they differ at the neurocomputational level. Or, to be more realistic, suppose there are numerous levels of abstraction at which we can provide neurocomputational models, and octopuses differ on some but not all of these. At which level must an octopus resemble us to be a candidate for consciousness? Elsewhere, I raise the same question for machines (Prinz, 2003). How close to the human brain must a computer be before we can say that it is probably conscious? I think there is no way to answer this question. To do so would require identifying the level of analysis that was crucial for consciousness in ourselves. That would require that we be able to look at the contribution of different levels individually. We have no way to do that, and if we did it wouldn't help. Suppose we could keep the psychological level of analysis constant while changing the neural mechanisms in a human volunteer. We might do that by replacing brain cells with microchips that work differently at a low level of analysis. The problem is that the human volunteer would report being conscious no matter what. Preserving function at the psychological level would guarantee it (see Prinz, 2003, for more discussion). So we cannot tell if the low-level alteration makes a difference vis-à-vis consciousness. I call this level-headed mysterianism.

There may also be another mystery that faces anyone trying to answer the Who Problem. Suppose we find a creature that exhibits signs of intelligence but is radically unlike us functionally. Suppose it has nothing that we can identify as a working memory store (all memories are always equally accessible and long-lasting) and no attention capacity (all perceived stimuli get processed equally). Suppose further that its perceptual systems are not hierarchical and that it perceives through senses unlike our own. Now we can ask whether this creature has any conscious experiences. We can reasonably speculate that it does not have conscious experiences like ours. But we cannot, it seems, conclusively speculate that it has no conscious life at all. The creature may have very complex sensory systems, and it may respond to the environment intelligibly. We would certainly want to call it a perceiver. The difficulty is in deciding whether its perceptions have phenomenal character. We cannot demonstratively rule out the possibility that experience is realized in radically different ways. This might lead one to an agnostic conclusion, which one might call radical realization mysterianism. I cannot think of a reason to take such a position especially seriously. We know of no such creatures, and if we met one, its radically different psychology would be grounds for holding off ascriptions of consciousness. Moreover, I am inclined to interpret the discovery of a material basis for consciousness in us as the discovery of an identity. Consciousness just is the operation of a neurofunctional mechanism found in us. To hold otherwise is to regard consciousness as an emergent property of such a mechanism – a property that could emerge in other ways. I suspect that this kind of emergentism isn't true, but I can think of no decisive way to rule it out. The evidence for the material basis of consciousness in us is not decisive evidence against consciousness arising in different ways. And, of course, this is the road to the hard problem.

If we are saddled with either of these two kinds of mysterianism (I am much more confident about the level-headed variety), then the theory of consciousness will remain forever incomplete. We will be able to discover conditions that are sufficient for consciousness, such as those found in our brains, but we will never have a perfectly settled list of the conditions necessary for consciousness. We will not know what level of physiology is essential, and perhaps we won't even know what psychological processes are essential. That is a humbling discovery.

But, lest we get too humble, we should take heart in the enormous progress that has been made. Discovering the material basis of consciousness in us is no mean feat. We are not quite there yet. A full account would include a correct neurocomputational theory of attention. That goal is certainly within reach. When it has been achieved, I think we will be in a position to say that we have cracked problems of profound importance.

Finding the neurocomputational basis of consciousness will be like finding the material basis of genetic inheritance. It will be equally worthy of an invitation to Stockholm. There may be residual mysteries pertaining to realizations of consciousness in other creatures, but that should not distract us from the success that may already be at hand.

Acknowledgments

I want to thank the audience of the McDonnell Philosophy and Neuroscience Conference and, especially, Andrew Brook for helpful comments.

References

Baars, B. J. (1997). *In the Theater of Consciousness.* New York: Oxford University Press.

Bar, M., and Biederman, I. (1998). Subliminal visual priming. *Psychological Science, 9,* 464–469.

Baylis, G. C., and Driver, J. (2001). Shape-coding in IT cells generalizes over contrast and mirror reversal, but not figure-ground reversal. *Nature Neuroscience, 4,* 937–942.

Bisiach, E. (1992). Understanding consciousness: Clues from unilateral neglect and related disorders. In A. D. Milner and M. D. Rugg (eds.), *The Neuropsychology of Consciousness* (113–139). London: Academic Press.

Buckner, R. L., Kelley, W. M., and Petersen, S. E. (1999). Frontal cortex contributes to human memory formation. *Nature Neuroscience, 2,* 311–314.

Corchs, S., and Deco, G. (2002). Large-scale neural model for visual attention: Integration of experimental single-cell and fMRI data. *Cerebral Cortex, 12,* 339–348.

Crick, F., and Koch, C. (1995). Are we aware of activity in primary visual cortex? *Nature, 37,* 121–123.

Dell'Acqua, R., and Grainger, J. (1999). Unconscious semantic priming from pictures. *Cognition, 73,* B1–B15.

Diekamp, B., Kalt, T., and Güntürkün, O. (2002). Working memory neurons in pigeons. *Journal of Neuroscience, 22,* RC 210.

Driver, J. (1996). What can visual neglect and extinction reveal about the extent of "preattentive" processing? In A. F. Kramer, M. G. H. Coles, and G. D. Logan (eds.), *Convergent Operations in the Study of Visual Selective Attention* (193–224). Washington, DC: APA Press.

Farah, Martha J. (1990). *Visual Agnosia: Disorders of Object Recognition and What They Tell Us About Normal Vision.* Cambridge, MA: MIT Press.

Ffytche, D. H., Howard, R. J., Brammer, M. J., David, A., Woodruff, P. W., and Williams S. (1998). The anatomy of conscious vision: An fMRI study of visual hallucinations. *Nature Neuroscience, 1,* 738–742.

Friedman, D. P., Murray, E. A., O'Neil, B., Mishkin, M. (1986). Cortical connections of the somatosensory fields of the lateral sulcus of Macaques: Evidence of a corticolimbic pathway for touch. *Journal of Comparative Neurology, 252,* 323–347.

Fries, P., Reynolds, J. H., Rorie, A. E., and Desimone, R. (2001). Modulation of oscillatory neuronal synchronization by selective visual attention. *Science, 291,* 1560–1563.

Izaki, Y., Maruki, K., Hori, K., and Nomura, M. (2001). Effects of rat medial prefrontal cortex temporal inactivation on a delayed alternation task. *Neuroscience Letters, 315,* 129–132.

Jackendoff, R. (1987). *Consciousness and the Computational Mind.* Cambridge, MA: MIT Press.

Kaas, J. H. (1993). The functional organization of somatosensory cortex in primates. *Annals of Anatomy, 175,* 509–518.

Kaas, J. H., and Hackett, T. A. (2000). Subdivisions of auditory cortex and processing streams in primates. *Proceedings of the National Academy of Science, 97,* 11793–11799.

Leopold, D. A., and Logothetis, N. K. (1996). Activity changes in early visual cortex reflect monkeys' percepts during binocular rivalry. *Nature, 379,* 549–553.

Logothetis, N. (1998). Object vision and visual awareness. *Current Opinion in Neurobiology, 8,* 536–544.

Luck, S. J., Vogel, E. K., and Shapiro, K. L. (1996). Word meanings can be accessed but not reported during the attentional blink. *Nature, 383,* 616–618.

Mack, A., and Rock, I. (1998). *Inattentional Blindness.* Cambridge, MA: MIT Press.

Marr, D. (1982). *Vision: A Computational Investigation into the Human Representation and Processing of Visual Information.* New York: W. H. Freeman.

Marshall, J. C., and Halligan, P. W. (1988). Blindsight and insight in visuospatial neglect. *Nature, 336,* 766–767.

Mather, J. A. (1991). Navigation by spatial memory and use of visual landmarks in octopuses. *Journal of Comparative Physiology, A, 168,* 491–497.

Mendola, J. D., Dale, A. M., Fischl, B., Liu, A. K., and Tootell, R. B. H. (1999). The representation of illusory and real contours in human cortical visual areas revealed by functional magnetic resonance imaging. *Journal of Neuroscience, 19,* 8560–8572.

Menzel, R. (1984). Short-term memory in bees. In D. L. Alkon and J. Farley (eds.), *Primary Neural Substrates of Learning and Behavior Change* (259–274). Cambridge: Cambridge University Press.

Miller, E. K., Erickson, C. A., and Desimone, R. (1996). Neural mechanisms of visual working memory in prefrontal cortex of the macaque. *Journal of Neuroscience, 16,* 5154–5167.

Nakamura, K., and Kubota, K. (1995). Mnemonic firing of neurons in the monkey temporal pole during a visual recognition task. *Journal of Neurophysiology, 74,* 162–178.

Niebur, E., Koch, C., and Rosin, C. (1993). An oscillation-based model for the neuronal basis of attention. *Vision Research, 18,* 2789–2802.

Olshausen, B. A., Anderson, C. H., and van Essen, D. C. (1994). A neurobiological model of visual attention and invariant pattern recognition based task. *Journal of Neuroscience, 14,* 6171–6186.

Prinz, J. J. (2000). A neurofunctional theory of visual consciousness. *Consciousness and Cognition, 9,* 243–259.

Prinz, J. J. (2001). Functionalism, dualism and the neural correlates of consciousness. In W. Bechtel, P. Mandik, J. Mundale, and R. Stufflebeam (eds.), *Philosophy and the Neurosciences: A Reader* (278–294). Oxford: Blackwell.

Prinz, J. J. (2003). Level-headed mysterianism and artificial experience. *Journal of Consciousness Studies, 10,* 111–132.

Prinz, J. J. (2004). *Gut Reactions: A Perceptual Theory of Emotion.* New York: Oxford University Press.

Ramsden, B. M., Chou, P. H., and Roe, A. W. (2001). Real and illusory contour processing in area V1 of the primate: A cortical balancing act. *Cerebral Cortex, 11,* 648–665.

Rees, G. (2001). Neuroimaging of visual awareness in patients and normal subjects. *Current Opinion in Neurobiology, 11,* 150–156.

Rees, G., Kreiman, G., and Koch, C. (2002). Neural correlates of consciousness in humans. *Nature Reviews Neuroscience, 3,* 261–270.

Rees, G., Wojciulik, E., Clarke, K., Husain, M., Frith, C., and Driver, J. (2000). Unconscious activation of visual cortex in the damaged right hemisphere of a parietal patient with extinction. *Brain, 123,* 1624–1633.

Rolls, E. T. (1998). The orbitofrontal cortex. In A. C. Roberts, T. W. Robbins, and L. Weiskrantz (eds.), *The Prefrontal Cortex* (67–86). Oxford: Oxford University Press.

Sahraie, A., Weiskrantz, L., Barbur, J. L., Simmons, A., Williams, S. C. R., and Brammer, M. J. (1997). Pattern of neuronal activity associated with conscious and unconscious processing of visual signals. *Proceedings of the National Academy of Science, 94,* 9406–9411.

Savic, I., Gulyas, B., Larsson, M., and Roland, P. (2000). Olfactory functions are mediated by parallel and hierarchical processing. *Neuron, 26,* 735–745.

Seghier, M., Dojat, M., Delon-Martin, C., Rubin, C., Warnking, J., Segebarth C., and Bullier, J. (2000). Moving illusory contours activate primary visual cortex: An fMRI study. *Cerebral Cortex, 10,* 663–670.

Seguin, E. G. (1886). A contribution to the pathology of hemianopsis of central origin (cortex-hemianopsia). *Journal of Nervous and Mental Diseases, 13,* 1–38.

Sheinberg, D. L., and Logothetis, N. K. (1997). The role of temporal cortical areas in perceptual organizations. *Proceedings of the National Academy of Science, 94,* 3408–3413.

Simons, Daniel J., and Chabris, Christopher, F. (1999). Gorillas in our midst: Sustained inattentional blindness for dynamic events. *Perception, 28,* 1059–1074.

Vignolo, L. A. (1982). Auditory agnosia. *Philosophical Transactions of the Royal Society of London Series B: Biological Sciences,* 298, 49–57.

Vuilleumier, P., Henson, R. N., Driver, J., and Dolan, R. J. (2002). Multiple levels of visual object constancy revealed by event-related fMRI of repetition priming. *Nature Neuroscience, 5,* 491–499.

Vuilleumier, P., Sagiv, N., Hazeltine, E., Poldrack, R. A., Swick, D., Rafal, R. D., and Gabrieli, J. D. (2001). Neural fate of seen and unseen faces in visuospatial neglect: A combined event-related functional MRI and event-related potential study. *Proceedings of the National Academy of Science, 98*, 3495–3500.

Yamada, A., Sekiguchi, T., Suzuki, H., Mizukami, A. (1992). Behavioral analysis of internal memory states using cooling-induced retrograde amnesia in Limax flavus. *Journal of Neuroscience, 12*, 729–735.

Zeki, S. (1993). *A Vision of the Brain.* Oxford: Blackwell.

12

Making Consciousness Safe for Neuroscience

Andrew Brook

Work on consciousness by neurophilosophers often leaves a certain group of other philosophers frustrated. The latter group of philosophers, which includes people such as Thomas Nagel, Frank Jackson, Colin McGinn, Ned Block, and David Chalmers, believe that consciousness is something quite different from the brain circuitry or other processes that are active in cognition. They feel frustrated because work on consciousness by neurophilosophers usually ignores this view, yet proceeds from an assumption that it is wrong. This work tends to assume, simply assume, that neuroscience not only will identify neural *correlates* of consciousness (which virtually all parties to the current consciousness debate now accept), but also (perhaps with the assistance of cognitive science) will eventually tell us *what consciousness is.* That is to say, it assumes that consciousness simply is a neural/cognitive process of some kind. Even more, it assumes that consciousness is a neural/cognitive process similar in kind to the processes that underlie (other aspects of) cognition and representation. That is to say, it assumes that consciousness is an aspect of general cognition.

Consciousness has appeared to be weird and wonderful to many people for a very long time. Daniel Dennett captured the feeling very nicely many years ago: 'Consciousness appears to be the last bastion of occult properties, epiphenomena, immeasurable subjective states – in short, the one area of mind best left to the philosophers. Let them make fools of themselves trying to corral the quicksilver of "phenomenology" into a respectable theory' (1978a, p. 149). Consciousness no longer appears *this* strange to very many researchers, but the group of people just mentioned continue to hold that it is very different from any brain or other process active in cognition. By contrast, like most consciousness researchers now,

neurophilosophers simply take for granted that consciousness will be do-
mesticated along with the rest of cognition – indeed, that it will turn out
to be simply an aspect of general cognition.

I am sympathetic to the position assumed by these people. However,
I do not think that one can simply ignore the opposition. In my view,
one must confront their arguments and show where they fail. In addition
to the general desirability of not ignoring one's opponents, there is a
specific reason why this needs to be done in the case of consciousness.
If the opposition is left unchallenged, it can easily appear as though the
neurophilosophy of 'consciousness' is in fact not talking about *consciousness,* has subtly changed the topic. It can easily appear that these essays are
merely talking about correlates of consciousness, not the real McCoy, consciousness itself. The opposition does not have much to say about what
this something else might be like – a deep streak of what Owen Flanagan
(1991) calls mysterianism runs through this work – but they think that
they have strong arguments in favour of the claim that it *is* something
else. I think these arguments must be answered, not ignored – which is
what I will try to do in this chapter.

The arguments all have the same cast. They consist of thought-
experiments designed to show either that complete cognitive functioning, even cognitive functioning of the highest sort, could proceed as it is
in us without consciousness, or that it could proceed as it is in us even
if the contents of consciousness were very different from ours. The most
common arguments of the former kind are zombie thought-experiments:
Something could behave like us or function cognitively like us or even
be a molecule-for-molecule duplicate of us without being conscious. One
common form of argument of the second kind is exemplified by inverted
spectrum thought-experiments: Where something seems green to us, it
could seem red to another. However, because this other has been trained
to express its experience using the word 'green', and so on, its behaviour –
and even its cognitive functioning – could be just like ours. These are the
kind of arguments that I propose to go after.[1]

---

[1] Chalmers' well-known (1995) distinction between what he calls the easy problem and
the hard problem of consciousness starts from this distinction between the cognitive
role of representations and something appearing to be like something in them. Understanding the former is, he says, an easy problem, at least compared to understanding
the latter. The easy problem is to understand the inferential and other roles of such
states. The hard problem is to understand how, in these states or any states, something could appear as something to me, how certain stimulations of the retina, processing of signals by the visual cortex, application of categories and other referential

Broadly, there are at least two ways in which one might go after such arguments. One way would be to show that the opposing point of view is unproductive. We will do that. Indeed, I will suggest that it is so unlikely to shed light on most of the interesting aspects of consciousness that consciousness simply could not have the character that the view says it has. The other would be to show that the arguments do not succeed. I will suggest, indeed, that there is deep incoherence built into most of them.

Thus, the chapter has three parts. First, we will articulate the difference between the two points of view in detail. Along the way we will examine one respect in which most neurophilosophy of consciousness is different from even physicalist philosophy of consciousness, much of it at any rate, and we will examine and reject the old charge that most neurophilosophers are really eliminativists about consciousness. Then, we will do a (by the nature of the current science, necessarily provisional) assessment of the prospects of the antiphysicalist point of view as an explanatory theory of consciousness. Finally, we will sketch a general line of argument that, I will argue, undermines most of the thought-experiments invoked in support of antiphysicalism.

As will have become clear, this chapter won't have a lot of neuroscience in it (though some of it appears in the final section). One of philosophy's important roles historically has been clearing conceptual ground, removing confusions and confounding notions, and providing empirical researchers with concepts adequate to do their work with clarity and precision. There is still a lot of obscuring underbrush in consciousness research. The profusion of current terminology alone is enough to show this. There is access consciousness, phenomenal consciousness, self-consciousness, simple consciousness, background consciousness, reflective consciousness, creature consciousness, state consciousness, monitoring consciousness, consciousness taken to be coextensive with awareness, consciousness distinguished from awareness, qualia, transparency, consciousness as higher-order thought, higher-order experience, displaced perception ... and on and on. The aim of this chapter is to clear a small part of the underbrush out of the way.

---

and discriminatory apparatus elsewhere in the brain can result in an *appearing*, a state in which something *appears* a certain way. Chalmers says that the easy problem is easy because it is simply the problem of the nature and function of representation in general, while the hard problem is hard because it is sui generis, quite unlike any other problem about cognition that we face. If the first problem is easy, I'd hate to see what a hard one is like, but on anticognitivism, the two will at least be quite *different* problems.

1 The Two Approaches to Consciousness

Whatever their position on other issues, most researchers now take at
least one central form of being conscious of something to consist in its
being like something to represent that state, and conscious states to be
states that make us conscious in this way. When I consciously perceive the
words on my computer screen, it is *like something* to perceive those words,
in Nagel's (1974) now-famous phrase, and the perception is a conscious
state. If I shift attention to the perception itself ('How clearly can I see
print on this screen?'), it then (at least then) becomes like something
to have, too – which makes the case for calling it a conscious state even
stronger.[2] Without committing ourselves as to whether these are the only
forms of consciousness, let us take these phenomena and their common
core, it being like something to have them, as our target. As we have seen,
researchers divide over whether consciousness so understood (from now
on simply 'consciousness') is a kind or aspect of cognition or of the brain
processes active in cognition.

By contrast, as we have also seen, others say, or assume, that consciousness *does* consist of a kind or aspect of cognition and the brain processes
active in cognition. This, in fact, is the position of virtually all empirical investigators, whether cognitive or neuroscientists (Baars, Crick and
Koch, Goodale and Milner, Jackendoff, Mack and Rock, Newman, Posner,
Shallice, Zeki, and many others), and some philosophers (P. M. and P. S.
Churchland, Dennett, Dretske, Prinz, Rosenthal, Thompson, Tye, and
many others).

The two sides of this division are sometimes called representationalism
and antirepresentationalism. The name is not entirely apt. Those who
are hostile to the idea that consciousness is representational would be
equally hostile to an idea that it is some other cognitive property. And
on the other side, there are those who view consciousness as a perfectly
straightforward aspect of cognition but reject the idea that consciousness
is a *representational* phenomenon. Representations (or, if you reject the
idea of representations, your favourite surrogate for representations)[3] are

---

[2] Confusion and (often implicit) disagreement abound in the use of consciousness terminology, and there is some stipulation in what I have just said. I hope the result is
nonetheless not controversial.

[3] So long as you allow that we need some term or set of terms for talking about how things
appear to people, I don't care whether you call it representation or something else. Note
that nothing here hangs on a particular view of representation. Whether one views being
a representation as a matter of referring to, standing in for, reliably covarying with, being
semantically evaluable, or something else, anticognitivists will deny that consciousness

only one part of a cognitive system. There is also all the machinery for processing representations (and the emotions, and volitions, and other things). Theorists who hold that consciousness is an aspect of cognition but not a representational aspect include, for example, all the researchers who hold that consciousness is attention or a kind of attention. So let us instead call the two sides of the division *cognitivism* and *anticognitivism* about consciousness.

Cognitivism – the view that consciousness is a representational property of representations or a cognitive property of the system that processes representations,

and,

Anticognitivism – the view that consciousness is neither a representational property of representations nor a cognitive property of the system that processes representations.

2 Another Divide in Research on Consciousness

Another divide in recent work on consciousness has not been as prominent as the cognitivist/anticognitivist one. It will help us situate neurophilosophy vis-à-vis the opposition. Many philosophers of consciousness focus on individual psychological states – individual perceptions or feelings or imaginings (Tye 1995; Chalmers 1996) – or at most, tiny combinations of such states (a thought directed at an experience, for example; Rosenthal 1991). Let us call this the *atomistic approach* to consciousness.

Atomist approach to studying consciousness – the view that conscious states can be studied one by one or in small groups without reference to the cognitive system that has them.

Atomists about consciousness talk about conscious states one by one ('What is it like for something to look red?'), or at most in tiny groups, for example a thought directed at a perception (the so-called higher-order thought view [Rosenthal 1991]), and ask questions such as, 'When a state is like something to have, what is this aspect of the state like?' The answers to this question then split along the lines of the cognitivist/anticognitivist division. Almost the whole of the massive literature on *qualia* is atomistic

has that character. And likewise for those who view consciousness not as a property of representations but of whole cognitive systems.

in this way. (*'Qualia'*, a philosopher's term, is a term for the felt quality of an individual conscious state, 'what it is like to have it'.) But notice: Atomists of either stripe ignore the cognitive system whose states these states of consciousness are. They may say the words '...look red *to me*' but they do nothing with the addition.

Most experimentalists, by contrast, focus on properties of whole cognitive systems: global workspace (Baars 1988), intermediate level of processing (Jackendoff 1987), attention (Posner 1994; Mack and Rock 1998), phaselocked spiking frequency (Crick 1994), or something similar. Let us call this the *system approach* to consciousness.

System approach to consciousness – the approach to consciousness that views it as a property of whole cognitive systems, not individual or small groups of representations or properties of individual representations such as their *qualia* (their 'felt quality').

For Posner or Mack and Rock, for example, to be conscious of something simply is to pay attention to it. Here is Mack and Rock: "Attention [is] the process that brings a stimulus to consciousness" (Mack and Rock 1998); "if a...percept captures attention, it then becomes an explicit percept, that is, a conscious percept" (Mack 2001, p. 2). Posner (1994) captures the spirit of this line of thinking about consciousness nicely: '[A]n understanding of consciousness must rest on an appreciation of the brain networks that subserve attention, in much the same way as a scientific analysis of life without consideration of DNA would seem vacuous' (Posner 1994, p. 7398).

Now, what is interesting about this atomist/systems divide is that unlike a great many other philosophers, virtually all neurophilosophers are system theorists. Dennett's (1991) multiple-drafts model is a prime example. For him, consciousness is a matter of one or more of the multiple drafts of various descriptions and narratives in us achieving a certain kind of dominance in the dynamics of the Pandemonium-architecture of cognition. (Curiously, he says almost nothing about attention.) Paul Churchland is another example. Here is how Churchland summarized his approach recently:

[Consider] the brain's capacity to focus attention on some aspect or subset of its teeming polymodal sensory inputs, to try out different conceptual interpretations of that selected subset, to hold the results of that selective/interpretive activity in short-term memory for long enough to update a coherent representational 'narrative' of the world-unfolding-in-time, a narrative thus fit for possible selection and imprinting in long-term memory. Any [such] representation is...a presumptive instance of the class of *conscious* representations. (Churchland 2002, p. 74)

In this volume, Prinz clearly takes a system approach. Indeed, the system approach to consciousness once dominated in philosophy – think of Descartes and Kant.

Of course, the system theorists we have just considered are all system cognitivists. Anticognitivists, however, can also be system theorists. Indeed, of the five anticognitivists that we mentioned at the beginning, two of them are broadly atomistic (Chalmers and Block) but two of them take a system approach (Nagel and McGinn), broadly speaking.[4] For the atomist anticognitivist, individual conscious states are lined up one to one with individual representations. However, what makes a state a conscious state is nothing representational, not even when a representation is a conscious state. Representations could do the representational work they do in the absence of consciousness or in the presence of very different conscious content.

System anticognitivists come by their hostility to cognitivism in a variety of ways.[5] Many system 'anticognitivists' are philosophers. Nagel (1974), for example, argues that the only way to understand what a point of view is is to have one. The more one tries to adopt a third-person or impersonal point of view on a point of view, the more one moves away from what it is. There is no reason, however, to think that the same restriction to the first person governs our understanding of representation or other cognitive functioning. If so, there is reason to doubt that consciousness is representational. McGinn (1991) argues that we cannot know how consciousness is linked to the representational activities of the brain; consciousness is 'cognitively closed' to our kind of mind the way that physics is to field mice. If we expect a scientific understanding of cognition to be possible, this would be a reason to group him with the anticognitivists.[6] Among nonphilosophers, many different anti- or at least noncognitivist views can be found. R. Penrose's (1999) view that consciousness is a quantum phenomenon, rather than anything active in cognition, is perhaps typical.

With this, the issue is now clear. When neurophilosophers simply assume that consciousness is cognitive, some aspect of the machinery for

---

[4] Jackson is difficult to classify.

[5] I am not sure that their authors would all accept the label 'anticognitivist' but I think that it is a fair label.

[6] This consideration applies to Nagel, too. Since science is done from an impersonal point of view, Nagel is also arguing for a limitation on how far we are able to understand consciousness scientifically. What a point of view is like will elude science.

managing representations, it will strike anticognitivists of both stripes, both atomist and system anticognitivists, that whatever they are talking about, it isn't consciousness. Cognition and representation of all kinds could proceed without consciousness, and there is no reason to think that consciousness is anything like either of them.

We should note before we leave the system approach that there is enormous diversity in views that could be regarded as taking this approach. System approaches are far from a monolith. We cannot begin to explore the whole range of options here, but here are some of the leading views. Consciousness consists in a global workspace of a certain kind (plus some other things) (Baars 1988). Consciousness is an intermediate level of representation, a phonetic or similar level between acoustic or visual input and full-blown conceptual content (Jackendoff 1987). In understanding consciousness, attention should be singled out for special...attention (Posner 1994; Mack and Rock 1998). Consciousness is attention feeding working memory (Prinz, this volume). Consciousness is the result of multiple constraint satisfaction, a property or properties emergent on brain/world interactions. Consciousness is a form of self-organization in a dynamic system. Consciousness is a draft winning the competition for cognitive resources (Dennett 1991). Consciousness is, or is a product of, phaselock synchrony (Crick 1994). Consciousness is the result of a certain tensor phase-space processing (P. M. Churchland 1995). All this from people who broadly agree with one another! There is no agreement even on something as basic as whether consciousness is a biological property (P. M. Churchland 2002) or a cultural/information-engendered property (Dennett 1991). The system approach to consciousness is in a considerable mess.

3 Eliminativism: A Third Axis?

Of course, one reason for leaving consciousness out would be a belief that there is no such thing, and one form of the charge against some neurophilosophers is that they are urging that there is no such thing as consciousness when there patently is. Thus, it may seem that we need a third distinction, between eliminativists and noneliminativists about consciousness. Here is what I mean by 'eliminativism':

Eliminativism about consciousness – the view that the term 'consciousness' will prove not to be a theoretically useful term, that nothing exists that resembles what we take consciousness to be like.

Well, if the views of any neurophilosophers entail that consciousness should be eliminated in favour of something else – something infinitely less interesting and important, of course – such a result would be entirely inadvertent. There are very few deliberate eliminativists about consciousness.

At one time, Patricia Churchland (1983) and maybe Paul Churchland flirted with the idea. However, even at their most eliminativist, they never advocated wholesale replacement of our consciousness talk in the way that they did for our intentional talk. In recent years, they have backed away from eliminativism about consciousness almost entirely. As we saw in the previous section, Paul Churchland is now quite happy to talk about consciousness as a perfectly real phenomenon in need of scientific exploration.

Some think that Dennett's (1991) multiple-drafts model of consciousness is eliminativist. This would be quite wrong, in my view. Dennett certainly rejects a dominant way of thinking about consciousness, what he calls Cartesian materialism. But to reject a *theory* of consciousness is not to deny the existence of consciousness. To the contrary, Dennett has said repeatedly that consciousness is a perfectly real phenomenon (1998, pp. 135, 146). As he sees it, consciousness involves more interpretation by the cognitive system than has been thought, a system that in turn has less unity and stability, less universal general cognitive structure than has been thought, and the resulting conscious states have less determinability and temporal stability than has been thought. However, none of this is to deny that there is something appropriately called consciousness. Dennett himself says that he wants to be a deflationist about consciousness, not an eliminativist; he wants to deflate the pretensions of theories that insist on seeing consciousness as something weird, wonderful, and exotic (2000, pp. 369–370).

And there are good reasons why eliminativism about consciousness is extremely rare. Could consciousness turn out to be a theoretically useless term? That would require that the term be merely a misleading name for a variety of processes much better named and described by other terms, so that consciousness is a vague umbrella term for a diverse group of different things more perspicuously dealt with by giving each its own name (P. S. Churchland 1983). Or that nothing in us is much like what the term consciousness depicts pretheoretically; there is nothing in the brain that could be usefully labelled 'consciousness'. (We don't yet have anything remotely resembling a story about what consciousness is *supposed* to name, so exactly how one would determine this is a nice question.)

Both ideas mislocate the role of the term consciousness and cognates in our cognitive life. Unlike, say, 'intentionality', 'consciousness' is not a term of art. The notion of consciousness has deep roots in everyday discourse. We talk about losing and regaining consciousness. We talk about becoming conscious of this and that. We talk about being intensely conscious, for example of ourselves. It is unlikely on the face of it that all these modes of discourse rely on an implicit theory, or are describing nothing real or a bunch of things better discussed in a different vocabulary (though doubtless, some of the latter will turn out to be the case).

In short, neurophilosophers take consciousness to be a perfectly real phenomenon, just as real as their anticognitivist opponents take it to be, and they want their theories to explain *it*, not something picked out by some successor notion.

4 Prospects for Anticognitivism

As we said earlier, one way to undermine anticognitivism would be to show that it is so unlikely to shed light on interesting features of consciousness that consciousness could not have the character that the view says it has. So let us ask, does anticognitivism have any serious prospect of explaining interesting features of consciousness? When compared, for example, to system cognitivism, which has the better prospect?

What do we want a theory of consciousness to explain? As has often been said, consciousness:

- can be faint, full, and so on.
- can be independent of, indeed can continue in the absence of, sensory inputs.
- disappears in deep sleep, and . . .
- reappears in dreams.[7]

Then there is consciousness of self. On the face of it:

- Consciousness of oneself and consciousness of one's acts of representing, desiring, and so on seem to be two different things.

Moreover,

- Consciousness of self and the cognitive activities that yield it appear to have some unusual properties. Consciousness of self seems to use

---

[7] This list of four items and the items in the lists that follow are derived from Paul Churchland's (1995, pp. 213–214) list of the Magnificent Seven features of consciousness. I go beyond his list in a number of ways.

what S. Shoemaker (1968) called reference without self-identification, the resulting consciousness seems to have what he called immunity to error through misidentification with respect to the first person, and the use of first-person pronouns seems to be, to use J. Perry's (1979) term, essential.

Next, consider the conscious cognitive system. There has to be such a system; consciousness is a matter of *something being conscious* of something:

- Consciousness requires a conscious subject. (Atomism falls short right here.)

What is a system capable of consciousness like? Here are some features of such a system:

- Such a system has some general cognitive features:
  - Often how things appear to such a system is the result of cognitive activity, sometimes intense activity, on the part of the system.
  - Many of the global cognitive faculties of such a system are closely linked to consciousness, for example, memory, attention, and language.
  - For consciousness, a system simply having information as a result of representing this, that, or the other is not enough; the system must make cognitive use of the information.
- Consciousness requires a system that is capable of representing; there is a representational base to consciousness.[8]
- Usually, a cognitive system is conscious of whole groups of represented items in one 'act of consciousness' (Brook and Raymont, 2005).
- Usually when a cognitive system is conscious of groups of items in one act of consciousness, it is also conscious of representing them and of itself as the common subject of these representations.

Explaining these features of consciousness is a basic requirement on a theory of consciousness. How do the different approaches do? The answer is straightforward. When it comes to explaining these lists of features, both atomistic and system anticognitivists just claw the air.[9] By contrast,

---

[8] This statement is not the same as saying that consciousness is representational, or even that consciousness of something requires that we be representing it. All it says is that consciousness requires *a system* that can represent. As we noted earlier, some kinds of conscious states may not be representations, mood states for example, or mystical states. (Actually, I think that all such states are representational but cannot argue the point here.)

[9] To be sure, for some mysterians, this is a cause for rejoicing, not regret.

cognitive system approaches at least hold out a hope of being able to explain some of these features of consciousness. If so, it is unlikely that anticognitivism is talking about consciousness.

Anticognitivism is impotent in another way, too. An adequate theory of consciousness should be able to explain all the main kinds of consciousness. Two of them are

Consciousness of the world – being conscious of the world around us,

and,

Consciousness of oneself and one's states – the consciousness that we have when, for example, we are conscious of *representing* items in the world or conscious of *ourselves* representing items in the world.

Most (maybe all) variants of anticognitivism have nothing to say about consciousness of the world. In zombie thought-experiments, the organism's relationship to its world is meant not to change. The question of whether zombies would still be conscious of the world is seldom asked, but once asked, it is far from clear that they would not be. For Nagel, it is what it is like to have a point of view that is forever beyond the reach of science, not what a point of view is. Although inverted-spectrum thought-experiments are about how the world appears to me, not how something in me appears to me, they are not really about consciousness of the world, either. What is supposed to appear inverted is something purely internal. Ex hypothesi, the discriminations and comparisons that I make, the actions I launch, all this and everything else to do with my colour relationships to the world around me remain the same. These thought-experiments take it that appearings can change while nothing else changes, and so they too have to be focussing on a property of consciousness of representations, not consciousness of the world. And so on.[10] (So much for the idea of a unified theory of consciousness.)

In this section, we have shown the potential of anticognitivism as an explanatory theory of consciousness to be fairly dismal. But there are still the arguments for anticognitivism.

---

[10] Some theorists make a distinction between state consciousness, those states of ours that are conscious, and creature consciousness, which is at least something like our consciousness of the world. They then focus in their theory of consciousness on the former. This illustrates the deficiency we are describing very nicely.

5 The Frustration and Arguments of Anticognitivists

The Frustration

We have said that for the anticognitivist, cognitivist system approaches to consciousness are not talking about *consciousness* or they miss the most interesting and central features of it. A passage from Dennett illustrates how the frustration can arise. Says Dennett, 'We are beginning to discern how the human brain achieves consciousness. [I and others] see convergence coming from quite different quarters on a version of [Baars'] global workspace model' (2001). Statements like this tend to make anticognitivists crazy! Why? Because it seems perfectly easy to imagine a global workspace grinding away doing its thing with no consciousness at all. The point can be generalized. For any form of representation and any representing system, couldn't one imagine such a system doing all the wonderful cognitive things that it does without consciousness? If there is anything to this challenge, then consciousness is not representational or anything cognitive. So it cannot simply be ignored.[11] Instead, we need to examine whether there is anything to it.

The Arguments

Here is one way in which an argument for this anticognitivism can get going. When something appears to us to be a certain way, the representation in which it appears can play two roles in our cognitive economy. The contents of the representation (or even the representation itself) can connect inferentially to other representations: If the stick appears to have two straight parts with a bend in the middle, this will preclude representing it as forming a circle. The representation can also connect to belief: If the stick appears straight with a bend in it, I will not form a belief that it bends in a circle. And to memory: I can compare this stick as it appears to sticks I recall from the past. And to action: If I want something to poke into a hole, I might reach for the stick. In all these cases, so long as I am *representing* the stick in the appropriate way, it would seem to be irrelevant whether I am *conscious* of the stick or not. But I am also *conscious* of the stick – it *appears* to me in a certain way. Now, it seems at first blush plausible to say that my representation could do the representational jobs just delineated whether or not I was conscious of the stick

---

[11] Dennett has been fighting in the consciousness wars for too long to neglect the opposition himself. Indeed, in the very paper just cited, he says that he will "diagnose some instances of backsliding and suggest therapeutic countermeasures."

or of my representation of it. If so, we might begin to suspect that being conscious plays no independent representational role.

The anticognitivist then advances arguments aimed at turning this suspicion into a conviction. The best known is the zombie thought-experiment introduced earlier. There could be creatures just like us behaviourally, cognitively, or even physically who nevertheless are not conscious.[12] Though they are built and behave in ways wondrously like us, all is 'dark' inside.[13] Zombie thought-experiments seek to establish that a representation is one thing; what makes it a conscious state (its *qualia*) is another. If this thought-experiment establishes that a split between cognition and consciousness is so much as possible, then all forms of cognitivism about consciousness are in trouble.[14]

A second familiar argument for this 'neo-dualist' conclusion, as Perry (2001) calls it, is the old thought-experiment about inverted spectra: The way in which colours appear to me could be inverted with respect to how they appear to you without changing how our respective representations of colour function as representations?[15]

A third argument could be advanced, too, though it has seldom been used. We will call it the argument from *imprisoned mind*s. An imprisoned mind is a mind that is working perfectly well but cannot express itself in behaviour (for many people, a scenario of utter horror). Unlike zombies, imprisoned minds actually occur. Curare can produce imprisoned minds, for example.

Curare is a muscle paralysant and in sufficient doses produces total paralysis. Not even an eyebrow moves. It is added to some anaesthetic

---

[12] For a good sample of this literature, see the *Journal of Consciousness Studies* target article by Flanagan and Polger (1995) and the remarkable array of comments that it generated. Inverted spectrum and a host of other thought-experiments (including dancing qualia, inverted earth, shrinking brain, and expanding brain) raise similar questions, but here we will confine ourselves to zombie thought-experiments. See also Polger's (2000) follow-up article and Dennett's (2000) reply. Note that zombie thought-experiments have extremely broad scope; they aim to establish that consciousness could be absent from *anything* to which a theory of consciousness could tie it.

[13] 'Dark' here is a highly misleading metaphor. Given the opacity of the skull, all is dark, indeed pitch black, in the brain of all conscious beings, too. Representing light and giving off light are two entirely different things. (Dennett 1991 makes very good use of this distinction.)

[14] Exactly what kind of possibility a zombie thought-experiment would have to establish is much debated: logical possibility, natural possibility (possible on some set of laws of nature), nomological possibility (possible on our laws of nature). Fortunately, I do not need to go into this exquisitely scholastic literature because I don't think that zombie thought-experiments can even be coherently thought.

[15] I say 'old' because it goes back as far as John Locke (1690).

mixes to keep surgery patients from twitching and moving. For a time, more curare as a ratio of the whole and lower doses of central nervous system (CNS) suppressors were used, especially with children. Upon regaining consciousness, some of these patients appeared traumatized, and so a surgeon who had to have a small procedure done volunteered to undergo the surgery with the same anaesthetic mix. To his horror, he did not lose consciousness. He just lost all capacity to express what he was feeling. So he felt every slice of the scalpel, the insertion of every stitch – and he could do absolutely nothing about it![16]

Imprisoned minds also arise as a result of strokes. Such a stroke can completely paralyse the body but leave the patient conscious for a few days. Many of these patients retain control over eye movements, but it is suspected (and could be only suspected; this is a real-life problem of knowledge of other minds) that some patients lose even this minimum channel of communication and are completely unable to communicate. Fortunately, such patients die after only a couple of days (so far as we know) (Zeman 2003; see also Calvin's 2003 review).

There are also cases in which victims of accidents who are in a persistent vegetative state have been assessed for therapy on the basis of differences detectable by MRI, in what the researchers call an N400 response to semantically usual and semantically odd sentences ('the pizza was served with piping hot cheese' vs. 'the pizza was served with piping hot socks'). Upon finding a reaction to the latter kind of sentence, therapy has been ordered – and patients have eventually recovered enough to walk and talk. Without the reaction, the plug could easily have been pulled on their ventilators (Colin Herein, personal communication, reporting on work in Halifax, NS).

Does the phenomenon of imprisoned minds have any implications for cognitivism about consciousness? To have such implications, I think that one would have to add a further condition, one similar to the condition built into zombie and inverted-spectrum thought-experiments and just as unrealistic. Here, the required further condition would be that not just behaviourally imprisoned minds but also imprisoned minds whose brains had ceased to function in the relevant ways are possible: not just no difference in N400 response, but no N400 response. Let us call an imprisoned mind that is not making a difference even to the brain a *radically*

---

[16] Some commentators suspect that this story is an urban legend, but Dennett, in one of the most scientifically literate treatments of pain ever written by a philosopher (1978b, p. 209), gives (now rather dated) references to medical accounts of it.

imprisoned mind (RIM). There could not be the *slightest* reason to think that an RIM existed, of course. However, all anticognitivists think they need is the bare possibility, in some sense of 'possibility' (see note 14).

Another, more exotic argument against cognitivism about consciousness flows from externalism about representational content. Externalism is the view, in Hilary Putnam's famous (1975) saying, that 'meaning ain't in the head'. The content of representations consists of some relationship between what is in the head and the world. Philosophers who accept this view then go one or the other of two ways about consciousness. Some continue to hold the commonsense view that the element of representations or represented objects of which we are conscious is in the head. They then argue that since representational content is not in the head, qualia are not representational content. Others hold that if representational content ain't (entirely) in the head, then how something appears (or anything else that the element of representations of which we are conscious consists in) ain't gonna be entirely in the head, either (Tye, 2003, plato.stanford.edu/entries/qualia, p. 10). The former abandons cognitivism. The latter defends cognitivism – but often at the price of considerable implausibility, depending on which external factor is invoked. (Most of our conscious contents remain exactly the same over even radical change of location of our body, for example, which would seem to rule out direct causal links as an external factor playing a role in conscious content.)

To support the claim that a separation of consciousness and cognition is possible, theorists often appeal to Levine's (1983) explanatory gap. According to Levine, one way to understand the connection between a phenomenon and a mechanism is to understand why, given the mechanism, the phenomenon has to exist. With consciousness, not only do we not know of any mechanism or causal process whose operation has to bring about consciousness, but we cannot even imagine what such a mechanism might be like. There is nothing like the same explanatory gap with respect to cognitive functioning, and so consciousness is radically unlike cognitive functioning, epistemically at least.[17]

Sometimes such arguments go so far as to conclude that what is distinctive to consciousness is not just not cognitive, it is not even physical. One way of arguing for this is to think of a zombie that is a molecule-for-molecule duplicate of oneself. If a zombie such as this is possible,

---

[17] Levine's explanatory gap is part of what makes the hard problem appear to be so hard, too.

then the conscious aspect of things is not molecular, that is, not physical. Another is F. Jackson's (1986) famous thought-experiment concerning Mary, the colourblind colour scientist. Mary knows everything there is to know about the experience of colour, therefore everything *physical* there is to know about the experience of colour, but she has never experienced colour herself. Then her problem is corrected and she experiences colour! Clearly, she gains something she did not have before. However, she knew everything physical about colour. Therefore, what she gains must be something nonphysical.

The people we have called system anticognitivists agree with other system theorists in viewing consciousness as a property of cognitive systems as a whole, not individual representations, but they have their own arguments hostile to cognitivism. We have already seen an argument of Thomas Nagel's (1974). Nagel argues that the only way to understand what a point of view is to have one. The more one tries to adopt a third-person or impersonal point of view on a point of view, the more one moves away from what it is. There is no reason, however, to think that the same restriction to the first person governs our understanding of representation or other cognitive functioning. If so, there is reason to doubt that consciousness is cognitive.

McGinn (1991) argues for a similar conclusion, that we cannot know how consciousness is linked to the representational activities of the brain, by urging that because our only direct, noninferential access to conscious states (introspection) is so different from our direct, noninferential access to brains (perception), we will never be able to find laws bridging the two domains.

R. Penrose's (1999) argument is indirect, going via computationalism. Relying on Gödel's theorem that a system of arithmetic axioms cannot prove intuitively true propositions that can be generated by the system, he argues that consciousness (and cognition) cannot be computational and, therefore, must be something weird and wondrous to do with quantum phenomena. Since the only halfway worked-out picture of cognition that we have is the computational one, the conclusion Penrose reaches is so different from any usual cognitive account that we might as well group it with the anticognitive approaches.

J. Searle's (1980 and many subsequent publications) Chinese room is meant to establish the same conclusion that consciousness and cognition are not (just) computational. The direction in which he goes in response is different from Penrose's, however. He argues that consciousness *and* cognition must both be some noncomputational biological element, at least in part. Unfortunately, he has never been able to identify a plausible brain phenomenon to play this role, but his conclusion is still clearly hostile to cognitive system theory.

Why These Arguments Don't Work

In my view, zombie, inverted-spectrum, and imprisoned-mind thought-experiments are the most serious challenge to cognitivism about consciousness (together maybe with externalism), but let us take a quick look at some of the other arguments first.

There are familiar, powerful difficulties with each of them. Mary acquires something new when she first experiences colour, to be sure, but there is no reason to think that what she acquires is anything more than a new mode of access to facts she already knows. ('Ah, so that is what light of 640 angstrom units processed through V1 to V4 and integrated in the XYZ cortex is like!') Nagel seems to be just plain wrong. By studying the brain and relating our findings to what people say about their experience (for reflections on this method, see Thompson, Lutz, and Cosmelli, this volume), we are rapidly finding out precisely what he says we cannot know: what kind of brain structure a point of view is. McGinn's cognitive closure claims are wildly premature – and again, seem to be seriously threatened by recent work using the method just sketched. Penrose seems to have given little or no thought to the role of heuristics in cognition. Not everything we do cognitively has to be governed by deductively closed rules. And Searle has never responded effectively to the so-called System Response: Maybe it is plausible to say that the Chinese room contains no consciousness or content in isolation, but if one hooked it up to a visual system and gave it exquisitely detailed control of a body, then the situation is no longer at all clear.

So there does not appear to be a serious threat to cognitivism about consciousness in any of this. Next, zombies, inverted spectra, and imprisoned minds (specifically, RIMs). The worry we are addressing is that for any cognitive system, we seem to be able to imagine that system behaving as it does, functioning as it does, even being built as it is, without it being conscious or while having conscious contents very different from what we would have in the same situation or while being conscious without that showing in brain or behaviour. A requirement common to all three situations is that the difference in consciousness be compatible with complete similarity in cognition, or cognition and brain, and ensuing behaviour, or that the presence and absence of consciousness be compatible with complete similarity in brain and behaviour. This is a crucial requirement because if the difference in consciousness goes with any difference in structure, cognition, or behaviour whatsoever, the desired separation of consciousness and cognition will not have been achieved and no argument that the two are radically distinct will have be given.

Put this way, zombie, inverted-spectrum, and RIM thought-experiments entail (and perhaps rely on) a particularly vigorous form of scepticism about other minds, other conscious minds at any rate.[18] Zombie thought-experiments purport to show that no behaviour, no cognition, even no neural structure is conclusive for the presence of consciousness. Inverted-spectrum thought-experiments purport to show that no behaviour, no cognition, even no neural structure is conclusive for the presence of conscious content of a particular kind. RIM thought-experiments purport to show that nothing about brains or behaviour is conclusive for the absence of consciousness. This connection with a deep-dyed and highly suspect form of scepticism should get our suspicions up.

The basis of scepticism about other minds (especially other consciousness) is a lack. *We do not know what behaviour, cognitive activity, or brain phenomena would constitute or justify an ascription of consciousness.* This is the explanatory gap, one of them anyway. Cross it successfully and the threat of zombie and inverted-spectrum thought-experiments is removed. RIMs are not quite so easy, but the first two first.

Traditionally, philosophy has tried to cross the behaviour/consciousness gap with a single mighty bound, to no great success. Not being a very strong jumper, my approach is to try two bounds instead. The first bound is to justify the ascription of representations. This is easy to do – we have to postulate the possession of representations to explain everything from how people (and other animals) react to the Müller-Lyer arrowhead illusion to how children acquire the capacity to recognize false beliefs in others as they develop. Moreover, nothing in the first bound puts us in conflict with the anticognitivist. Zombies and inverted-spectrum subjects would have *representations*. It is just that they would not be conscious, or would not have the conscious content that we would have when we had the same representations.

In the second bound, we argue that concerning consciousness, once representations of the right kind are in place, there ain't nothin' left

---

[18] Many statements of the problem of knowledge of other minds focus on knowledge that others are conscious. Some extend the sceptical worry to intentionality in representation and behavioural control, too, but, as we will argue shortly, intentionality is not the problem here that consciousness is.

over to be left out. This second move works more straightforwardly with
zombies than in inverted-spectrum cases, so let us explore it with zombies
first. Here is what a philosopher's zombie will be like. It will talk to us about
how proud it feels of its kids. It will describe the stabbing pain of arthritis
in its hip. It will say how pleasant it is to stretch out on a hammock on a
hot summer day. It will report its memories, hopes, dreams, fears just as
we do. In short, there will be absolutely nothing in its behaviour, *including
its self-reporting behaviour*, that distinguishes it in any way from a normal,
conscious human being. Even though, supposedly, it is not conscious, it
will *represent itself to itself* as conscious, *feel* pleasure, *have* pain, and so on.
And it will do all these things, ex hypothesi, *on the basis of representations and
representations of the right kind.* Even the zombie itself will be convinced that
it is conscious. So what could possibly be missing? The supposition that
it is possible that the zombie might nonetheless not be conscious seems
to be a truly perfect example of a supposed distinction that makes –
and could make – no possible difference.

Is my argument here verificationist? Don't know. Don't care, either.
Whatever you want to call it, the principle here is that for a sentence to
be an assertion (intuitively, for a sentence to be meaningful), the difference between its being true and being false must mark some difference.
The difference need not be detectable – but it must be stateable. In the
case of the zombie, we should ask not just what is but what *could be* missing.
Can we so much as form an idea of something that could be missing? If
the answer is no, the philosopher's zombie is not fully imaginable, hence
not possible, not in any relevant sense of possibility.

Possible objection: 'A thing representing itself to itself in some belief-
inducing way is not the same as representing itself to itself in the it-
is-like-something-to-have-the-state way'. Response: No? What could this
difference consist in? For notice, it would not only be a difference that
no one else could detect; it would be a difference that the organism
itself could not detect! We shouldn't simply *assume* we can imagine cognition without sneaking consciousness in by the back door. *Show* us
that you can do it! As I argued long ago (Brook, 1975), we must at
least imagine the relevant difference in enough detail to be able to
assert plausibly that if what we imagine came to pass, we would actually have the one without the other. To see the silliness of this 'notion'
of a zombie, reflect on a point that Dennett makes: Even if we
could make some sense of the idea of such a zombie, why on earth
should we *care* about what would be missing? It would have nothing
to do with consciousness as we encounter it, in oneself or in others

(2000, pp. 381–382). (This observation also supports the conclusion reached in section 4.)

The same demolition job can be done on inverted-spectrum thought-experiments. It must be admitted, however, that some people find this application less persuasive. We can imagine, it is said, that Santa Claus' suit seems blue to Suma when it seems red to me, without anything else changing – without, therefore, this difference showing in her behaviour or in her cognition (inferred from behaviour in any case) or her brain physiology. That means that she will engage in the same additive and subtractive exercises with the colour as it appears to her as we do with the colour as it appears to us. When she combines, for example, a chip of her apparent colour with yellow, she will get orange, not green – and she will say that the combination seems orange to her, not green. She will behave as though the suit appears red to her in all other respects, too. In particular, she will say that the suit seems red to her (she is trained to call apparent blue 'red').

A telling story, again from Dennett. Suppose that a particular shade of blue reminds you of a car in which you had a bad accident and so is a colour to be avoided. Now your colour spectrum is inverted. At first, things are fine. The things that used to look blue to you now look yellow and you are not averse to them. In time, however, you adapt to the inversion. (Evidence for this is that you again call 'blue' the shades of colour that others call 'blue', fit these shades into a colour wheel the way others do, and so on.) Suddenly, you start avoiding the things that you again call blue *and it is because they remind you of the car in which you had the bad accident.* That is to say, the shade of colour in front of you strikes you as the same as the shade of colour of the car as you remember it. Does the shade of colour in front of you and/or the remembered colour of the car now seem to you as yellow seems to others or as blue does (Dennett 1991, p. 395)?

Here is what I am inclined to say: 'Given that everything, *everything*, is exactly as it would be if the colour now appeared blue to you, what could appearing yellow *consist in* here?' Given that you will say that the colour in question appears blue to you and you will react in every respect as though it does, what could the additional state of affairs of it appearing yellow *consist in*? What difference would correspond to 'X appears yellow to you' being true and it being false?

As we said earlier, many people find the strategy just adopted more persuasive in the case of zombies than in the case of inverted spectra. One reason might be this. We can distinguish a macro- from a microproblem of knowledge of other minds. The macroproblem is whether a being that behaves like us has a mind or is conscious at all. The microproblem is whether, granting that the macroproblem is solved, it has a mind or consciousness like ours. In real life, the macroproblem almost never arises. By contract, we face the microproblem a million times a day. ('She says nice things but what does she really think of me?') Thus, while many people are prepared to allow, on analysis, that, well, yes, they can't make sense of the idea of a (philosopher's, i.e., behaviourally, cognitively, and even neurally indistinguishable from us) zombie, a lot of people are by no means as willing to allow this about the idea of a (behaviourally, cognitively, neurally invisible) inverted spectrum. For that reason, it takes more therapy (as Wittgenstein called it) to wean people away from the idea of a (philosopher's) inverted spectrum than from the idea of a (philosopher's) zombie. For example, for inverted spectra of the right kind to be possible, subjects would have to be able to make distinctions that are impossible to make. (For the argument for this claim, see Brook and Raymont, forthcoming, chapter 3.)

Now RIMs. A way to infer consciousness from behaviour and cognition is not going to help us with whether RIMs are possible because, ex hypothesi, there is no behaviour and cognition. It would be nice to have an argument that showed that RIMs are impossible. But I do not. Instead, I will allow the logical possibility of RIMs and try to show that this possibility is not a problem for cognitivism.

The 'nothing left over to be left out' move does not work for RIMs. We know perfectly well what would be left over when the brain stopped while a RIM continued – exactly the same kind of mind aware of itself in exactly the same kind of way as you and I have. Granted, since the idea of a RIM disconnects conscious minds from everything detectable by others, it leads to some bizarre questions: If RIMs are possible, what argument do we have, for example, that the Esc key on my keyboard could not be conscious?[19] All we have is no evidence that it is conscious – but that is what we would have with RIMs, too. (So can we ever be sure that dear old granny is not still with us, trapped inside the lifeless body that we are about to put in the ground?) Still, bizarre is not the same as incoherent.

Indeed, a more radical form of anticognitivism than we have seen so far would be in business. In order for something to be conscious, things must be like something to it, and we have argued that this requires some

---

[19] When Wittgenstein canvasses the idea of someone in pain turning into a stone and asks, "What has a soul, or pain, to do with a stone?" (§283), he may be invoking a similar idea.

modes of cognitive functioning. So a RIM would have to continue to have such modes, and cognition would have to be just as detachable from the brain as consciousness. We would be pushed to what on our own terms could only be called an anticognitivist account of cognition![20]

Curiously, this link is helpful to us. On the RIM story, consciousness and cognition at least still go together. Which means that cognitivism about consciousness would remain an open possibility. Which would be progress.

However, it is progress in the wrong direction. The whole point of wanting to show that consciousness is a form of cognition is to lend support to the project of developing a unified account of the mind and brain. Such an account has to be physicalist – has to argue that both consciousness and cognition are properties of a physical system, the brain. RIM stories thus appear to threaten physicalism. But this appearance may be deceiving.

Physicalists need not be committed to the view that consciousness *must* be physical. They can accommodate the mere logical possibility not just of RIMs but also ghosts in the machine, ectoplasmic consciousness, and all manner of things that are conscious without being physically realized. How? Physicalists accept the possibility of multiple realizability of conscious and of other cognitive states. If so, physicalists can insist that all *actual* conscious states are realized or implemented in purely physical systems, even that all possible conscious and cognitive states *compatible with the laws of nature* are likewise implemented, and yet allow for the possibility of mental states that are not physically realized.

Before the physicalists among us start to gasp and groan, I hasten to add that this concession doesn't concede very much. The issue before us concerns the physical status of consciousness and cognition. If they do not *have* to be physical, this need be nothing more than an artefact of conscious and cognitive types being anomalous, that is, not being reducible to physical (in this case, brain-state) types. Since that is also true of clocks and radios and hundreds of other types of things whose types are multiply realizable physically and therefore not reducible to any physical types, the possibility of RIMs does not establish that consciousness or cognition are any more nonphysical than clocks are.[21]

---

[20] John Kulvicki pointed this out to us.

[21] For more on multiple realizability and its implications, see Brook and Stainton 2000, ch. 5, especially the discussion of Descartes' indivisibility argument for mind/brain dualism.

RIM and zombie thought-experiments are asymmetrical. If zombie thought-experiments worked, something could be missing even though everything physical is there. If so, the missing element would have to be nonphysical. But the possibility of a nonphysical realization holds no implications for what something is like *as realized in our world*. Whatever the general reducibility of clock types, in our world clocks are composed entirely of matter. The possibility of RIMs holds no implications that anything else is true of consciousness or cognition. Consciousness and cognition *could* be nonphysical. So could clocks. But there is no reason to think that any of them are nonphysical, or even that they could be nonphysical in any world with our laws of nature.

Conclusions: Zombie thought-experiments cannot be done as specified. Inverted-spectrum thought-experiments cannot be done as specified. RIM thought-experiments do establish a possibility of consciousness and cognition being nonphysical but not in any way that is not also true of clocks and radios. If anticognitivism in its various forms does not succeed, there is no reason to think that consciousness is anything more than an aspect of representing and/or of the cognitive machinery for managing representations. We need not be ignoring consciousness or changing the subject when we look to give a cognitive account of it. Consciousness is safe for neuroscience.

Final note: Whatever the merits of the specific moves made in this chapter, I think that it shows how traditional philosophical techniques still have a role to play in neuroscience. These techniques are certainly not neuroscience, but they clear away confusions and lay out better ways of thinking about things. If I am really lucky and the particular moves succeed, what would I have accomplished? I would have set things up so that when the next generation puddle about in their hippocampi and thalamuses and V1 and MT, they will know that the cognitive functions and neural implementations of these functions that they find are, or least could be, constituents of *consciousness*, not merely something correlated with consciousness.

Acknowledgments

Thanks to Kathleen Akins, Dan Dennett, Zoltán Jakab, Jerzy Jarmasz, Luke Jerzykeiwicz, Jamie Kelly, Christine Koggel, Kris Liljefors, James Overall, Don Ross, Sam Scott, Rob Stainton, Edina Torlakovic, Chris Viger, Tal Yarkoni, and especially Paul Raymont, who has written a book with me on these topics (forthcoming). This chapter is derived from chapters 1 and 3.

References

Baars, B. 1988. *A Cognitive Theory of Consciousness*. Cambridge: Cambridge University Press.

Brook, A. 1975. Imagination, possibility, and personal identity. *American Philosophical Quarterly* 12, 185–198.

Brook, A., and Raymont, P. 2005. Unity of Consciousness. *Stanford Encyclopaedia of Philosophy*. http://plato.stanford.edu.

Brook, A., and Raymont, P. forthcoming. *A Unified Theory of Consciousness*. Cambridge, MA: MIT Press.

Brook, A., and Stainton, R. 2000. *Knowledge and Mind*. Cambridge, MA: MIT Press/A Bradford Book.

Calvin, Wm. 2003. Review of Zeman (2003). *New York Times Book Review*, Sept. 28, 2003.

Chalmers, D. 1995. Facing up to the problem of consciousness. *Journal of Consciousness Studies* 2, 200–219.

Chalmers, D. 1996. *The Conscious Mind*. Oxford: Oxford University Press.

Churchland, P. M. 1995. *The Engine of Reason, the Seat of the Soul*. Cambridge, MA: MIT Press/A Bradford Book.

Churchland, P. M. 2002. Catching consciousness in a recurrent net. In Andrew Brook and Don Ross, eds., *Daniel Dennett*. New York: Cambridge University Press, pp. 64–80.

Churchland, P. S. 1983. Consciousness: The transmutation of a concept. *Pacific Philosophical Quarterly* 65, 80–95.

Crick, F. 1994. *The Astonishing Hypothesis*. New York: Scribner's.

Dennett, D. 1978a. Toward a cognitive theory of consciousness. In his *Brainstorms*. Montgomery, VT: Bradford Books, pp. 149–173.

Dennett, D. 1978b. Why you can't make a computer that feels pain. In his *Brainstorms*. Montgomery, VT: Bradford Books, pp. 190–232.

Dennett, D. 1991. *Consciousness Explained*. Boston: Little, Brown.

Dennett, D. 1998. Real consciousness. In his *Brainchildren*. Cambridge, MA: MIT Press, pp. 131–140.

Dennett, D. 2000. With a little help from my friends. In D. Ross, A. Brook, and D. Thompson, eds., *Dennett's Philosophy: A Comprehensive Assessment*. Cambridge, MA: MIT Press, pp. 327–388.

Dennett, D. 2001. Are we explaining consciousness yet? *Cognition* 79, 221–237.

Flanagan, O. 1991. *Consciousness Reexamined*. Cambridge, MA: MIT Press.

Flanagan, O., and Polger, T. 1995. Zombies and the function of consciousness. *Journal of Consciousness Studies* 2, 313–321.

Jackendoff, R. 1987. *Consciousness and the Computational Mind*. Cambridge, MA: MIT Press/A Bradford Book.

Jackson, F. 1986. What Mary didn't know. *Journal of Philosophy* 83, 5, 291–295.

Levine, J. 1983. Materialism and qualia: The explanatory gap. *Pacific Philosophical Quarterly* 654, 354–361.

Locke, J. 1690. *Essay Concerning Human Understanding*. London: Basset.

Mack, A. (2001). Inattentional blindness: Reply to commentaries. http://psyche.cs.monash.edu.au/v7/psyche-7-16-mack.htm.

Andrew Brook

Mack, A., and Rock, I. 1998. *Inattentional Blindness*. Cambridge, MA: MIT Press.

McGinn, C. 1991. *The problem of consciousness: Essays towards a resolution*. Oxford: Basil Blackwell.

Nagel, T. 1974. What it is like to be a bat? *Philosophical Review* 83, 435–50.

Penrose, R. 1999. *The Emperor's New Mind: Concerning Computers, Minds, and the Laws of Physics*. Oxford: Oxford University Press.

Perry J. 1979. The problem of the essential indexical. *Noûs* 13, 3–21.

Perry J. 2001. *Knowledge, Possibility, and Consciousness*. Cambridge, MA: MIT Press.

Polger, T. 2000. Zombies explained. In D. Ross, A. Brook, and D. Thompson, eds., *Dennett's Philosophy: A Comprehensive Assessment*. Cambridge, MA: MIT Press, pp. 259–286.

Posner, M. 1994. Attention: The mechanism of consciousness. *Proceedings of the National Academy of Science USA* 91, 7398–7403.

Putnam, H. 1975. The meaning of 'meaning'. *Mind, Language and Reality: Philosophical Papers*, Vol. 2. Cambridge: Cambridge University Press, pp. 215–272.

Rosenthal, D. 1991. *The Nature of Mind*. Oxford: Oxford University Press.

Searle, J. 1980. Minds, brains, and programs. *Behavioral and Brain Sciences* 3, 417–458.

Shoemaker, S. 1968. Self-reference and self-awareness. *Journal of Philosophy* 65, 555–567.

Tye, M. 1995. *Ten Problems of Consciousness*. Cambridge, MA: MIT Press.

Tye, M. 2003. Qualia. plato.stanford.edu/entries/qualia.

Wittgenstein, L. 1953. *Philosophical Investigations*, trans. G. E. M. Anscombe. Oxford: Basil Blackwell.

Zeman, A. 2003. *Consciousness: A User's Guide*. New Haven, CT: Yale University Press.

Index